CONTEMPORARY TECHNOLOGY

Innovations, Issues, and Perspectives

by

Linda Rae Markert
State University of New York at Oswego
Oswego, New York

Patricia Ryaby Backer
San Jose State University
San Jose, California

Publisher
The Goodheart-Willcox Company, Inc
Tinley Park, Illinois

Copyright 2003

by

The Goodheart-Willcox Company, Inc.

Previous editions copyright 1997, 1993, 1989

All rights reserved. No part of this book may be reproduced, stored in a retrieval system, or transmitted in any form or by any means, electronic, mechanical, photocopying, recording, or otherwise, without the prior written permission of The Goodheart-Willcox Company, Inc. Manufactured in the United States of America.

Library of Congress Catalog Card Number 2002066840

International Standard Book Number 1-56637-982-2

3 4 5 6 7 8 9 10   03   06 05

Library of Congress Cataloging in Publication Data

Markert, Linda Rae.
        Contemporary technology: innovations, issues, and perspectives / by Linda Rae Markert and Patricia Ryaby Backer.

                p. cm.

        Includes index.

        ISBN 1-56637-982-2

        1. Technology.   I. Backer, Patricia Ryaby   II. Title

T45.M36 2002

600—dc21   2002066840

                                                CIP

Introduction

Contemporary Technology: Innovations, Issues, and Perspectives is written for college and university courses related to technology and society. It is intended to introduce you, as one of its readers, to the pervasive nature of technological innovations and to increase your awareness of both the promises and the uncertainties associated with the use of technology as a creative human enterprise.

The chapters are organized under three subheadings. Section I, "Innovations," contains five chapters discussing recent developments and breakthroughs in the following technical disciplines: biotechnology, information systems, space exploration, medicine, and manufacturing. Section II, "Issues," includes four chapters focusing on some of the public controversies associated with significant international and global issues in technology transfer, appropriate technology, environmental policy, and defense strategy. The final three chapters make up Section III, "Perspectives," and address a variety of diverse sociological viewpoints regarding humankind's effective use of technology and the projected stability of our "pseudo traditional" social institutions.

Contemporary Technology presents technical facts and figures and scientific theories in a general way to provide you with an overview of the innovations, issues, and perspectives on the topic of technology in your everyday life. We acknowledge that it is impossible, in these few pages, to cover all the bases, investigate all the angles, or delve into all of the topics with the thoroughness everyone might desire. We believe, however, this book will prove to be a useful starting point in your study of technology and its impact on society. The text presents cases, both pro and con, to be considered in the application or eventual use of technology. We provide explanations for many key technical concepts and periodically define words with which you may be unfamiliar. The lengthy list of suggested readings at the end of each chapter will enable you to further study areas of particular interest.

It is the authors' desire and optimistic expectation that the course or activity for which you are reading *Contemporary Technology* will "teach" you to actively think, analyze, question, criticize, and ultimately search out new sources of information. It is the authors' primary objective to motivate you to take a strategic role and participate in current and future technological ventures.

Linda Rae Markert

Patricia Ryaby Backer

About the Authors

About the Authors

Dr. Linda Rae Markert and Dr. Patricia Ryaby Backer are well qualified to write this newest edition of *Contemporary Technology.* Dr. Markert is the author of the first three editions of this textbook, and we are pleased to welcome Dr. Backer to the writing team for this release.

Dr. Markert is presently the Dean of the School of Education at the State University of New York (SUNY) at Oswego. Prior to this administrative post, she chaired the Department of Technology at SUNY Oswego, where she continues to hold a professorship. As chair, she taught courses and supervised technology education student teachers. She held a professorship in and was the Graduate Coordinator for the Division of Technology at San Jose State University (SJSU). Employed at that institution for fifteen years, she taught a variety of graphics and professional courses, including a graduate seminar in International Technology. Dr. Markert holds her doctorate in educational administration from the University of the Pacific. She received an appointment as a Visiting Scholar in the Science, Technology, and Society Program at the Massachusetts Institute of Technology, and participated in the Institute for Management and Leadership in Education at the Harvard Graduate School of Education. Dr. Markert is published in a number of professional journals, serves on editorial boards, and is a sought after speaker.

Dr. Backer is presently the Chair of the Department of Aviation and Technology at SJSU. She holds an Associate Professorship in the Department of Aviation and Technology and has taught courses in many technical areas, several of which include technical illustration, multimedia production, manufacturing organization and communication, and statistical quality control. Prior to joining the faculty at SJSU, she held instructor appointments at Tennessee Temple High School, Tennessee Temple University, and Thomas Jefferson High School for Science and Technology, a Governor's Magnet School in northern Virginia, where she taught courses in geometry, computer science, physics, business statistics, and philosophy of education. At Tennessee Temple University, she also evaluated the pre-student teaching observation program for preservice elementary and secondary teachers. Dr. Backer holds her doctorate in neurocognition from The Ohio State University. She received a Fulbright Scholar lecturing award in Peru, where she taught and consulted on the topics of computer-based multimedia and web teaching materials. She also received fellowships from the National Endowment for the Humanities and the Army Research Institute. Dr. Backer has given numerous national and international presentations on many topics, all of which have been published in professional proceedings, and she is a successful research grant recipient.

Both authors have taught courses for which this textbook is selected and used. Working and teaching in Central New York and in the heart of Silicon Valley keeps Dr. Markert and Dr. Backer in focus with emerging trends in technology and their impacts on society.

Contents

Chapter 1

Science, Technology, and Society

DATELINE

Ybor City, Florida

Super Bowl XXXV was played in Tampa, Florida, and it provided the venue for the local police to try out a new high-tech surveillance system. Cameras were strategically placed throughout the stadium to scan every face in the huge crowd of energized fans. Using Visionics Corporation's FaceIt® computer software, police officers were able to search the crowd for faces matching those on its database of known criminals.

In nearby Ybor City, the nightlife district of Tampa, a similar system was deployed to search for faces of wanted criminals. Essentially, the software installed in the FaceIt system receives a video signal from the cameras looking into the public streets, and it detects the people's faces from the video. It converts the images of people's faces to a mathematical equation, and if there is a match in the equation between an entry in the database (wanted criminals) and the live image, the police are alerted with a known level of confidence.

The use of cameras and computers to look at every citizen attending a sporting event or going to dinner while on vacation is a bit unsettling. The FaceIt system has worked. It identified nineteen people attending Super Bowl XXXV who were wanted for various crimes....

Introduction

What better way to open the first chapter of a book about contemporary technology than to relate an incident about the highly pervasive nature of technological systems in our everyday lives? In all likelihood, most people attending football games or enjoying a restaurant meal are entirely unaware of the fact that they might be under investigation or being recorded. See Figure 1-1. Reasonably, most of us have become accustomed to being filmed while using a bank's automatic teller machine (ATM), but we are less keen on having a private dinner engagement videotaped without our knowledge or consent.

In recent years, we have begun to take greater notice of the importance of privacy, since a significant portion of it has been or is threatened to be taken away. Privacy is viewed as a personal right we assume we have, yet we take it for granted until something or someone infringes on it. In reality, our current technological surveillance capacity makes personal privacy, in many aspects of our lives, a thing of the past. At the same time, however, our society guards against antisocial conduct through these same surveillance technologies, in order to protect personal and group rights and enforce social rules and regulations.

Philosophically, *privacy* can be viewed or defined in any number of ways. *Privacy* may be regarded as an entitlement of a person to determine what personal information about oneself may be communicated to others, or *privacy* may be defined in relation to control over access to information about oneself and the intimacies of personal identity. Another definition of *privacy* identifies it with a state or condition of limited access to a person, the intimacies of life, one's thoughts, or one's body. In each of these

Figure 1-1.
On an ordinary Saturday afternoon in early winter, thousands of spectators are casually watching a basketball game inside a large college dome stadium. Unbeknownst to them, surveillance cameras may have been installed. (Michael Poirier)

instances, control over information about oneself is relative, as is control over one's bodily security and personal property. The more we venture into the public realm to pursue interests with others in our lives, the more we inevitably risk privacy invasion and perhaps identity theft. It stands to reason that if we want medical doctors to cure us, or if we choose to conduct our private business over the Internet, then we must be willing to relinquish what might be considered private or personal information about ourselves.

Without question, as these ideas illustrate, we are living in an age of cyberspace where computerized information storage and retrieval systems have become the rule, rather than the exception. Many of us have actually become quite accustomed to the idea of record keeping redundancy and various other forms of technological surveillance. All kinds of personal statistics, credit references, academic records, annual earnings figures, health reports, and even grocery store purchases are stored electronically by a number of agencies around the world. These bits and bytes of what was once considered to be private information are transmitted back and forth between banks, universities, government offices, and detective agencies in a matter of a few seconds.

For a while, it seemed as if the grand advancements in communications technology and electronic database management would result in what literary observers called a "paperless society." To date, not many people seem willing to completely relinquish their right to have a hard copy of any document being reviewed. In fact, the advent of computerized information systems, coupled with the ever present electrostatic copier, has led to a society more full of paper, as opposed to one that uses less of it. The advent of electronic mail (e-mail) has made its mark here as well; many people routinely print their e-mail messages along with an array of e-mail attachments, in order to read them at a more convenient location away from the computer screen (such as while riding the subway home from work or while sitting in a lounge chair at the beach).

More often than one might expect, most people insist on having multiple copies of everything they write or transcribe for someone else. These oversimplified examples of contemporary daily living illustrate the point that technology pervades every business transaction and every individual's life. For this reason, it is no longer possible, if indeed it ever was, to be casual about science and technology. See Figure 1-2. Large and small, banal and impressive, scientific innovations and technological improvements pour out at an increasingly accelerated rate of speed. A great portion of such technological change, modification, and improvement is hardly even noticed these days. Many of us have become numb to novelty unless it is formidable, majestic, or entertaining in its outward appearance.

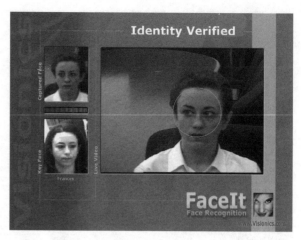

Figure 1-2.
FaceIt computer software allows police officers to search for the faces of wanted criminals in a crowd.

The types of breakthroughs receiving the greatest amount of media attention and the subsequent interest of the public seem to take place most often under the guise of exotic new high-tech industries. Among these disciplines one might list biotechnology, computers, robotics, artificial intelligence, lunar mining, fiber optics, pharmaceuticals, supersonic transport, lasers, glass ceramics, satellites, and space shuttles. Behind all the glitter of high-tech, we find a gamut of new, but somewhat unremarkable, technologies that cut across a wide range of businesses, including sporting goods, convenience foods, household appliances, tools and gadgets, clothing fabrics, children's toys (often also enjoyed by adults), video games, soft drinks, music, and art. Contemporary scientific and technological advances range from tiny improvements in devices that we take for granted (such as coated frying pans or the shape of the buttons on the remote control devices linked to an entertainment system), to heroic leaps forward that have far-reaching impacts on the status of life on the planet (for example, nuclear weapons, space stations, the Human Genome Project, artificial organs, and global communication networks).

Who has the authority to say any one of these changes is any more or any less important than any other? That is one of the most rhetorical questions of the era. Technologies one social group favors are simultaneously abhorred by other groups. Some people's lives are hanging in the balance of whether or not a medical breakthrough will be made before it is too late. Others believe too much money is already being spent on medical research only benefiting those who can afford it. Some people profess to be antitechnologists, while others are avid supporters of technology. Still more people find themselves somewhere along this continuum of extremes.

Key Concepts

As you review the material contained in this chapter, you should learn the meanings of the following terms.

science
technology

The Nature of Science and Technology

Over the past several decades, academic leaders around the world have developed a well-regarded discipline given the label "Science, Technology & Society" (STS). Through curricular proposals, scholarly papers, annual conferences, and hundreds of articles and books, these professionals have garnered a great deal of respect for an area of study that is, in every sense of the word, *interdisciplinary.* Its principles, theories,

and objectives pervade the many circles of physical scientists, life scientists, humanists, business analysts, educational specialists, engineers, technology educators, and industrial technologists. Across the nation, schools of engineering have started to develop required courses seeking to achieve a balance between technical, social, and ethical issues throughout the design process. These courses help ground the education of university science, technology, and engineering students, with the ideals of social and ethical responsibility as an integral part of their intended professions. Through interdisciplinary projects, students are taught the relevance of social, environmental, ethical, cultural, and political factors to their work. One of the aims of STS research has been to investigate the relationships (both historical and contemporary) between science and technology, in an attempt to reveal how the developments in both fields influence the nature of social institutions. STS researchers encourage us to think about the social, political, and environmental influences on and consequences of science and technology.

Can Technology Be Defined?

The word *define* is a variant of the Latin term *definire,* which translates to mean "to set bounds to." In order to define a word or concept, one must therefore be able to limit its meaning through the establishment of boundaries. A definition not only explains the true or accepted meaning of a term, but it also delineates what is *not* true about it. In this day and age, *technology* is one of those words (like *engineering*) that has been used, overused, and misused to the extent that it almost defies definition. If one has difficulty explaining what technology is *not*, then it must be everything else. By definition, a definition should *exclude* something, since it cannot include everything.

In many cases, the word *technology* has become so diluted that a secondary modifier is demanded for further clarification. Thus, we have expressions such as *medical technology, military technology, household technology, manufacturing technology, biotechnology, information technology, computer technology, agricultural technology,* and *industrial technology.* The list continues as a greater number of specialized permutations become evident. From this perspective, technology seems to make reference to the procedures and artifacts that are aligned with a particular discipline. On the other hand, the degree to which the research foundations for each discipline are interrelated makes it hard to say exactly where an exclusive product of technology fits. For example, is a solar heated grain elevator part of agricultural technology, or does it belong under the energy technology heading or even in the manufacturing technology category? Modifiers enable us to focus on a certain discipline, but they do not really define the term *technology* very effectively.

It has also become fashionable to abbreviate the term *technology* to its shortened *tech* version. There is *high-tech, low-tech,* and *no-tech.* We commonly use the prefix *techno-* to refer to a host of social conditions such as technophobia, technostress, technocrat, technotoys, technoworld, technodummies, technowizards, and technolifestyle. Once again, since we have become conditioned to these modern day expressions, they are implicit in their use, and we say these expressions without giving them a second thought. Each of these terms has something to do with either the *level* of technological development or society's *behavioral* or *psychological response* to technology.

Perhaps it is easier to explain the meaning of technology by generating a list of its profile characteristics. As you review the following inventory, place a check mark next to any features you honestly believe are *not* true about this concept.

❏ Technology creates new economic opportunities and social benefits and, at the same time, produces new social problems.

❏ Technology is a powerful force that improves human productivity.

❏ Technology is an extension of human physiological capabilities and biological potential.

❏ Technology involves inventing new things and modifying the old ones to make them more efficient.

❏ Technology is evident in every culture, regardless of its level of sophistication or stage of development.

❏ Technology enables humans to exert control over the natural environment.

❏ Technology liberates us from demeaning and demanding labor and, therefore, creates more leisure.

❏ Technology has increased the human life span by conquering many debilitating diseases.

❏ Technology incorporates human knowledge into physical hardware that will eventually respond to some human need or desire.

❏ Technology is intrusive in our lives and threatens our right to privacy.

❏ Technology is a process for transforming raw materials into useful goods and services.

❏ Technology is human made and desire-driven.

❏ Technology creates uselessness by displacing people and trivializing their work.

❏ Technology has made many people apprehensive about the future.

❏ Technology is fundamental to the survival of civilization.

❏ Technology is destructive to nature.

❏ Technology is making the world increasingly incomprehensible.

❏ Technology is future-oriented and, therefore, progressive.

❏ Technology is motivated by a pragmatic or instrumental interest.

❏ Technology utilizes the methods, tools, and skills typically characteristic of the process we call *innovation*.

Now, if you made a number of check marks in the margin as you scanned through this very partial list of modern technology's features, go back to those items and read them over a second time. While there is definitely a thread of deception to each of these ideas, convincing examples can also be cited to validate every one of them. Consider for a moment the FaceIt software technology described in the opening *Dateline* story, as aligned with the first item in the list above. Surveillance cameras might be viewed as an invasion of individual privacy (social problem), but this technology has the potential to also be used to locate missing children (social benefit).

Technology is not a simple word to define. It is used over and over again in this book within a whole series of contextual arrangements. The following simple definition for *technology* is not any more correct or useful than any other, but it can be used to establish boundaries.

Is Science Different from Technology?

The terms *science* and *technology* are so routinely combined to form a single phrase that some people find it nearly impossible to distinguish between the two. What is scientific is also perceived to be technological,

and vice versa. It was not always this way. For centuries, people regarded science and technology as distinct fields of study, entirely separate from one another. *Science* was understood to be the pursuit of knowledge that would enhance human understanding of the natural environment. *Technology* simply entailed making and inventing things humans could use to control or cope with the natural environment. Early scientists were not often concerned with practical applications of the knowledge they secured, and early technologists did not have an interest in understanding the reasons things worked the way they did.

> **TECHNOLOGY**
> The cumulative sum of human means developed in response to society's needs or desires to systematically solve problems.

Therefore, one of the early distinctions between science and technology made reference to the world of "know-what" and "know-why" on the one hand, and the world of "know-how" on the other. Throughout the twentieth century and into the twenty-first century, the line between scientific inquiry and technological application has become quite obscure. In fact, one of the chief characteristics of contemporary technology is it is based more on the scientific knowledge of the nature of things— why materials react in certain ways and which laws are in force at any given point in the process of development. Some excellent examples of science-based technology are found in atomic energy, fiber optics, biotechnology, antibiotics, and microengineering. The traditional language clearly distinguishing science from technology began to break down several decades ago and has nearly vanished.

In the mid-1990s, shortly after the author Michael Shermer founded the

Skeptics Society and published his first issue of *Skeptic* magazine, we were introduced to another novel concept—something he labels *pseudoscience*. In much of today's media, numerous claims are presented so they appear scientific, but they lack supporting evidence and plausibility. These claims are what pseudoscience is made of (such as hypnosis). It is often difficult for many of us to recognize the point at which valid science leaves and pseudoscience begins. To some degree, this level of confusion underlies the negative reactions or skepticism some people have toward contemporary scientific breakthroughs.

The authors maintain that contemporary scientists are still motivated by curiosity and the desire to improve social well-being, but they also recognize the need for the results of their research to be industrially relevant. In this regard, a recent strategy within the scientific community has been an attempt to establish that a close link exists, via technology, between pure science and its application to economic ends (entrepreneurial science). Thus, science is what is, while technology is what can be, as a result of science. Scientists have been anxious to demonstrate that their apparently curiosity-oriented work would, ultimately, lead to critical profit making technological innovations and business ventures. This book reveals numerous instances in which this strategy has been quite successful and illustrates cases in which technology is actually used to "legitimize" science. In sum, it is difficult to separate science from technology. To simplify the incomprehension, the following explanation for the term *science* is proposed.

> **SCIENCE**
> A stream of human events involving a mathematical or systematic study of nature resulting in a body of knowledge that is practical, as well as theoretical.

Overview of the Text

The following twelve chapters are organized under three subheadings. Section I, titled "Innovations," contains five chapters discussing and explaining recent developments in the following technical fields: biotechnology, information technology, space exploration, medicine, and manufacturing. These areas have been selected to illustrate the dynamic interaction between seemingly disparate fields of specialization.

Section II, "Issues," includes four chapters focusing on the dilemmas and controversies commonly associated with technological change. The subject areas receiving a significant amount of public attention during recent years are presented here: technology transfer, appropriate technology, environmental protection and degradation, and defense strategy.

The final segment of the text is Section III, "Perspectives," and it contains three chapters addressing several diverse sociological viewpoints regarding the effective utilization and management of technology and the projected stability of our traditional social institutions.

The format for each chapter is consistent throughout the book. As you found with this first chapter, each chapter opens with a *Dateline* incident that is, in some way, pertinent to the subject matter covered. These brief vignettes are designed to illustrate that technology marches on throughout the world at all levels, without the least regard for time. They are meant to show you that a simple glance through today's newspaper or last week's national magazine will reveal some event (major or minor) in technology that is already or will soon be shaping or reshaping the way we live each day. Be alert as you read through each chapter, and you will note the authors make reference to the *Dateline* story in order to tie it into the material being discussed.

Scientific and technological facts and figures are presented in a general way in this book. Highly respected publications, entire series, and comprehensive CD-ROM databases have been written or developed on every topic contained here. This fact alone makes it a challenge to provide a brief overview making any sense. Informed readers will unquestionably react to our method of content coverage by saying something like "the authors left this out," "the authors forgot to mention that important historical fact," or "the authors don't provide enough examples to back up their ideas." For this very reason, a lengthy reference list of suggested readings for further study is provided at the end of each chapter. Your ongoing perusal of technical subject matter, either directly related to your career objectives or indirectly supportive of them, is strongly recommended.

Several publications to which you might want to consider opening a subscription in the future include *Technology Review, Computers and Society, Invention & Technology, The Wall Street Journal, American Scientist, Science, The Futurist, The Technology Teacher, Skeptic, Science Digest, Scientific American, Harvard Business Review, Fortune,* and *IEEE Technology and Society Magazine.* While this list is not exhaustive of the titles available, it is a representative sampling that

is highly readable and quite informative. In addition to these printed media, we can strongly recommend you listen to numerous programs broadcast on National Public Radio (NPR), a few of which include *Morning Edition, Public Interest, Talk of the Nation, Science Friday,* and *All Things Considered.*

The discussion questions found at the end of each chapter (with the exception of this first chapter) are designed to stimulate further creative thinking about the innovations, issues, and perspectives presented under each of the major subject headings. We encourage you to work through these discussion questions, which are essentially inquiry-based and tied to authentic learning experiences. They may be completed individually, but we recommend using cooperative study groups and seminar discussions that will facilitate most effective interdisciplinary learning and dialogue. Since it is impossible for any one author or solitary instructor to cover all the bases or investigate all the angles, you must take some personal responsibility for the learning endeavor.

Toward this end, it is essential to recognize that the development of a technological vocabulary goes far beyond a rote memorization of terms and their meanings. If the formal study or informal activity for which you are using this book teaches you anything, it should teach you how to think, learn, and search out new sources of information. No single reference is ever completely infallible. A technological vocabulary changes and expands on a regular basis. In just a few years, a whole new set of terms will be "on the books," but not included on the lists of Key Concepts found at the beginning of each of our chapters. You will see a written segment at the beginning of each chapter where a number of key concepts are introduced. Collectively, throughout the textbook, they provide the foundation for a contemporary technological vocabulary.

The list of technology terms you will have studied when you finish reading our textbook is by no means exhaustive. Furthermore, between the time of this writing and your reading of the material, any number of new terms will have appeared in all technology-related disciplines. As you read through this textbook, please make an effort to expand your own technology vocabulary—endeavor to learn new words and avoid simply "glossing over" key terms in your reading. We hope our inclusion of definitions of scientific and technological terms both in chapters where they are discussed and in a comprehensive glossary at the end of the textbook will be helpful as you learn about contemporary topics and issues in technology. Moreover, it is our expectation that our methods of introducing you to an array of science and technology innovations, issues, and perspectives will enhance and improve your personal level of technological literacy.

Suggested Readings for Further Study

Alcorn, P. A. 2000. *Social issues in technology.* 3d ed. Upper Saddle River, N.J.: Prentice Hall.

Allen, T. J. 1977. *Managing the flow of technology: Technology transfer and the dissemination of technological information with the R & D organization.* Cambridge, Mass.: Massachusetts Institute of Technology.

Barrett, D. 2001. In dreams begin technologies: How innovations can come during sleep. *Invention & Technology* 17 (2): 30–34.

Barry, John A. 1991. *Technobabble.* Cambridge, Mass.: MIT Press.

Brod, Craig. 1984. *Technostress: The human cost of the computer revolution.* Reading, Mass.: Addison-Wesley.

Brosan, M. J. 1998. The impact of psychological gender, gender-related perceptions, significant others, and the introduction of technology upon computer anxiety in students. *Journal of Educational Computing Research* 18 (1): 63–78.

Etzkowitz, H. 2001. The second academic revolution and the rise of entrepreneurial science. *IEEE Technology and Society Magazine* 20 (2): 18–29.

Florman, S. C. 1996. *The introspective engineer.* New York: St. Martins.

Hallinan, K., M. Daniels, and S. Safferman. 2001. Balancing technical and social issues: A new first-year design course. *IEEE Technology & Society Magazine* 20 (1): 4–14.

Hayes, D. 1989. *Behind the silicon curtain: The seductions of work in a lonely era.* Boston: South End.

Hazen, R. M. 1997. The great unknown. *Technology Review* 100 (8): 38–45.

Norman, D. A. 1990. *The design of everyday things.* New York: Doubleday.

Pacey, A. 1983. *The culture of technology.* Cambridge, Mass.: MIT Press.

Rheingold, H. and H. Levine. 1982. *Talking tech: A conversational guide to science and technology.* New York: William Morrow.

Rosen, J. 2000. *The unwanted gaze: The destruction of privacy in America.* New York: Random House.

Shermer, M. 2001. *The borderlands of science: Where sense meets nonsense.* New York: Oxford Univ. Press.

Teich, Albert H., ed. 1993. *Technology and the future.* 6th ed. New York: St. Martins.

Toffler, A. 1980. *The Third Wave.* New York: Bantam.

Walters, G. J. 2001. Privacy and security: An ethical analysis. *Computers and Society* 31 (2): 8–23.

Wenk, E., Jr. 1986. *Tradeoffs: Imperatives of choice in a high-tech world.* Baltimore: John Hopkins Univ. Press.

Westrum, R. 1991. *Technologies and society: The shaping of people and things.* Belmont, Calif.: Wadsworth.

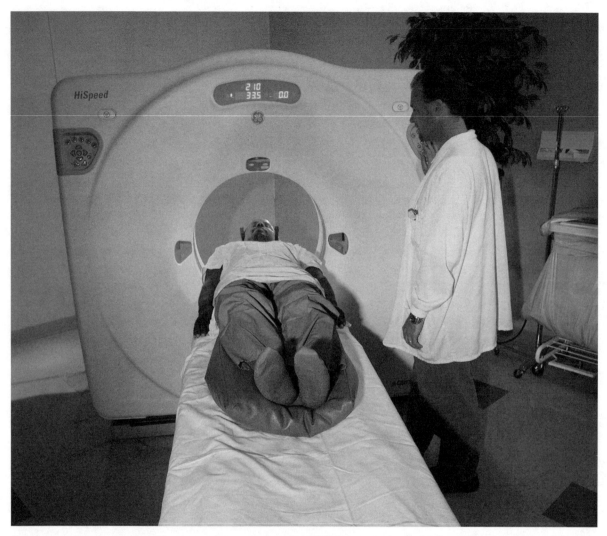

Science and technology affect every aspect of society. With increasingly high-tech equipment, doctors are better able to understand the human body. Advancements in digital imaging allow health professionals to view more complete images of the body, thus ensuring better health care. (NASA)

Chapter 2

Biotechnology

Reykjavik, Iceland

Iceland has become the first country in the world to sell the rights to its citizens' genetic codes to a biotechnology company. The data gained from Iceland, because of the uniformity of its people—a largely blue-eyed, blond hair populace—is expected to provide an invaluable resource for studying human genetics, leading to fundamental insights into many diseases, proponents say. Additionally, because Iceland has a strong health care system, scientists can correlate diseases with environmental and social causes.

Not all the scientists in Iceland support this plan, however. Some fear the database could make private details of medical histories available. Also, patients might be less willing to divulge information to their doctors. There are fears that the anonymous nature of the data will be compromised when it begins to be analyzed by researchers.

The supporters of this plan point to the goal—to improve delivery of health services. "On balance, I think the potential advantages will outweigh the risks involved," noted Solveig Petursdottir, a member of Iceland's Parliament....

Introduction

The practice of *biotechnology* has existed in science for thousands of years, but it has acquired a new definition and perception in the minds of the public over the last twenty years. The first use of the word *biotechnology* can be traced to the late 1910s, when it was used to refer to the industrial production of materials, such as acetone used to make cordite, an explosive.

The traditional uses of biotechnology were mostly in the manipulation of agricultural products through the process of artificial selection. Different strains of plants and animals were crossed in order to produce particular genetic traits. Many previously wild species of plants and animals were domesticated to suit the demands of the human population. Through this artificial selection, the genes of these species were irrevocably changed. Although we generally forget about these early uses of biotechnology,

the results can be seen in the diversity of animals today. For example, thousands of years ago, all dogs were wolves. Today, however, there are thousands of breeds of dogs. Many of these breeds developed as dogs were crossbred to maximize certain traits. This process of breeding and crossbreeding a species is the essence of traditional biotechnology.

The modern era of biotechnology began in the early 1950s, when James Watson and Frances Crick discovered the structure of DNA. The actual birth of the biotechnology industry is generally marked in the mid-1970s. Modern achievements of biotechnology include the transfer of specific genes from one organism to another, human stem cell research, cloning, and the fusing of different types of cells to produce medical products, such as monoclonal antibodies (MAbs). *Biotechnology* has become accepted as the umbrella term that encompasses all of these techniques. We define *biotechnology* as follows:

> ### BIOTECHNOLOGY
> The manipulation of biological organisms to make products benefiting human beings.

Biotechnology has triggered a great deal of public scrutiny and interest over the years. Consequently, it represents many things to a variety of professional disciplines. Molecular biologists view it as a powerful scientific tool, while entrepreneurs and venture capitalists recognize its versatility as a novel and potentially lucrative technology. For agronomists (scientists who specialize in field crop production and soil management), biotechnology holds the promise of another "Green Revolution," like the one of the 1960s, in which there were dramatic increases in the production of cereals and grains in developing countries because of improved seeds and production techniques. For medical scientists, biotechnology harkens a future that will bring cures to dreadful genetic diseases. Of course, there are numerous critics of biotechnology who focus on the potential for its abuse in the environment, among people, and in society as a whole.

The widespread use of biotechnology today promises abundance for humanity, but simultaneously brings with it a series of mysterious potential risks. Biotechnology has an additional difficulty for laypeople in that the technologies involved are difficult for nonscientists to understand. It is always easier to fear something one does not understand. As biotechnology affects us more in our daily lives, it will be subject to more controversy and discussion. Genetically engineered microorganisms in any application present an array of complex problems for the environmental ecosystem and the regulatory agencies commissioned to protect it.

The commercialization of biotechnology started in the mid-1970s, when a number of small entrepreneurial firms were founded in the United States. These companies were specifically formed to expand the ever-growing body of fundamental knowledge in molecular biology and exploit it toward a profitable end. The market for biotechnologically engineered products has grown exponentially and continues to do so. Large, well established corporations in the United States, Japan, and Europe have expanded their research and development (R & D) programs to investigate new techniques in biotechnology. Along with this expansion of research has come a renewed focus on the patenting of biotechnological products, including those derived from experimentation with human cells. It appears that a struggle is being waged to dominate a fledgling market whose lifeblood is found in basic scientific research.

Biotechnology is different than many other technologies. Its growth and development have occurred under a climate of intense public scrutiny. This scrutiny has brought increasing pressure to biotechnology companies to justify their product developments and reduce the risks to people and the environment. This chapter describes a variety of biotechnology uses today. The broad spectrum of disciplines involved in biotechnology research is presented with an eye for both the associated promises and perils. The following questions will be addressed in this chapter:

❑ What is the science behind the biotechnology movement?

❑ What types of commercial breakthroughs will be made in the health care field and in gene therapy?

❑ Will biotechnology use and agriculture lead to another "Green Revolution"?

❑ What are the ethical and business aspects related to cloning?

❑ Why has the question of regulation in the biotechnology industry been so complicated to address?

Key Concepts

As you review the material presented in this chapter, you should be able to explain what is meant by each of the following terms and phrases:

    biomining
    biotechnology
    blastocyst
    cell fusion
    clone
    DNA chip
    Escherichia coli (E. coli)
    gene therapy
    genome
    genetically modified (GM) food
    monoclonal antibody (MAb)
    pharming
    recombinant DNA (rDNA)
    transgenic
    xenotransplantation

The Critical Technologies

Politically, as well as economically, the United States is well suited to the continuing and extensive development of the biotechnology industry. Generally speaking, the laws and policies of our government have enabled industrialists and scientists to capitalize rapidly on biotechnology and apply it to many areas. Much of the basic research on biotechnology has been conducted within the university system and in government laboratories. In the United States, there is a symbiotic relationship between R & D efforts. It is usual, even expected, that a researcher will commercialize any discovery. This entrepreneurial spirit and the availability of financing for risky ventures have led the country to the forefront in the commercialization of biotechnology. Additionally, the U.S. industrial sector has exemplified its diversity and flexibility, with reference to making efficient use of numerous laboratory achievements in biotechnology. All of these factors have facilitated the rapid commercialization of biotechnology in the United States.

What Is Deoxyribonucleic Acid (DNA)?

To understand biotechnology, a basic knowledge of deoxyribonucleic acid (DNA) is indispensable. The actual structure of the DNA molecule itself is identical among all living things. See Figure 2-1. It appears as a three-dimensional twisted ladder, with two strands twisted into a long spiral—a double helix. The steps of the ladder are pairs of bases: adenine (A) with thymine (T) and guanine (G) with cytosine (C). In the process of reproduction, the two strands of DNA unwind and serve as the foundation for generating a new strand. This process results in the duplication of DNA. The structure of the DNA determines what the organism will be.

The set of instructions for making an organism is called its *genome.* The genome contains the master blueprint for the organism and consists of DNA and the associated protein molecules, organized into structures called chromosomes. You may recall from previous biology courses that the nucleus of each living cell hosts a number of chromosomes. Different species have different numbers of chromosomes: for example, humans have forty-six, corn has twenty, and goldfish have ninety-four. Strung along the chromosomes are the units of the genetic code, the genes that are composed of DNA. If all the DNA in a single human cell was uncoiled, it would stretch out over five feet (about two meters). In these five feet of human DNA, there are

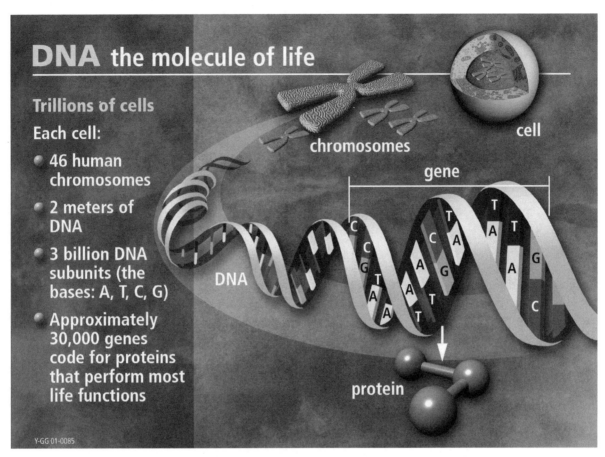

Figure 2-1.
This illustration shows the composition of DNA in a human being. (U.S. Department of Energy Human Genome Program)

some one hundred thousand genes, which vary in length, often extending over thousands of base pairs. Overall, the human genome consists of 3 billion units (the bases A, T, C, and G). In each DNA molecule there is the blueprint (genetic code) for bones, muscle, tissue, and blood. Every living being has its own unique genetic makeup. The DNA present in every cell of every living organism has the capacity to direct the functions of that cell.

Directly altering an organism by changing its genetic code requires that the specific gene (or set of instructions) along the DNA coil be modified and that this same set of instructions be changed in *every* cell of the organism. This task is not too difficult to accomplish in the contemporary research laboratory with single-cell organisms, such as bacteria, but remember that the human organism is one hundred trillion times as complicated, in terms of cell structure.

It is for the reason of simplicity that the most obvious candidate for the pioneer research in biotechnology was a single-cell bacterium. The organism selected by scientists in the early 1970s was a bacterium commonly found in the human intestinal tract, *Escherichia coli* (**E. coli**). Besides chromosomes, the *E. coli* nucleus contains plasmids, which are small strands of DNA. A plasmid exists and reproduces as a circular molecule. The benchmark genetic manipulation research conducted by Herbert Boyer (University of California—San Francisco), Stanley Cohen (Stanford University), and their colleagues provided the information needed to engineer *E. coli*.

Recombinant DNA (rDNA)

Recombinant DNA (rDNA) consists of a DNA sequence artificially produced by joining pieces of DNA from different organisms. The rDNA experiment is considered successful if the desired traits are passed on to succeeding generations. This technique can also be used to develop useful microorganisms having the ability to degrade toxic wastes. New strains of agriculturally essential plants, such as wheat, corn, and rice, represent by-products of rDNA technology. Recombinant DNA technology can be used in a wide range of industrial enterprises to develop the following:

❑ Microorganisms that produce new products.

❑ More efficient methods of producing currently available products.

❑ Large quantities of products that are scarce in the natural environment.

The biological principle underlying rDNA technology is simple, but staggering. If genes from a higher organism are implanted in bacteria, the bacteria will follow the genetic instructions of the higher cell. An *E. coli* bacterium's cells divide and reproduce every twenty-five to forty minutes. They continue to do this even after the genes from another organism have been spliced into them.

Bacterial enzymes (called restriction enzymes) are used to transfer DNA from one organism to another. Restriction enzymes are used like scissors to cut the DNA at specific points so new DNA can be inserted. There are well over a hundred restriction enzymes used in biotechnology, and each restriction enzyme cuts a specific and different part of the DNA molecule. After the restriction enzyme cuts the DNA, the result is a set of double stranded DNA fragments with single stranded ends, called sticky ends. The bases of the sticky ends can then form a bond with the foreign DNA molecules. The result is an edited, or recombinant, DNA molecule. The rDNA molecule then takes on the characteristic of the inserted gene. The first product made by rDNA technology was human insulin in the early 1980s, produced by Eli Lilly & Co.

The diagram in Figure 2-2 illustrates the procedures used to clone the human protein interferon using rDNA technology. Human cells invaded by viruses naturally produce small quantities of interferon. This protein is effective in fighting many viral diseases, including some forms of cancer. The gene for interferon production is mapped using sophisticated laboratory equipment and removed from a human chromosome. It is next inserted into the *E. coli* bacterium, which acquires the capacity to make interferon.

If the biotechnical engineer's objectives are fulfilled, a new organism has been created. The transformed bacterium possesses mixed, or recombinant, DNA, in which human and bacterial genes are present side by side. The bacterium is cultured under carefully controlled conditions in order to allow the gene for producing human interferon to work in its new site with the *E. coli*. Once the bacterium has been correctly transformed, interferon synthesis can commence using huge vats or fermentation chambers. The rapid reproduction cycle of the bacterium speeds this process. Finally, substantial quantities of interferon are extracted, primarily for use in the health care field.

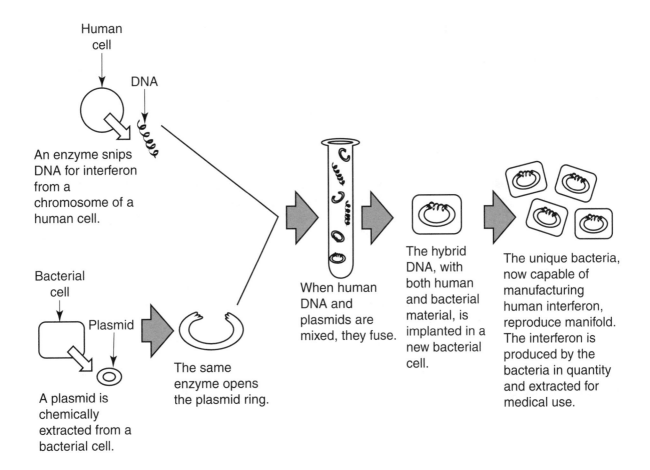

Human cell

DNA

An enzyme snips DNA for interferon from a chromosome of a human cell.

Bacterial cell

Plasmid

A plasmid is chemically extracted from a bacterial cell.

The same enzyme opens the plasmid ring.

When human DNA and plasmids are mixed, they fuse.

The hybrid DNA, with both human and bacterial material, is implanted in a new bacterial cell.

The unique bacteria, now capable of manufacturing human interferon, reproduce manifold. The interferon is produced by the bacteria in quantity and extracted for medical use.

Figure 2-2.
This diagram illustrates the procedures used to clone the human protein interferon, using rDNA technology.

Human interferon is so potent that the body produces it in minuscule amounts. For this reason, its use in the medical field was extremely limited until recently (in the past, it has reached costs of seventy to one hundred dollars per millionth of a gram!). During the late 1970s, a Finnish laboratory, supervised by Dr. Kari Cantell, processed ninety thousand pints of blood and produced only one-tenth of a gram of pure interferon. Subsequently, to attack a common cold, doctors treating one patient in a clinical study needed to use about two thousand dollars worth of interferon. Using rDNA, these newly engineered bacteria produce interferon as a metabolic by-product, and the same amount is acquired in the laboratory for about a nickel. Four different strains of lab grown bacteria and one strain of yeast are now used commercially to produce human insulin, human growth hormone, bovine growth hormone (used to increase the amount of milk a cow produces), and chymosin (a protein from calf stomachs used for cheesemaking).

Recombinant DNA is not the only form of biotechnology used to transmit DNA from one organism to another. In fact, using a variety of techniques, DNA has been transferred to bacterial, yeast, animal, and plant cells. In any transfer technique, the plasma membranes or cell walls must be penetrated without permanently damaging the cell. The removal of the cell wall allows the scientist to introduce foreign DNA directly into the host cell. When used with plants, scientists produce protoplasts (plant cells minus the cell wall) before introducing DNA through the plasma membranes. After the creation of the protoplast, the DNA can be introduced into the cell via infection with a bacterium; by microinjection of foreign DNA, using a fine needle; or by electroporation, using brief pulses of high voltage electricity to form holes in the membrane of the cell. Protoplast fusion is another technique used in biotechnology (with limited success), in which the external walls of two different types of cells are removed, allowing their specific genetic materials to combine and mix. This fusion causes a natural (but random) genetic recombination. Once the new DNA, consisting of both foreign DNA and the inserted gene, is inserted into the cell, the foreign DNA becomes part of the genetic makeup of the host cell. The host cell now contains the gene transmitted by insertion, and it will pass this gene to future generations.

Monoclonal Antibody (MAb) Technology

Monoclonal antibodies (MAbs) are produced by **cell fusion,** using cells capable of forming antibodies and myeloma (a tumor of the bone marrow) cells capable of sustained division and growth. When myeloma cells were fused with antibody-producing mammalian spleen cells, it was found that the resulting hybrid cells, or hybridomas, produced large numbers of MAbs. The term *monoclonal* indicates the offspring, or clones, of a single hybridoma cell produce them.

The production of antibodies in higher life-forms is one aspect of a complex series of metabolic processes referred to as the immune response system. This system is located in the spleen, lymph nodes, and blood of the organism. Specialized sets of cells, known as B-lymphocytes, are organized to detect and destroy foreign substances posing any sort of threat (for example, bacteria, viruses, and pollen molecules). As these foreign substances, called antigens, enter the body, the special cells churn out large numbers of complex molecules, called antibodies. The structure of these molecules enables them to lock on to particular antigens. More importantly, specific antibodies will only fit one kind of antigen. The foreign substances are thus blocked in their actions. The end result of this aspect of the immune response is the antigen's removal from the body.

Separate from their natural functions in the protection of life-forms via this immune response, antibodies have long been used in clinical research to identify particular molecules or cells. They also play a major role in the diagnosis of diseases. Antibodies that recognize known antigens are commonly used to detect the presence and level of drugs, hormones, bacterial elements, and viral products in sensitive blood samples.

Antibodies are actually large proteins that are specified genetically. Molecular biologists have devoted a great deal of research toward considering how the genes in antibody-producing cells are switched on and off to produce the correct molecule at the right moment. The conventional method to produce antibodies was to inject a laboratory animal with an antigen, allow a period of time for antibodies to form in the animal, and then collect the antibodies from the blood serum of the animal (antiserum). This method had problems. Since the antiserum was collected from animals, it contained undesirable animal extras. Also, this technique produced a relatively small amount of usable antiserum per animal. It was this type of problem that led to the development of the MAb technique.

In contrast to the conventional method, MAb technology allows scientists to produce large amounts of pure antibodies. A mouse is immunized with a foreign substance, or antigen, to simulate the production of antibodies. A few weeks later, the spleen of the mouse is removed, and the antibody-producing cells are isolated from the spleen. Fusing the antibody-producing cells with myeloma (tumorous) cells produces MAbs. Antibody-producing cells cannot survive on their own in artificial cultures. They must be fused with another type of cell and cultured in a medium. The resulting cell is called a hybridoma.

The hybridomas resulting from this procedure are cloned, and each clone is tested for the production of the antibody desired. These clones may be established either in an in vitro (outside a living body or in an artificial environment) culture system or injected into mice, where the hybridomas grow in abdominal cavity fluids. Either of these methods allows for the preparation of large quantities of very specific MAbs to counteract almost any available antigen. By allowing the hybridoma to multiply in a culture, it is possible to produce a population of cells, each of which produces identical antibody molecules. These antibodies are called MAbs because they are produced by the identical offspring of a single, cloned antibody-producing cell.

Although prohibitively expensive at first, the costs of producing MAbs have continually dropped. New production techniques under testing include using transgenic plants and animals to produce antibodies more efficiently. In the health area, more than eighty MAbs are now in clinical trials to obtain Food and Drug Administration (FDA) approval (most for cancer imaging or therapy).

MAbs have been used for many purposes, including the diagnosis and treatment of disease and the purification of proteins. In the late 1990s, Roche Laboratories was approved to market a MAb to help prevent acute kidney transplant rejection, Zenapax® immunosuppressant. This drug uses the body's own immune system to prevent the acute rejection that can make kidney transplants fail. Herceptin® pharmaceutical, a MAb manufactured by Genentech, has been used to treat metastatic breast cancer. Since the FDA approved it in 1998, over twenty-five thousand women have been treated with the Herceptin pharmaceutical, which can dramatically improve the effectiveness of chemotherapy against some advanced breast cancers. MAbs are not limited to medical uses. The Agricultural Research Service, a unit in the U.S. Department of Agriculture, designed a MAb protein to screen new varieties of tomatoes, potatoes, and eggplants to ensure they do not exceed safe levels of bitter tasting natural compounds called glycoalkaloids.

Polymerase Chain Reaction (PCR)

Polymerase chain reaction (PCR) is a technique for making many copies of a specific DNA sequence—we can think of this technique as photocopying DNA. In nature, most organisms duplicate their DNA in the same way. PCR mimics this duplication process in a test tube. There are three steps to the PCR process. The first step is to separate the two chains of DNA by heating. The second step is to cool the test tube and add a large quantity of the four bases (A, C, G, and T) and a primer to bind to the ends of the DNA strands. The primer is derived from the *Thermus aquaticus* bacterium, discovered in Yellowstone National Park. The final step is to make a complementary copy of the DNA strand by using a pair of short primer sequences matching the ends of the sequence to be copied. The three steps in the PCR reaction take less than two minutes, and since the cycle is repeated thirty or more times, 1 billion copies of DNA are produced in about three hours!

Following the discovery of the PCR techniques in the early 1980s, its use in molecular biology exploded. PCR has become a quick method for detecting mutations associated with genetic diseases. One diagnostic PCR test was developed to screen for adenovirus, one of the causes of the common cold. Adenovirus can lead to left ventricular dysfunction (LVD), which is a rare disease of the heart muscle. This disease often afflicts young athletes in their prime and has caused high school athletes to die suddenly while playing sports. PCR has made its greatest impact on the field of DNA fingerprinting for genealogical studies and forensics. PCR allows the forensic researcher to take a minute portion of blood or semen and grow enough copies of the DNA to analyze. The use of PCR technology has revolutionized the analysis of data from crime scenes. It is possible that by licking an envelope, you deposit enough DNA to trace your identity from the seal of the envelope. Police crime labs use DNA analyses to determine the likelihood that a suspect is guilty. Although DNA analysis cannot prove someone is guilty of a crime, it can indicate the odds of someone else having the same DNA markers. This type of evidence is being used more frequently in court cases.

Bioprocess Technology

Modern bioprocess technology is an extension of ancient techniques using natural biological processes for developing usable products. The production of beer is a classic case of bioprocess technology. In beer production, the combination of yeast cells and cereal grains undergoes a fermentation process to make the beverage. Modern bioprocess technology is not exactly a genetic technique, but it allows the biological methods of production to be adapted to large-scale industrial use. Bioprocesses are systems in which complete living cells or their components (such as enzymes and chloroplasts) are used to create desired physical or chemical changes. Examples of industrial products created by bioprocess technology include polymers, plastics, chemicals, fuels, lubricants, newsprint ink, pharmaceuticals, cosmetics, and construction materials. Advances made in bioprocess technology could ultimately determine the commercial success of selected industrial applications for both rDNA and MAb techniques.

Commercially oriented industrial biological syntheses are currently executed in single batches, allowing only a small quantity of the product to be recovered from large quantities of cellular components, nutrients, wastes, and water. In large-scale fermentation tanks, genetically engineered microbes are placed in a liquid solution of catalytic nutrients required by the microorganisms to propagate. One method used to separate the desired end product (the protein interferon) from the fermentation solution entails running the mixture through membranes collecting the larger protein molecules and allowing the smaller molecules to pass through. Another technique uses bioreactors, where either the catalytic enzymes or the cells containing the rDNA genes are immobilized on a substrate. Improvements made in techniques for immobilizing cells or enzymes, as well as in the design of bioreactors, have helped to increase production and facilitate the recovery of many substances (such as proteins, hormones, and wastes). More research in this area is still needed, however. Techniques are also being developed that can assist in the design of more efficient bioreactors, sensors, and recovery systems.

Current applications of bioprocess technology include the following:

❑ Production of cell matter, such as baker's yeast and single-cell proteins.

❑ Production of cell components, such as enzymes and nucleic acids.

❑ Production of chemical products, including ethanol, lactic acid, and antibodies.

❑ Catalysis of single substrate conversions, such as glucose to fructose.

❑ Treatment of waste to allow recycling of garbage.

Present research in biotechnology covers a wide range of interests. Progress is continually being made in both solving the engineering problems unique to genetically modified (GM) organisms and identifying ways in which genetically altered organisms may be used to enhance the efficiency of certain bioprocesses.

The remainder of this chapter will consider applications of biotechnology in use today in different areas: agriculture, industry, and medicine. Also, we will cover international and regulatory issues related to the use of biotechnology today. It is evident from the amount of money and effort expended in this effort that the effects of biotechnology will only increase in the future. How these technologies will affect the quality of life is an open, and unresolved, question.

Biotechnology and Agriculture

Old style plant breeding was a slow and unreliable process. The biologist would cross plants with desirable traits, such as resistance to drought or pests, and hope these desirable traits would be passed to the offspring. However, as with human genetics, where two brown-eyed parents can have a blue-eyed child, not all characteristics are readily apparent. Now, with modern techniques in biotechnology, scientists can identify traits by their genes and move those genes (and the traits) to create new and improved organisms.

Genetically Engineered Crops

The first two products of rDNA technology in agriculture were not actual food products; rather, they assisted in the production of food. In 1990, Pfizer introduced an enzyme (Chy-Max® chymosin) that was an artificial substitute for natural chymosin, an enzyme used by cheesemakers to convert milk into curds and whey when making cheese. The traditional source of chymosin was found in the stomachs of slaughtered suckling calves. The use of the rDNA Chy-Max chymosin allowed cheesemakers to have a consistent supply of this enzyme. It also had another unexpected effect. Since traditional cheese was made with a product from a calf's stomach, certain vegetarians and kosher Jews would not eat cheese. The new cheese could now be eaten, making the vegetarians, and certainly the cheesemakers, happy.

There was little controversy with the chymosin enzyme. That was not the case with the second rDNA food product—recombinant bovine somatotropin (rBST). BST occurs naturally in cows' pituitary glands and is related to the control of milk production. It was well-known for many years that injections of BST into cows increased milk production. In the early 1980s, Monsanto developed an rDNA-derived version of BST (called rBST), and the company received approval in the early 1990s to market rBST to dairy farmers. Recombinant BST can increase milk production by 20 percent when farmers inject it into a cow, starting at sixty days into the milking period.

The road to FDA approval for rBST was a long one and one fraught with controversy still existing today. To approve food additives, the FDA uses a standard called Generally Regarded as Safe (GRAS). Since rBST mimics a naturally occurring protein (except for the addition of one amino acid at the end), it easily met the GRAS standard. What was unexpected was the widespread opposition to this drug. During the late 1980s, there was extensive criticism of rBST and the FDA approval process from dairy-producing states, Congress, and the general public. Some critics were concerned that a drug was being added to the milk supply at all, while other critics based their concerns on economic reasons—rBST would further reduce the number of small dairy farms. Wisconsin even banned the use of rBST for one year in 1990.

After rBST was in production, additional criticism arose over how ***genetically modified (GM) foods*** should be labeled. As we will see in this chapter, this criticism of biotechnology occurs again and again as new biotechnologies are developed and applied by companies. For rBST milk, the FDA concluded that since rBST did not change the composition of milk, no extra labeling was required. Dairies can, however, indicate their milk comes from non-rBST cows by adding a label.

After the debates over rBST, there was considerably more scrutiny by the public and food activists about the next developments in GM foods. Interestingly enough, the next GM food was an actual food product—the Calgene Flavr Savr® tomato.

The Flavr Savr tomato was genetically engineered by Calgene to allow the tomato to remain on the vine longer to ripen before harvesting. Most other tomatoes are harvested while green and firm so they are not crushed when shipped. After shipment, the green tomatoes are treated with ethylene gas to ripen them. The Flavr Savr tomato reduces the level of a ripening enzyme, polygalacturonase (PG), that normally causes tomatoes to soften as they ripen. Since Flavr Savr tomatoes have lower levels of PG than normal tomatoes, they can remain on the vine longer before they are picked, and they will be firmer when sold in supermarkets.

As the first GM food product sold, the Flavr Savr tomato received extensive scrutiny by the FDA. In the early 1990s, the FDA instituted a new policy requiring that biotechnology-derived foods meet the same safety standards required of all other foods. This was a critical juncture for the biotechnology companies. By its ruling, the FDA established a precedent that GM foods were generally going to be evaluated the same as normal foods. Therefore, if the new GM food is significantly the same as the normal food, the FDA does not conduct a comprehensive scientific review of the GM food. Instead, the FDA requests that the company provide a summary of the GM food safety and nutritional distribution to the FDA and discuss its results with the FDA's scientists prior to commercial distribution. The FDA calls this a consultation process rather than a review process. By any name, the effect was to make it easier for U.S. companies to market GM foods in the United States.

The consultation process includes a safety assessment of new GM foods and includes evaluations of the effects of the genetic modification, the source and function of the introduced genetic material, analytical studies to determine whether the genetic modification had any effects on the composition of the food, and the safety of new or modified substances in the food. Using these consultation criteria, the Flavr Savr tomato was reviewed by the FDA in the mid-1990s and found to be as safe as normal tomatoes.

Since then, there has been an explosion of GM foods. In fact, it is estimated that GM seed accounts for 40 to 60 percent of all food planted in the United States. In 1999, approximately 40 percent of corn, 50 percent of soybean, and 33 percent of cotton in the United States were genetically modified. A total of forty-six GM foods had finished the FDA consultation process by the middle of 2000. See Figure 2-3. In addition to foods like the Flavr Savr tomato, there are GM foods that include genes resistant to pesticides. Some of the most controversial GM foods are those whose seeds are resistant to Roundup® chemical herbicide (created by Monsanto). Since these plants (soybeans, cotton, corn, and canola) have the Roundup herbicide–resistant gene, farmers can spray their fields with the pesticide, killing all plants (including weeds) except for the Roundup herbicide–resistant plants. Also, this way, farmers can use no-till agriculture, which reduces soil erosion by 70 percent. However, some scientists worry this resistance to Roundup herbicide can spread to the weeds and using these seeds will lead to the creation of "super weeds."

Other biotechnology efforts have tinkered with the nutritional makeup of plants. A consortium of scientists from Germany and Switzerland created a new rice strain to improve the nutritional value of rice. Almost half of the world's population eats rice daily and is dependent upon this food for much of its caloric intake. Dependence on natural rice, however, can lead to a vitamin A deficiency that is already a serious health problem in at least twenty-six countries, including highly populated areas of Asia, Africa, and Latin America. Vitamin A deficiency, which can lead to vision impairment, affects around 70 percent of children under age five in Southeast Asia. The new rice, called Golden Rice®, has three extra genes that increase the synthesis of beta-carotene, which can be used by people to produce Vitamin A. Golden Rice, with its distinct yellow-orange hue, will be distributed free to government-run breeding centers and agriculture institutes in India, China, and other rice-dependent Asian nations. Local farmers in these areas will each be allowed to earn an annual ten thousand dollars, without paying royalties. The International Rice Research Institute in Los Baños, Philippines is evaluating the safety and use of Golden Rice. After this evaluation is completed, this GM food will be distributed to developing countries.

For hundreds of years, humans have been changing the biological makeup of their food. Now, with biotechnology, these changes are coming faster. A new generation of GM foods will include traits that will improve a plant's taste, size, or nutritional content. Overall, GM foods hold the promise of increased yields with less pesticide use and more nutritional value. They promise a new "Green Revolution" to double production and nutritional content of key crops, like corn, wheat, and rice.

Transgenic Animals and Plants

GM foods are considered to be *transgenic* if they contain genes outside their species. An example of a transgenic plant is Monsanto's NewLeaf® Plus potato. The NewLeaf Plus potatoes include genes that

| Foods Derived from rDNA Technology May, 2000 | |

U.S. Food and Drug Administration Center for Food Safety & Applied Nutrition Office of Premarket Approval

2000	
Aventis	Male sterile corn
1999	
Agritope Inc.	Modified fruit ripening cantaloupe
BASF AG	Phytaseed canola
Rhone-Poulenc Ag Company	Bromoxynil tolerant canola
1998	
AgrEvo, Inc.	Glufosinate tolerant soybean
	Glufosinate tolerant sugar beet
	Insect protected and glufosinate tolerant corn
	Male sterile or fertility restorer and glufosinate tolerant canola
Calgene Co.	Bromoxynil tolerant/insect protected cotton
	Insect protected tomato
Monsanto Co.	Glyphosate tolerant corn
	Insect and virus protected potato
	Insect and virus protected potato
Monsanto Co./Novartis	Glyphosate tolerant sugar beet
Pioneer Hi-Bred	Male sterile corn
University of Saskatchewan	Sulfonylurea tolerant flax
AgrEvo, Inc.	Glufosinate tolerant canola
Bejo Zaden BV	Male sterile radicchio rosso
Dekalb Genetics Corp.	Insect protected corn
DuPont	High oleic acid soybean
Seminis Vegetable Seeds	Virus resistant squash
University of Hawaii/Cornell University	Virus resistant papaya
1996	
Agritope Inc.	Modified fruit ripening tomato
Dekalb Genetics Corp.	Glufosinate tolerant corn
DuPont	Sufonylurea tolerant cotton
Monsanto Co.	Insect protected potato
	Insect protected corn
	Insect protected corn
	Glyphosate tolerant/insect protected corn
Northrup King Co.	Insect protected corn
Plant Genetic Systems NV	Male sterile and fertility restorer oilseed rape
	Male sterile corn
1995	
AgrEvo Inc.	Glufosinate tolerant canola
	Glufosinate tolerant corn
Calgene Inc.	Laurate canola
Ciba-Geigy Corp.	Insect protected corn
Monsanto Co.	Glyphosate tolerant cotton
	Glyphosate tolerant canola
	Insect protected cotton
1994	
Asgrow Seed Co.	Virus resistant squash
Calgene Inc.	Flavr Savr® tomato
	Bromoxynil tolerant cotton
DNA Plant Technology Corp.	Improved ripening tomato
Monsanto Co.	Glyphosate tolerant soybean
	Improved ripening tomato
	Insect protected potato
Zeneca Plant Science	Delayed softening tomato

Figure 2-3.
This table lists the foods derived from rDNA technology, as of 2000.

give the plants resistance to infection by the potato leaf roll virus and the Colorado potato beetle. The NewLeaf Plus potatoes contain the natural pesticide *Bt*. *Bt* is the common abbreviation for *Bacillus thuringiensis,* a group of naturally occurring soil bacteria, found everywhere on Earth. Results of the last few years of planting showed the NewLeaf Plus potato is less susceptible to infection by the potato leaf roll virus. Also, farmers use fewer pesticides with this transgenic food. Since the NewLeaf Plus potato includes a gene from another species, it was evaluated and approved by both the FDA and the U.S. Department of Agriculture (USDA).

Since the discovery of *Bt* (and its genetic transfer into food), most GM foods have included transgenic genes. By far, the most important trait introduced into foods is pesticide tolerance (led by Monsanto with its Roundup Ready® crops). See Figure 2-4. The next most inserted trait is for insect resistance. Papaya is a significant crop in Hawaii, but the yields have been dropping over the last ten years because of the papaya ring spot virus (PRSV), which is rapidly transmitted by a number of aphid

Figure 2-4.
This is a field trial in Dahnsdorf, Germany, with herbicide-resistant seed. The surrounding fields serve as a catch crop for transgenic pollen. (BBA/B. Hommel/ H. Baier/2001)

species. From 1993 to 1997, Hawaii's papaya production fell from 58 million pounds to 36 million pounds. Researchers began to investigate how to transform the papaya so it would resist PRSV. They decided to inoculate the papaya with a gene from the virus, with the hope that the new papaya would become resistant. They used rDNA techniques to isolate and clone a gene in the virus that then was injected into the cells of the papaya plant. This plant inoculation worked.

All is not green for transgenic foods. There is a small, but vocal, segment of consumers actively opposing all use of GM foods. There are three kinds of concerns. The first is that pollen from GM crops could contaminate organic crops. The second concern is that GM crops might negatively impact biodiversity. On this point, there is uncertainty even among scientists as to whether bioengineered traits can pass to other plant species. The concern is that, for example, insect-resistant *Bt* crops will crossbreed with weeds and other undesirable plants and cause an explosion of insect-resistant plants that, in turn, would cause more damage to the ecosystem. The third area of concern relates to terminator seeds— seeds sold by biotech companies that result in sterile seedless plants, thus requiring the farmer to buy new seed each year. To differentiate GM foods even more in the minds of the public, they are acquiring the negative label of "Frankenfoods."

In 1995, when the first transgenic seed was available, farmers planted 4.3 million acres of the new bioengineered varieties. By the late 1990s, over 70 million acres of crops were planted with transgenic seeds. In monetary terms, the global sales of transgenic crops grew from $75 million in 1995 to $2.2 billion in 1999, a thirty-fold increase. This trend may not continue, however. The results of surveys of American farmers taken in 2000 show that corn growers, at least, planned to reduce this planting of GM crops in 2000 by almost 25 percent. Why has this slowdown in GM food happened?

Most scientists claim that eating GM foods is harmless because they are not significantly different from traditional foods. There is, however, growing concern about the effects of GM foods on animals and humans. International public reaction to a recent study of monarch butterflies trumpets this concern. At Cornell University, researchers in a laboratory fed monarch butterfly larvae milkweed leaves dusted with pollen from *Bt*-modified corn. Almost half the larvae died. This news was used in some countries to ban the development of new GM foods. Recently, a new study completed in an actual cornfield, analyzing the effects of *Bt* corn pollen on black swallowtails, contradicted the famous monarch butterfly report, but the news contradicting the monarch butterfly report was not featured on CNN. Why? Many within the industry and government believe that opposition to GM foods stems mainly from a fear of the unknown, with most of the reasoning based on emotion, not fact. Coupled with the normal news tendency to highlight bad news over good news, it is easy to see how this story was overlooked. Whether any scientific data, no matter how credible, could sway the opponents to GM foods is certainly debatable.

Consumer response to GM foods in the United States, although significantly less negative than the response in Europe, has led some U.S. food companies to stop using GM products. Two large U.S. baby food companies, Gerber and H. J. Heinz, announced in the late 1990s they would switch to non-GM food ingredients in baby food. McDonald's asked its suppliers to stop sending GM potatoes, and Frito-Lay, the nation's largest snack producer, pledged to stop using GM corn in chips.

Current agricultural research will stretch the content of GM foods. In the pipeline are vegetables containing medicine to cure a range of medical problems. Imagine eating a banana or tomato to get vaccinated for hepatitis, skipping a pill and taking your morning's vitamins through your breakfast of eggs, or taking your contraceptive through a side dish of corn on the cob! All these GM foods are currently being developed, but fruits and vegetables are not the only GM foods being developed. Recombinant DNA techniques also are being used to transfer foreign genes to animals and to turn plants into manufacturing factories.

Why transfer foreign genes to animals? The simple answer is to produce drugs for human beings. Researchers realized they could use the normal functions of a goat, sheep, or cow to produce drugs during the normal production of milk. The milk can then be purified to obtain the pure drug, or the drug-enhanced milk can be used directly by the consumer. Although no transgenic animal product has been approved yet, there are a number of products in clinical trials.

Even more lucrative than GM foods are plants that could make materials for industrial products as they grow. Plant biologists at Monsanto are working on a plastic that could be grown inside the plant in the field. Although the Monsanto work is still in the development stage, there already are biotech poplar trees planted in Oregon. If the bioengineering works, these trees will grow to full height in just six years, ten times the global rate for poplars.

Despite the concerns and negative publicity, there is no doubt there will be an increase in transgenic crops and animals. According to the Biotechnology Industry Organization, the growth potential for GM foods is enormous and will lead to new developments in energy, food, and chemicals. The political climate for transgenic plants and animals is harder to predict.

International Concerns with Genetically Modified (GM) Foods

It is no secret that more of the food we eat contains genetically modified (GM) products, and there is a real opportunity for biotechnology to lead to a second "Green Revolution" and produce more food for a growing world. The expansion of GM foods seems unstoppable, at least in the United States, but the United States is not alone in producing GM crops, despite being, by far, the largest producer of GM food. In 1999, 72 percent of transgenic crops were planted in the United States, while Argentina had 17 percent, and Canada had 10 percent.

The production levels of GM foods are linked to their acceptance levels in these countries. In the United States, there is a high level of use of GM foods and a higher level of acceptance than in most of the world. Why Americans are more likely than Europeans to eat GM foods is not clear; however, surveys indicate that Americans have more trust in government and the regulatory process. So, when a product is approved for sale by a government agency such as the FDA, Americans generally believe it is safe. In contrast, since the mad cow disease scare in the mid-1990s, European consumers have been more distrustful of government experts. During that crisis, public health officials in the United Kingdom gave people false assurances about the safety of eating beef from diseased cows. Although mad cow disease has nothing to do with GM foods, the scare led to increased food anxiety. The memory of unsafe food is very fresh in the minds of European consumers.

Overall, European countries are generally opposed to the use of genetically engineered food—this opposition is the cause of a trade dispute between the United States and the European Union. Despite assurance from the FDA as to the safety of GM foods, many consumers in the European Union remain wary of the new technology, and this normal wariness was heightened by the widespread press coverage of the monarch butterfly experiment that seemed to justify all the fears of using biotechnology in food. In addition to banning GM crops, there have been many incidents, particularly in the United Kingdom, in which activists destroyed GM crop trials.

In the late 1990s, the European Union halted the authorization of new GM foods until it approved new regulation. Also, the European Union mandated labels on all foods containing GM ingredients. The de facto ban on the commercialization of genetically engineered plants in the European Union will be in effect for at least a couple of years. Bans of GM food are also being considered in Thailand, Ethiopia, and India.

GM foods are more warmly received in other parts of the world. China already uses one GM crop: pest resistant cotton developed by Monsanto. The Philippines is conducting field experiments on a new strain of rice resistant to bacteria and new *Bt* maize resistant to local pests.

The new Biodiversity Protocol, finalized in 2000, will affect the transport of GM foods across international boundaries. The protocol allows individual countries to decide whether or not they are willing to accept imports of GM foods, and more critically to allay consumer fears, all products containing GM organisms must be clearly labeled. Whether the labeling of GM foods will end this debate is difficult to see. There are significant economic benefits to companies developing new GM foods. Also, export of GM foods can enhance a country's world trade. The future of GM food is difficult to predict.

Industrial Uses of Biotechnology

When we think about biotechnology, we forget it is more than new food and medical products. The industrial uses of biotechnology extend much further than those isolated areas. New developments in biomaterials have led to products reducing their environmental impacts. Next time you are at your local grocery store, look at the plastic bag you carry to your car. Chances are, if the bag is recyclable, the bag was made from potatoes using new biomaterials.

Recent developments in genetic engineering have expanded the industrial role of biotechnology. In this section, we will discuss three different industrial applications of biotechnology: waste management, biomining, and DNA chips.

Waste Management

The first method of sewage treatment was to let the contaminated water flow along open ditches to clean it. Next came forced aeration—the insertion of air into wastewater to speed up its purification. Then came the use of biotechnology in waste treatment. Composting and wastewater treatment are some examples of older environmental biotechnologies. Today, the treatment of waste has moved beyond traditional cleaning techniques and uses biological microorganisms to remove pollutants from the environment. This biologically based treatment is called bioremediation.

The history of bioremediation starts more than three decades ago. During the 1970s, microbiologists developed anaerobic biodigesters to help clean waste. These biotechnology products ate the pollutants from the waste and allowed more thorough cleaning.

Most current biotechnology waste treatments are remedial; that is, they attempt to clean what is already dirty. These waste treatments use special microbes that selectively eliminate certain pollutants. Interestingly, the use of microbes for waste treatment happened by accident. During World War II, in addition to the massive loss of human life, there was tremendous environmental damage. In particular, because of the many sea battles, there was a large amount of oil spilled into the ocean. When scientists began to study the worst oil spill areas after the war, they discovered there were petroleum-eating microbes that had broken the oil down into harmless compounds. This discovery was soon applied to other waste treatment problems.

Bioremediation can be applied in one of two ways. Nutrients can be added to stimulate native microbes to clean the pollution. A famous example of this was the cleaning of the beaches along the Prince William Sound after the Exxon Valdez disaster that occurred in the 1990s. Workers noticed petroleum was removed more quickly from beaches that received an application of fertilizer than from beaches that were steam cleaned. The fertilizer stimulated the oil-eating microbes and quickened the removal of oil. See Figure 2-5. A second method has been used in the United States and Europe in the decontamination of toxic and hazardous waste sites. In this technique, samples are taken to a laboratory, where nutrients are added to enrich the selected microbes to remove the contamination. Then, the samples are returned to the waste site to clean them.

Bioremediation has been successfully used to remove DDT, mercury, kerosene, and oil, among other pollutants, from soil. In one use of bioremediation in Great Britain, in just twelve months, most of the oil and heavy metals pollutants at a former tar factory were eliminated after being polluted for more than one hundred years!

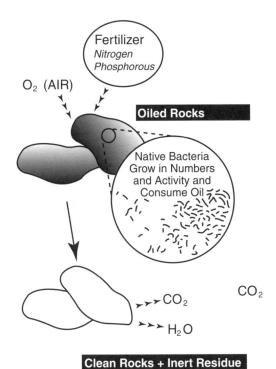

Figure 2-5.
Fertilizer acting to balance the amounts of nitrogen and phosphorus with the available carbon can enhance the natural biodegradation of oil. When present in optimal amounts, the three elements can be fully utilized by microorganisms to form carbon dioxide and water from oil.

Plants are also used to treat environmental problems. In this technique, plants act as filters for the contaminants by collecting and accumulating the waste in their roots and shoots. In another method, plants can also act to stimulate microbe cleaners in the soil by releasing enzymes into the soil. Transgenic technology can enhance the ability of plants to function as soil cleaners. Recently, this technique was used to clean methylmercury, a highly toxic chemical, from soil.

Biotechnology applications can even change the end products of waste. A North Carolina State University scientist has developed a process for changing animal waste into various products, including methane gas for fuel, liquid nutrients for aquaculture, and feed additives for poultry and livestock. When demonstrated in China and North Carolina, it was estimated that a farm with fifty thousand chickens could produce nearly 10 million BTUs of energy a day—enough energy to supply two hundred homes!

Biomining

The pursuit of precious metals for jewelry, coins, and other finery has been the trademark of man since the beginning of time. In tombs traced to ancient civilizations such as the Incas, Mayans, and Egyptians, we find many examples of gold and silver ornaments. Most of these metals were obtained by mining; obviously, the first metals mined were the easiest to get. The techniques used to extract minerals from the earth have not changed significantly over the past few centuries. Ore containing the mineral is dug up from the earth. Then, the ore is crushed, and the precious minerals are extracted either by extreme heat or by using toxic chemicals, such as cyanide.

Over the past few years, mining techniques have been changing, and this change can be attributed to the use of new *biomining* techniques. There are two main biomining techniques used today: bioleaching and biooxidation. Both of these techniques rely on the behavior of certain bacteria that, at very high acidity levels, can eat away any unwanted iron and minerals, leaving the valuable mineral (silver, copper, or gold) behind.

Bioleaching uses the natural behavior of certain bacteria to drain (leach) precious metals from raw ore. In recent years, it has been used to successfully recover a variety of metals, including copper, nickel, zinc, lead, cobalt, gold, and uranium. Bioleaching is used widely in the extraction of copper, with approximately 25 percent of all copper worldwide, worth more than $1 billion annually, being produced through bioprocessing. This technique is particularly successful with lower quality copper ore. After removal from the mine, the ore is treated with sulfuric acid. The acid stimulates the growth of *Thiobacillus ferooxidans* bacteria, which literally eat the ore. As the ore is eaten, the copper is released and collected in solution. To make the process even more environmentally sound, the sulfuric acid can be recycled.

Another technique used in biomining is bio-oxidation. Bio-oxidation is a type of priming technique used on gold ore imbedded in sulfur. Bacteria are added to the crushed ore to separate the gold from the sulfur. Then the gold can be extracted using a traditional cyanide process. Since the ore containing gold is pretreated, more gold can be extracted with less use of cyanide. The first full-scale bio-oxidation gold mine was the Youanmi mine in Australia. Now, there are at least six more bio-oxidation mines operating around the world.

Integral to this use of biotechnology are bacteria and microbes assisting in mineral processing. There is a need to find bacteria that can resist heat because the bioprocessing of ore releases a tremendous amount of heat. Biotech researchers are now searching for new bacteria under the ocean near volcanic vents and geysers. The goal is to find bacteria that can be used for biomining temperatures up to one thousand degrees Celsius. One mine, Gold Mines of Australia, is using high temperature (thermophilic) bacteria in their processes. The mine was able to process 98 percent of the available gold because of the efficiency of the thermophilic bacteria. This is compared to a gold recovery rate of only 40 to 60 percent with traditional cyanide processing.

It isn't only new bacteria that scientists are trying. A new technique in biomining involves using plants to separate gold from ore. Researchers at Massey University in New Zealand found that chicory and some mustard plants would absorb the gold from the ore. After the gold was accumulated, the plant is burned, and the miner collects the gold from the ashes. See Figure 2-6.

DNA Chips

The future is DNA. At least, that is what many companies hope. DNA's promise extends much beyond medical diagnosis and drug treatment. DNA chips can also screen for water quality, determine the identity of endangered species, and allow rapid identification of crime suspects in the field. A **DNA chip** is a wafer of glass (sometimes a wafer of nylon) onto which DNA strands are placed. The DNA is placed in distinct spots, and each spot contains a unique DNA

Figure 2-6.
Eric Mathur, Senior Director of Molecular Biodiversity for Diversa Corporation, collects a sample from a geothermal vent in Yellowstone National Park, Wyoming. (Diversa Corporation)

sequence. Hundreds of DNA sequences are assembled in parallel rows onto the DNA chip.

When the test sample, such as a drop of blood, is washed over the DNA chip, a change occurs in only some of the DNA sequences. A machine then reads the DNA pattern and converts the results. DNA chips are attractive because they can be specially designed for a specific purpose. One chip, manufactured by Affymetrix, allows fast and easy screening for a variety of contaminants in water, including bacteria, viruses, and parasites. Not only does the DNA chip reduce the testing time of the water from forty-nine to four hours, it is also ten times cheaper!

It is clear from even these three examples that industry has only begun to explore the possibilities from biotechnology research. The future holds the promise of more specialized products, "greener" and cleaner technologies, and cheaper medical diagnostics. If this promise is kept, the potential applications will be impossible to predict.

Biotechnology for Health

Genomics is the branch of biotechnology dealing with the structure and organization of DNA. Today, scientists have identified the DNA of many organisms, including human beings. Also, as human DNA is further defined and analyzed, the genes responsible for a number of inherited diseases, such as Huntington's disease and sickle cell anemia, will be identified. The largest initiative in genomics is the Human Genome Project (HGP).

The Human Genome Project (HGP)

The Human Genome Project (HGP) is an international research effort to map the DNA of human beings. In 2000, the then-President of the United States, Bill Clinton, announced the completion of the first draft of the survey of the entire human genome. This ten-year effort included scientists from the United States and several other countries. Decoding the human DNA is not the end of the HGP—it is really the beginning. In order to understand the importance of the HGP, we must look back to what motivated this project. In the 1980s, there was a widespread belief among doctors and scientists that understanding the human DNA would allow for better medical treatment and diagnosis.

The HGP is an international fifteen-year, $3 billion effort that formally began in 1990. The goal of the HGP is to discover all of the estimated one hundred thousand human genes and make the information about these genes accessible to researchers for further study. Another goal is to determine the complete sequence of the 3 billion DNA subunits (for an extended explanation of DNA, see the previous section in this chapter, "What Is DNA?").

The U.S. HGP, managed by the National Institutes of Health (NIH) and the Department of Energy (DOE), coordinates the work on the human genome done in the United States. The work is being completed at three DOE national laboratories: Lawrence Livermore, Lawrence Berkeley, and Los Alamos. Also, research is being conducted at numerous colleges, universities, and other laboratories throughout the United States. On average, the U.S. HGP funds about twenty separate researchers each year.

The United States is not alone in the HGP. There are at least eighteen other countries that have human genome research programs. Some of the larger programs are in Australia, Brazil, Canada, China, Denmark, France, Germany, Israel, Italy, Japan, Korea, Mexico, the Netherlands, Russia, Sweden, and the United Kingdom. The Human Genome Organisation (HUGO) helps to coordinate international collaboration in the genome project.

The original goals of the HGP were to identify and describe the complete sequence of the human genome; develop technologies for genome analysis; train scientists to use the tools of the HGP to improve human health; and examine the ethical, legal, and social implications of human genetics research. The last goal of the HGP is particularly important and represents a shift in traditional scientific funding. Ethical analysis is imbedded in the HGP, and this type of analysis has received extensive support and attention. In fact, the DOE and the NIH have set aside 3 to 5 percent of their annual budgets for this project to study ethical, legal, and moral issues.

In addition to studying the human genome, the HGP has completed the genome for four other organisms: the bacterium *E. coli,* the yeast *Saccharomyces cerevisiae,* the fruit fly *Drosophilia melanogaster,* and the roundworm *Caenorhabditis elegans.* Also, a new goal added in the late 1990s focuses on identifying individual variations in the human genome and seeing how these variations affect an individual's predisposition to disease, drugs, and environmental conditions.

The commercial potential for the information derived from the HGP is great. As more of the human genome is deciphered, there will be more specific genetic tests developed to identify individuals with genetically caused diseases. Also in the future is the development of specialized drugs with a sales projection exceeding $45 billion of all DNA-based products and technologies by the year 2009.

Genetic Testing

One of the likely by-products of the HGP will be an increased availability of genetic tests to screen for various diseases. Thus far, HGP research has led to the location of genes responsible for a wide range of human diseases, including cystic fibrosis, Huntington's disease, some forms of breast and ovarian cancers, and some forms of Alzheimer's disease. Consumers now have the option of taking a test to determine if they have a genetically related condition.

Genetic testing usually involves taking a person's cells (frequently from a blood sample) and examining the DNA. The genetic test will usually focus on a particular portion of the DNA in order to determine whether the person has the medical condition. However, a positive DNA test does not mean one will ever get the disease. In some cases, it just means the person has an increased risk of eventually getting the disease, and the risk varies depending on the disease. One breast cancer test determines whether a woman has a BRCA1 or BRCA2 mutation. A BRCA1 carrier has a 50 to 90 percent chance of developing breast cancer in her lifetime; however, the inherited BRCA1 mutated gene is only responsible for about 5 to 10 percent of breast cancers. So, a woman might find out that she does not have the BRCA1 mutation and still get cancer, or she might have the BRCA1 mutation and not get cancer.

A genetic test can give a false sense of security, and at the same time, it will become a part of a person's medical history. Once an individual is tested, the result is a genetic blueprint. If the person has children (and, for example, a genetically linked disease), this information can lead to finding genetic disorders for that person's children. This information can become part of their medical history, too. As the *Dateline* excerpt indicates, the people of Iceland are confronting these issues today. A company called Decode Genetics has studied the gene pool of 275 thousand Icelanders to find links between inherited illnesses and DNA. The company uses the medical records of the Icelanders as well as the country's genealogies, which have been recorded for more than one thousand years. Two other countries, Tonga and Estonia, recently have signed gene agreements similar to Iceland's. Researchers hope one of these populations could be a gold mine for DNA-related medical information. For the citizens, however, there is an increased risk of gene discrimination.

The confidentiality of test results is a problem in the United States. Since many people have employer-provided health insurance, there is a chance the results of genetic testing will be identified with an individual. Genetic testing is one area in which the technology is outpacing government regulation. In the 1990s, there was legislation introduced in the U.S. Congress to address the issue of genetic discrimination in the workplace. As of now, no federal legislation has been passed relating to genetic discrimination in insurance coverage or the workplace. Federal employees are protected from genetic discrimination by an Executive Order, signed by then-President Clinton in 2000. Individual states have a variety of genetic nondiscrimination laws that differ significantly. Generally, the state laws prohibit employers from requiring genetic testing as a condition of employment, but overall, genetic discrimination has not been tested in the courts.

So, the consumer is left with a dilemma. Is it better to know or not know? Should one take the genetic test to determine whether one has a DNA-linked disease and risk having the information made available to his insurance company or employer? Or should one avoid the test and hope for the best? These tests are still a matter of choice in the United States, but whether they will be in the future, no one knows.

Human Gene Therapy

Gene therapy is a method to treat, cure, or prevent human disease by changing a person's DNA. It differs from most other therapies that attempt to fix a problem after it has occurred. Gene therapy attempts to prevent a problem from happening by modifying a person's DNA.

Gene therapy has been heralded as the technology of the future for a long time, but in reality, its development has had a long and twisting path. Scientists have had to overcome many hurdles in maximizing the potential of gene therapy. The first such hurdle was the gene delivery tool. Since our DNA is present in every cell of our bodies, billions of cells have to be modified in order for gene therapy to be effective. A second hurdle relates to the complexity of our DNA and the causes of many diseases. Most diseases involve the interaction of several genes; therefore, most genetic diseases are related to more than one gene. This fact makes it more difficult to use gene therapy.

There are two different techniques in gene therapy. In an ex-vivo approach, cells are taken from the patient, and the new gene is introduced into the cells in a laboratory. The modified cells are then grown in culture. When the cells have multiplied, they are reinjected into the patient. This technique was used to successfully treat three babies born with the rare immune system disease called severe combined immune deficiency, or SCID (you might know it as the bubble boy syndrome). Bone marrow was removed from the babies, and the marrow cells were treated in a laboratory with healthy copies of marrow cells. Then, the healthy cells were reinjected into the children. Ten months later, their immune systems were normal.

Most gene therapy treatments, however, take an in-vivo approach, in which the corrected gene is injected directly into the patient using a virus as the delivery mechanism. See Figure 2-7. The theory is that instead of the virus making the person sick, the virus will replace the defective gene with the corrected copy. There is no guarantee, however, that the virus will replace the defective gene; instead, the virus could insert the "new" gene into another position in the cell's DNA. The "new" gene might actually harm the cell and cause it to function incorrectly.

Although gene therapy has the potential to be a powerful tool for medicine, there have been few successes thus far. Since 1990, according to the NIH, more than 390 gene therapy studies were begun, with over four thousand patients and more than a dozen medical conditions. One promising gene therapy for heart disease generates the growth of new blood vessels to the heart, allowing patients to grow their own heart bypasses. It has proven successful in twenty-four patients.

The future of gene therapy is uncertain. While it is true this method has come a long

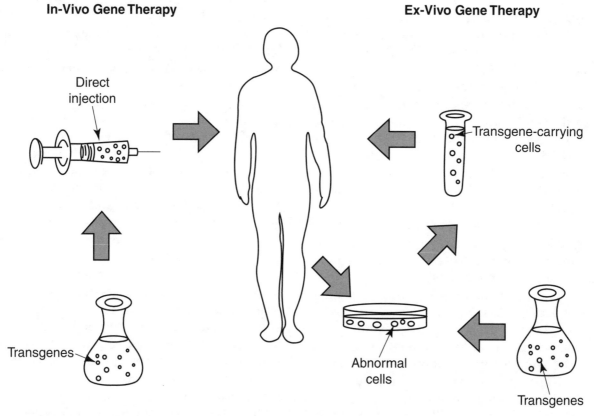

Figure 2-7.
The route of gene delivery is categorized into in-vivo and ex-vivo gene transfer methods. For in-vivo gene therapy, the transgene incorporated in genetic material is directly delivered into the patient's body. For ex-vivo gene therapy, the transgene is transferred in vitro into abnormal cells removed from a patient, and the transgene-carrying cells are relocated into the patient.

way, scientists still face many challenges in perfecting it. Gene therapy has successfully changed the lives of some people, but the downsides of using this form of biotechnology are still quite apparent. Considering all of the benefits and risks involved, gene therapy is a promising, but terribly unpredictable, technology. Undoubtedly, scientists will continue to research this complex technique and develop even more unimaginable applications for the future.

Xenotransplantation

Xenotransplantation is the transplantation of animal cells, tissues, or organs into human beings. Xenotransplantation first emerged into the news in the mid-1980s, when cardiac surgeon Leonard Bailey transplanted a baboon heart into a newborn girl known as Baby Fae, who had congenital heart disease. The baby lived for twenty days after the operation.

Since the mid-1980s, doctors have used baboon livers and marrow as transplant material, but a problem exists with organ rejection. Every time an organ is transplanted from one organism to another (even from human to human), there is a great risk of rejection because the host body recognizes the donor organ as foreign and attacks the foreign cells. To allow normal human-to-human transplants to succeed, organ recipients must take drugs that suppress their immune systems. New methods in xenotransplantation might eliminate or reduce these organ rejections.

Xenotransplantation in the past has used the animal version of the organ in transplantation. For example, pig heart valves have been used in humans for more than twenty-five years without negative effects,

but what if the pig organ was genetically identical to the human organ? The chance of organ rejection would be much less. Recently, scientists have been able to insert human genetic material into pigs so a pig organ would be acceptable by the human body.

Pigs are considered to be the best animal for xenotransplantation because pig organs are similar in size to human organs. Since over sixty-two thousand people in the United States wait every year for organs, the potential health benefits for pig-to-human transplants cannot be underestimated.

So far, xenotransplantation has had some wild ups and downs. Despite the number of animal-to-human transplants done, not one was successful (that is, the human patient lived). Also, research has shown it is possible for viruses to transfer from animals to humans along with the organ. On the other hand, people who received pig tissue as skin grafts and to repair parts of the pancreas did not pick up any pig viruses.

In the United States, there have been no trials so far with whole animal organs and few medical trials with the transfer of animal cells and tissue. Diacrin, a biotech company in Massachusetts, developed treatments for Parkinson's disease and Huntington's disease using brain cells from pig fetuses. In another medical trial, ten Swedish diabetic patients received transplants of pig pancreas cells that manufacture insulin.

From gene therapy to xenotransplantation, biotechnology holds the promise of a healthier tomorrow. What technologies will emerge from this quickly changing field is anyone's guess. Whatever the future, one thing that is certain is the increased use of biotechnology in the health field.

Cloning

Cloning jumped into the headlines in the mid-1990s when Scottish researchers presented Dolly, the first mammal to be cloned from an adult. See Figure 2-8. She is different because she is an identical copy of another adult sheep, and she has no biological father. Although cloning experiments with frogs and toads date back to the 1970s, and experiments involving plants and animal embryos have been performed for years, the news of a cloned sheep led the news worldwide.

CLONE

An exact genetic copy. It is an individual grown from a single somatic cell of its parent. Clones are genetically identical to their parents.

Figure 2-8.
Dolly, the first fully-grown mammal to be cloned, is an identical copy of another adult sheep. (Roddy Field, Roslin Institute)

How exactly was Dolly created? A team of Scottish scientists, led by Ian Wilmut, scraped skin cells from the udder of an adult sheep (sheep 1). They then removed an unfertilized egg from another sheep (sheep 2) and removed the nucleus of the unfertilized egg. The skin cells from sheep 1 were fused with the unfertilized egg of sheep 2. The composite egg was cultured in a petri dish for a period of time and then placed into the uterus of a surrogate mother (sheep 3). When the lamb, Dolly, was born, she was an exact genetic copy, or clone, of sheep 1, the sheep that provided the transferred cell nucleus. Dolly was not genetically linked to sheep 2, the sheep that provided the egg.

Since the birth of Dolly, researchers have successfully cloned calves, pigs, bulls, monkeys, cats, and mice. In New Zealand, scientists were able to clone the last surviving member of a rare breed of cow. The cloning process itself is difficult, however. The success of Dolly followed 276 other attempts to clone a sheep by the Scottish team. Many fetuses were rejected, aborted, or deformed. Now, for every one hundred attempts, just two to five live offspring typically are born. Why the success rate is so low is hard to determine. Some scientists believe that cloning—since it is irregular and asexual—may be inherently flawed. Others wonder if the problem lies in the in vitro technology used in the process. In any case, the low success rate of cloning currently limits its widespread use. In addition to the low success rate, it is not clear whether a clone is actually older because it is a copy of an adult. The evidence suggests Dolly's DNA is older than her age. Particularly, her chromosomes were shorter than a normal sheep of the same age. Dolly's chromosomes were the same length as those of the six-year-old donor sheep.

Scientists believe cloning will allow the recovery of endangered or extinct species of animals. Chinese scientists have successfully produced a cloned giant panda embryo from an adult animal. The San Diego Zoo has been collecting cells from endangered species since the mid-1970s, and many believe this frozen zoo could be used to create these animals by cloning. Other cloning initiatives have a different motivation. In 2002, Texas scientists announced the first successful clone of a pet cat named CC (which is short for Copy Cat). The cat clone is actually not the intended result for these researchers—their aim is to clone a dog, a more complicated procedure. It appears that if this dog cloning is successful, it will cater to a willing public. The company funding this research has had thousands of calls from devoted pet owners willing to pay twenty thousand dollars to clone their animal companions.

Medical Uses of Cloning

Ian Wilmut did not decide to create Dolly for purely scientific purposes. Instead, Dolly was created as part of a plan to make transgenic animals that would produce drugs in their milk (to review transgenic animals, see the section "Transgenic Animals and Plants"). Transgenic animals contain a foreign gene allowing them to produce proteins. See Figure 2-9. For example, a cow could be genetically engineered to produce milk containing human insulin. This way, the cow could produce a large quantity of the drug insulin in its milk. The problem in traditional breeding of transgenic animals is that, because of genetics, often the offspring will lose the transgenic gene. A company could invest millions of dollars to create the insulin-producing cow, and none of the cow's offspring could have the same insulin-producing gene. In contrast, if the insulin-producing cow was cloned, the clone would have the exact genetic makeup and, therefore, would also be able to produce insulin in its milk. The use of

Figure 2-9.
The world's first cloned transgenic female calf, Victoria, was produced using Advanced Cell Technology's proprietary cloning technology and is carrying a copy of the green fluorescent protein (GFP) marker gene. (Advanced Cell Technology)

genetically altered livestock to produce drugs is called *pharming.* One product of pharming that has been produced from a transgenic flock, alpha-1-antitrypsin to treat cystic fibrosis, is in the FDA approval process.

A combination of cloning and genetic engineering can alter the nutritional content of milk. For example, a significant group of people is lactose intolerant. Presently, these people must either avoid milk or drink milk that has had lactose removed. Through genetic engineering and cloning, a herd of cows could be bred to produce milk free of lactose. The tinkering with milk does not have to stop there. Many children cannot drink cow's milk when they are young, but if human proteins replaced the cow proteins, then the milk would be closer to human milk.

Besides using cows, sheep, and other animals as factories, cloning can also provide animals that would donate organs to people (xenotransplantation). Transgenic pigs have been genetically engineered to contain an additional human protein to reduce the rejection of their organs once they are transplanted into humans. The cloning of these pigs would allow the transmission of this human protein to their offspring.

Cloning also brings a promise of quickly increasing the quantity and quality of various foods. Current breeding methods require that the animal scientist choose the desired trait and breed two individuals with the most desirable characteristics, but the use of biotechnology and cloning would shorten this process. A cow could be genetically engineered to be leaner and then that cow could be cloned to produce beef having little fat. Consider the other by-products of animals. We get wool from sheep. What if we could get twice the amount of wool, or even colored wool, from a sheep? The possibilities seem endless.

Obviously, the widespread use of cloning is not without risk. Cloning, by its nature, will reduce the genetic diversity of plants and animals. This goes against the processes of evolution and natural selection. Also, improving the quantity and quality of food could lead to population growth that, in turn, would cause more stress on Earth. What about the danger of developing new diseases that could be spread from animal to man? The answers to these questions are not clear yet. The best possible alternative would be for scientists to investigate the risks and benefits of cloning at the same time.

Stem Cell Research

One application of biotechnology is in the area of stem cell research. Stem cells are cells from early in the human development cycle. After a sperm fertilizes an egg, a single cell is formed. This single cell then divides into a hollow sphere of cells, called a *blastocyst.* The blastocyst has an outer layer of cells and an inner group of cells, called the inner cell mass. In normal fertilization inside a woman's body, the outer layer of cells begins to form the placenta. The inner cell mass goes on to form virtually all the other tissues of the body—these are called stem cells.

Stem cells, presently obtained from frozen embryos created during in vitro fertilization, have the ability to divide without limit and evolve into specialized cells. For example, a blood stem cell will specialize and become red blood cells, white blood cells, and platelets. Since stem cells have the ability to divide and specialize, they hold the possibility of a renewable source of replacement tissue to treat diseases such as Alzheimer's disease and Parkinson's disease. For example, the production of insulin in the pancreas is disrupted in people who suffer from type I diabetes. Transplanting healthy pancreas stem cells into a person with type I diabetes could reverse this process and cure the diabetes.

Ethical Concerns

The most common ethical and moral arguments against cloning originate from various religious groups. Many religious philosophies believe life is sacred and is created and controlled by a deity. Generally, organized religion takes an anti-cloning stand, based upon a belief that humans should not take over the divine power of creation, but there is a wide range of opinions on cloning and biotechnology.

Many religious groups separate the cloning of animals from the cloning of human beings. It is no surprise to note that the cloning of human beings is much more controversial. Pope John Paul II made a statement, on behalf of the Catholic Church, specifically condemning the cloning of human beings. In contrast, Abdelmo'ti Bayyumi, a theologian from Al-Azhar University, the supreme authority on Sunni Islam, stated that neither animals nor humans should be cloned. Israel's chief rabbi, Meir Law, stated the cloning of any creature is against Jewish law and recommended a ban on all animal cloning. In contrast to these leaders, however, some American Jewish and Muslim religious leaders testified before the National Bioethics Advisory Commission and stated that they believe cloning has a legitimate purpose in society.

Cloning has also struck a chord in the political world. Human cloning is illegal in Great Britain and Norway, but not in the United States. U.S. federal funds cannot be used for any human cloning research though, at least for the next five years. The debate over human cloning echoes previous concerns on human embryo research. Politicians, at least in the United States, are hesitant to allow federal funding for any research involving the manipulation of humans. In addition to ethical concerns, the business viability of cloning is different across the world. In the United States, it is possible to patent both the cloning process and GM organisms. This is not the case in the European Community, where one cannot patent a GM organism, but can patent the cloning process.

Regulatory Considerations

Many industrialists and business leaders grimace at the mention of regulation or regulatory agencies. Regulation often brings to mind endless government bureaucracy and increased costs directed toward legal compliance. This does not seem the case among biotechnology firms, however. Because the success of biotechnology products depends so strongly upon their perceived safety, to some extent regulation provides that security for companies.

Imbedded in the biotechnology regulation processes in the United States is the belief that GM organisms are not inherently risky. This underlying principle underlies the regulation process even today, and this positive perspective toward GM organisms probably accounts for their acceptance in the United States, at least in the scientific community. At the onset of the twenty-first century, the National Research Council, which provides science, technology, and health policy advice under a congressional charter, published a report on biotechnology. This report analyzed the potential effects of biotechnology and noted that the modern process did not appear to introduce new risks compared to older methods.

The United States is not alone in its regulation of biotechnology. International companies must also deal with a wide range of laws and regulations from other countries and governmental entities. Are new biotechnology products worth the trouble? How will regulation affect the developments in this field? The answers to these questions, now, are impossible to know.

Regulation of the Biotechnology Industry in the United States

Regulation of agricultural biotechnology in the United States began in the late 1980s, after more than a decade of mainly laboratory-based R & D. GM foods are regulated by three federal agencies in the United States: the FDA, the Environmental Protection Agency (EPA), and the USDA. Broadly, the FDA is responsible for the safety and labeling of all foods and animal feeds derived from plants. This mandate includes foods derived from GM crops. The EPA regulates pesticides. So, if a GM food contains a gene for pesticides, that food product would fall under its regulation. An example of this type of food is Monsanto's NewLeaf Plus potato that contains the natural pesticide *Bt*. The third agency, the USDA's Animal and Plant Health Inspection Service (APHIS), has the responsibility to oversee the agricultural and environmental safety of planting and field testing genetically engineering plants. In addition, the Department of Health and Human Service's NIH has developed guidelines for the laboratory use of GM organisms. These guidelines cover GM food, as well as other biotechnology products. The regulatory process includes many steps and can take up to ten years to finish. See Figure 2-10. Each agency is responsible for a particular aspect of regulation, but all the agencies work together to assure that all food reaching the public is safe.

Food and Drug Administration (FDA)

The Food and Drug Administration (FDA) is responsible for the safety of all food and drugs; this includes GM organisms. Since the early 1990s, the FDA declared that biotechnology-derived foods must meet the same safety standards required of all other foods. By this, the FDA established that no special rules were going to be created for GM food. Instead, GM foods would be evaluated the same as regular foods. The FDA calls this a consultation process, rather than a review process. The consultation process requires that the developers notify the FDA of their intent to market a GM food or animal feed at least 120 days before marketing. Companies must send the FDA documents summarizing the data they have gathered to prove that a GM food is as safe as conventional food. The documents must describe the genes they use, whether the genes are from a commonly allergic plant, how the plant functions biologically, and how many of the genes will be found in the new GM food. Also, the company must provide a summary of the GM food safety and nutritional distribution to the FDA and discuss its results with the FDA's scientists prior to commercial distribution.

To approve food additives, the FDA uses a standard called GRAS. After new genetic material is introduced into a crop, studies must be done to determine the safety and whether the new genetic material had any unintended effects. After the company submits the data to the FDA, FDA scientists review the information and generally raise questions. If the GM food passes, the new GM food is considered to be as safe as the regular food. The FDA will supply the company with a letter stating that it has completed the consultation process.

Most GM foods do not have to be labeled in the United States, and they appear at your neighborhood grocery store in the produce section without any notice. The FDA requires special labeling of GM food only if its composition is significantly different than its normal counterpart, or if its nutritional value has been altered. Also, if the GM contains a gene that might cause an allergy, then it would require special labeling.

Environmental Protection Agency (EPA)

The Environmental Protection Agency (EPA) gets involved in the review of a GM food if the plant has been altered to include pesticides—for example, insect protected corn or soybeans. Some GM plants are genetically engineered to resist pesticides. Monsanto has a range of GM plants that are resistant to Roundup chemical herbicide (created by Monsanto). Since these plants (soybeans, cotton, corn, and canola) have the Roundup herbicide–resistant gene, farmers can spray their fields with the pesticide, killing all plants (including weeds) except for the Roundup herbicide–resistant plants. The EPA, in the case of these types of

Seeking Federal Approval for a New Form of Biotechnology

1. The first opportunity comes almost immediately after a scientist discovers a potentially marketable product concept. Following guidelines established by the National Institutes of Health (NIH), developers of biotech products empanel an advisory group (Biosafety Committee) made up of employees and members of the general public. This panel reviews the environmental and health possibilities posed by developing the proposed idea. If the committee determines there is unacceptable risk, it will recommend that the concept not be developed.

2. If the concept passes initial considerations, a review must be conducted to determine if existing research facilities are adequate to conduct the research. The USDA must review and approve facility plans, including greenhouses where the plants will be developed and tested.

3. The developer must seek USDA approval in order to conduct field trials.

4. USDA must also give authority for the developer to ship seeds from a greenhouse to a field trial site.

5. Another formal interface comes after the developer has generated a full package of data, submits it to USDA and requests a "determination of non-regulated status," meaning the plant can be grown, tested, or used for traditional crop breeding without further USDA action.

6. If a developer plans to plant more than ten acres of a plant expressing a pesticidal protein in research or field trials, the EPA must grant an experimental use permit (EUP).

7. EPA reviews data on the human, animal, and environmental safety of the pest control protein to determine whether limits (tolerances) should be set on the amount of protein in food derived from the improved plant. In instances where there is substantial data on the safety of the protein and a history of safe use, the developer may request an exemption from the requirement of a tolerance, which may or may not be granted.

8. The final EPA step is a formal review of the data generated through years of study. During this final review, which typically takes approximately eighteen months, EPA considers whether or not to register the product for commercial use.

9. FDA meets with a developer of a biotechnology product early in the process and provides guidance as to what studies FDA considers appropriate to ensure food and feed safety. The recommended studies vary, depending on each product and the product's proposed use and function. The interactive FDA involvement in pre-market review of a biotech food spans several years. At the end of this process, the FDA provides a letter to the developer confirming that they have no more questions regarding the food and feed safety of the product. Even after a product is on the market, FDA has authority, under the Food, Drug, and Cosmetic Act, to immediately remove from the market any food that the FDA deems unsafe. FDA's authority is immediate and final.

Figure 2-10.
This excerpt, taken from the website of the International Consumers for Civil Society, lists the regulatory steps in the process of seeking federal approval for a new form of biotechnology. The entire report is available at http://www.icfcs.org/biotechreg.htm.

GM plants, would determine whether these crops pose risks to food or food safety. Also, the EPA would determine whether the GM food would require special labeling for consumer awareness.

In its review of new GM foods, the EPA examines several parameters. First, it looks at the type of genetic modification and where in the plant it will show. For example, if the genetic modification will only affect a plant's root, the EPA will focus on the effect of the modification on soil organisms. Next, the EPA is concerned about the toxicity of the added protein. Rodents are fed the protein produced by the introduced gene at levels at least one hundred thousand times the levels that a person would ingest. Also, the EPA requires that the GM company complete digestibility studies. These studies assess how long it would take for the protein to break down in a person's stomach and digestive system. Last, the EPA concerns itself with the toxicity of the added gene to other organisms such as birds, beneficial insects, and fish. After a complete analysis of the GM food, the EPA will either grant permission or deny the GM food.

United States Department of Agriculture (USDA)

Within the United States Department of Agriculture (USDA), APHIS has the job of protecting American farming against pests and diseases. APHIS reviews all GM plants to make sure these plants do not become a plant pest in the environment. A primary role for APHIS is the determination of possible environmental consequences. It analyzes whether the modified plant could cross-pollinate with other plants, thereby causing a change in the environment. This was a particular concern with Monsanto's Roundup Ready series of GM plants. Many scientists and environmental activists were afraid the pesticide resistance of the

Monsanto plants could spread to the weeds, thereby causing an ecological disaster.

APHIS is involved at two different stages: the field-testing of a GM plant and the commercial distribution. If a company wishes to field-test a GM plant, they must receive APHIS approval before proceeding. An APHIS scientific reviewer evaluates the environmental impact of the field test and assesses the impact of the new GM plant on endangered or threatened species. If the field test is approved, APHIS and state agriculture officials may inspect the site before any planting is done.

Before a GM plant can be sold commercially, the GM company must petition APHIS for a "determination of nonregulated status." The GM company must provide APHIS with extensive information about the development and field-testing of the GM food. After analyzing all of the data, APHIS will only allow commercialization of the plant if there is no significant risk to other plants in the environment.

The three main agencies involved in the regulation of GM foods work together to assess the human and environmental effects of biotechnology. This coordinated process of regulation has allowed the approval of forty new agricultural products since the late 1980s. A new initiative, begun in 2000, is to increase public awareness and education about the use of biotechnology in food products. It is hoped, with increased awareness, there will come an increased sense of security about GM foods in the mind of the average citizen.

International Efforts in Biotechnology Regulation

The main international effort in biotechnology regulation is the new Biosafety Accord, which was signed by over 130 countries in 2000. The new protocol will affect the transport of GM foods across international boundaries. The protocol allows individual countries to decide whether or not they are willing to accept imports of GM foods. A country can decide to ban imports of GM food if that nation believes there is insufficient evidence of the safety of GM food.

More critically, to allay consumer fears, all products containing GM organisms must be clearly labeled. Exporters must obtain the permission of the importing country for the first shipment of a GM organism. This process is called *advanced informed agreement*. When a GM crop is approved in a particular country, the information will be published and available via an Internet-based Biosafety Clearing House. If an exporting nation believes that a GM food is being blocked for no scientific reason, it can appeal to the World Trade Organization.

Summary

Biotechnology has emerged from secluded R & D laboratories to join the ranks of business and industrial enterprises around the world. Startling developments in rDNA procedures, GM food, and cloning are taking place almost daily. As these developments take place, there is a growing unease about the future role of biotechnology in our society.

Biotechnology is a powerful new technology characterized by a series of promises, as well as an accompanying list of threats. Decisions are being made that may spell either ultimate relief or certain disaster for the human species as we know it. A synopsis of the pros and cons associated with biotechnology research incorporates a wide range of disciplines including pharmaceuticals, agriculture, energy, defense, animal science, gene therapy, and environmental protection. As with other technological forces, society must manage and control the developments in biotechnology with foresight based upon convergent thinking and systematic decision-making.

Discussion Questions

1. Explain some of the reasons biotechnology has become a topic of technological controversy.
2. Why was the *E. coli* bacteria chosen for use in early rDNA research projects?
3. Consider your last trip to the grocery store. Which of the products you bought contained GM ingredients? What is your opinion of GM food?
4. Are you in favor of or opposed to xeno-transplantation? If you needed an organ, would your opinion change? Why or why not?
5. In an attempt to determine the degree to which your academic contemporaries are up-to-date on the developments taking place in biotechnology, you are about to play the role of a reporter. Randomly select five people on campus and record their responses to these basic questions. Be prepared to discuss your findings in a discussion to be assembled in class.
   a. Briefly explain what is meant by biotechnology.
   b. Are you in favor of further industrial research in this field?
   c. Provide at least one reason for your stance on this issue.
   d. What is your academic major?

Suggested Readings for Further Study

Bains, William. 1998. *Biotechnology from a to z.* New York: Oxford Univ. Press.

Barbour, Virginia. 1999. Gene-therapy advances greeted positively. *Lancet* 354 (9195): 2057.

Beardsley, Tim. 1996. Vital data. *Scientific American* 274 (March): 100–105.

Bonn, Dorothy. 2000. Genetic testing and insurance: Fears unfounded? *Lancet* 355 (9214): 1526.

Cook-Deegan, Robert. 1996. *The gene wars: Science, politics, and the human genome.* New York: W. W. Norton.

Ferber, D. 1999. GM crops in the cross hairs. *Science* 286 (26 November): 1662–1666.

Friedmann, T. 2000. Principles for human gene therapy studies. *Science* 287 (24 March): 2163–2165.

Hall, Stephen S. and James Watson. 2002. *Invisible frontiers: The race to synthesize a human gene.* New York: Oxford Univ. Press.

Hawkins, Dana. 1997. Dangerous legacies. *U.S. News & World Reports* Health Guide (10 November): 99.

Holden, C. 2002. Carbon-copy clone is the real thing. *Science* 295 (22 February): 1443–1444.

Hollon, T. 2001. Gene pool expeditions. *The Scientist* 15 (4): 1.

[Internet]. *Access Excellence.* <http://www.accessexcellence.org/>.

[Internet]. *Bio Online.* <http://www.bio.com/>.

[Internet]. *Genetic Engineering News Online.* <http://www.genwire.com/>.

[Internet]. *The Human Genome Organization (HUGO).* <http://www.gene.ucl.ac.uk/hugo/>.

[Internet]. *Human Genome Project.* <http://www.ornl.gov/hgmis/>.

[Internet]. *National Center for Biotechnology Information.* <http://www.ncbi.nlm.nih.gov/>.

Keller, Evelyn F. 1993. *A feeling for the organism: The life and work of Barbara McClintock.* New York: W. H. Freeman and Company.

Lewis, R. 2000. Porcine possibilities. *The Scientist* 14 (20): 1.

Mann, Charles C. and Mark L. Plummer. 2002. Forest biotech edges out of the lab. *Science* 295 (1 March): 1626–1629.

Miller, Henry. 1997. The EPA's war on bioremediation. *Nature Biotechnology* 15 (6): 486.

National Research Council. 2000. *Genetically modified pest-protected plants: Science and regulation.* Washington, D.C.: Author.

Olson, Joan. 1999. When crops save lives. *Farm Industry News* 32 (6): 4–5.

Paarlberg, Robert. 2000. The global food fight. *Foreign Affairs* 79 (3): 24–38.

Palevitz, Barry A. 2000. DNA surprise. Monsanto discovers extra sequences in its Roundup Ready soybeans. *The Scientist* 14 (15): 20.

Pires-O'Brien, Joaquina. 2000a. GM foods in perspective: Part one. *Contemporary Review* 276 (1608): 19–23.

———. 2000b. GM foods in perspective: Part three. *Contemporary Review* 276 (1610): 141–143.

———. 2000c. GM foods in perspective: Part two. *Contemporary Review* 276 (1609): 84–89.

Rotman, D. 1998. The next biotech harvest. *Technology Review* (September–October): 34–41.

Runge, C. Ford, and Benjamin Senauer. 2000. A removable feast. *Foreign Affairs* 79 (3): 39–51.

Shapiro, Robert B. 1999. How genetic engineering will save our planet. *The Futurist* (April): 28–29.

Stolberg, Sheryl Gay. 1999. Could this pig save your life? *The New York Times Magazine* (3 October).

Suzuki, David, and Peter Knudtson. 1989. *Genethics: The clash between the new genetics and human values.* Cambridge, Mass.: Harvard Univ. Press.

Watson, James D., and Francis H. C. Crick. 1953. Molecular structure of nucleic acids: A structure of deoxyribose nucleic acid. *Nature* 171.

Watson, James D., Michael Gilman, Jan Witkowski, Mark Zoller, and Gilman Witkowski. 1992. *Recombinant DNA.* 2d ed. New York: W. H. Freeman.

Wexler, Alice. 1996. Mapping fate: A memoir of family, risk, and genetic research. Berkeley: Univ. of California Press.

Information technologies are used in every area of science and technology.

Chapter 3

Information Technologies

DATELINE

Beijing, China
    China recently announced plans to issue new "smart" identity cards to all 1.26 billion citizens. The smart cards will have an imbedded microchip that will store personal information. These new smart cards will replace the existing paper identification (ID) cards and will reduce ID counterfeiting.
    China estimates it will take five years to implement this new ID system in the entire country; however, they plan on starting the implementation this year. First year college students will be among the first to get the new cards. It is expected that these cards will be required for the students to get college loans....

Introduction

Without question, information is the most basic resource driving social progress. Information is an essential ingredient for competitive economic growth. Many people in contemporary societies now view increased access to all types of information as the status quo. Some individuals may even experience e-mail withdrawal if they are away from a computer link for more than a few hours.

Nevertheless, an abundance of information can only be considered useful to the extent that recipients have the capacity to receive, assimilate, and utilize it in an efficient manner. The modes or methods we use to transfer information among individuals and machines are continually expanding and becoming eminently more powerful.

Information is the primary raw material of communication. Computer mediated communication (CMC) technologies have transformed the way individuals work together, consult with one another, hire personnel, and make new friends. In every sense, communication has become an incredibly complex industry that makes demands on all of us. For each convenience information technologies provide us, however, there is most likely a disconcerting side effect. Traditional modes of communication (such as face-to-face conversations, the telephone, and letters) are being largely displaced by the use of electronic messaging technologies (for example, electronic mail, or e-mail; computer conferencing; bulletin boards; and Internet relay chat rooms). Despite the variety among these forms of interaction, a salient feature common to each is the participants' lack of physical or real-time presence.

The term *CMC* has been around for many years; however, its meaning has gradually been expanded as new communication technologies have emerged. In its simplest form, CMC refers to communication between two people through the use of digital or computer means. The information technologies make CMC environments a reality all around us. Intercontinental messages bounce off satellites to connect people around the world. First grade children feel more at ease with computer terminals and keyboards than they do with writing paper and pencils. Massive quantities of information on just about any topic can be obtained via an assortment of databases and interactive on-line services. Much of the industrialized world is becoming overly dependent on the use of electronic information technologies to transmit messages, conduct business, look for employment, select a college to attend, store information, complete financial transactions, and seek companionship.

This phenomenon has created a world of virtual interaction devoid of much regulation or regulatory feedback, but replete with social anonymity. A whole new area of study regarding the ethics and etiquette of interpersonal interaction has evolved alongside the emergence of information technologies in all aspects of our lives. In today's information saturated society, individuals are commonly expected to respond to requests as quickly as the computers their coworkers and friends are using initiate them.

Typical key developments enabling us, and perhaps even forcing us, to communicate with one another at such a breakneck pace include wireless networks, e-mail, listserv systems, the Internet, the World Wide Web (WWW), subscriber networks such as AOL® or MSN® communication services, multimedia technologies, virtual reality, and intelligent software. Each topic represents an entire industrial enterprise of its own. A basic understanding of these areas or developments is therefore a critical starting point for any discussion of information technologies.

The long-standing marriage between computers and various communication tools has created and nurtured new conditions of employment. Society is being cajoled to rethink legislation dealing with telecommunication standards and the protection of intellectual property. More people are telecommuting and, as we proceed further into the twenty-first century, the number of telecommuters stands at 30 million in the United States. Numerous corporations are weighing the advantages against the potential deterrents to using videoconferencing or teleconferencing as an alternative to face-to-face meetings among employees who reside in various parts of the country. For the

geographically dispersed organization, all of the technologies associated with remote conferencing technologies may address the problems of distance, crisis management, personal time, and travel expenditures.

The linked assortment of technologies making up this ever-changing landscape of communication is both well established and still emerging. Fax numbers, cellular phones, and e-mail addresses are more the rule today than the exception, which they were just a few short years ago. Agent programs, homepages, websites, cyberspace, and virtual reality have become familiar topics of adult social conversations. Information technology terms also pervade the vocabulary of today's adolescents, who sometimes pride themselves in being able to break into computer files managed by schools, credit agencies, corporations, and government offices. Factor into this formula the evolution of digital ID technology, such as described in the *Dateline* of this chapter, and society has yet another source of concern. The proliferation of computers in our society can cause much alarm with regard to our lack of individual privacy and the security of information systems. The welcome mat to the information age has been at the threshold of our daily routines for quite a few years now. On the other hand, the horizon lying beyond the threshold is continually changing, both in its appearance and location.

The primary objective of this chapter is to briefly describe and discuss a few of the more prominent changes taking place in the information and communication sectors of our society. Information technologies continue to reshape the way we work, live, and entertain ourselves. There is no denying that we regularly come in contact with computerized goods and services—refrigerators and automobiles talk, miniature telephones are carried in shirt pockets, some people conduct all personal bank transactions from their homes or cars, and countless companies are doing business on the Internet. National governments around the world maintain huge databases on their citizens, in effect, spelling the end to individual *privacy,* in the conventional sense of the word.

Many individuals openly express fear and anxiety regarding their inabilities to keep up with various information technologies. Other individuals display an air of indifference and seem willing to ride out the waves of change in order to digest only those technological breakthroughs mattering the most to them. There are also those computer technophiles who cleverly keep abreast of every new development in media-related and communication-based systems. A fourth group of individuals insists it wants nothing at all to do with CMC technologies. Regardless of the segment of the population with which individuals feel most comfortable associating, the infrastructure of life and daily routine is often influenced and constricted by the success or failure of a distant computerized service or remote database, over which the individual has minimal, if any, control. This chapter provides a foundation for a continuing dialogue that can ultimately respond to the following questions:

❏ What effect has each of the following technological breakthroughs had on the profile of our information-driven culture?
wireless networks
computer mediated communication (CMC)
the Internet
intelligent software

❏ What are the pros and cons of telecommuting? Would you take a job that included a large percent of telecommuting? Why or why not?

❏ What is the difference between an intranet and an extranet?

❏ What impacts are leading edge information technologies having on electronic record keeping activities in our society?

❏ Is individual privacy an artifact of the not-so-distant past? Do we have any recourse to which we can turn for security?

❏ How will e-commerce change consumers' buying habits? How will it change organizations that produce goods for the marketplace?

Key Concepts

As you read through the various sections of this chapter, you should be alert for the following terms and phrases. It is expected you will be able to explain their meanings.

Advanced Research Projects Agency (ARPA)
computer blacklist
computer mediated communication (CMC)
cookies
e-commerce
e-finance
electronic mail (e-mail)
extranet
fiber optics
firewall
intelligent software agent
Internet
intranet
local area network (LAN)
optical network
telecommuting
teleconferencing
telematics
testbed
Trojan Horse
videoconferencing
virtual private network (VPN)
virtual reality
virus
web bug
wide area network (WAN)
wireless network
worm

Technologies to Sustain an Information-Driven Culture

Today's information-dependent culture has introduced phrases like "surfing the Net" and "cruising the information superhighway." The information superhighway and the *Internet* are constant topics in news programs, technical conferences, and journal articles. Drastic changes in competition and equipment are increasingly apparent, as telephone companies, cable television providers, and hardware and software vendors in the computer and communications industries form and, just as quickly, dissolve alliances and partnerships. New uses for existing technologies and new technologies to serve future demands continue to characterize a major transition in the industrialized world's communication infrastructure. Research and development (R & D) in information technologies is directed toward providing the average person open access to extensive computing power and an incredible amount of information through a networked fusion of communication equipment and distributed computers.

The U.S. government has articulated its commitment to providing computation and knowledge resources to its citizenry through various successive initiatives, including the National Information Infrastructure (NII), the Next Generation Internet (NGI), and Large Scale Networking (LSN).[1] The vast range of applications being envisioned for these projects requires an integration of computers and networks with, ultimately, wireless access. In the simplest sense, a network is a communication system connecting geographically dispersed computer users by means of circuits, switches, and control software. Networks provide the ability for multiple computers to communicate with one another. There are several important networks in use today. The nation's telephone network, built over the past century on an analog model, is the most mature segment of the communication infrastructure. Similarly, standards for the dissemination of television and radio programming over the air and, more recently, over cable are based on analog rather than digital transmission. Subsequently, many of us continue to live in our homes where telephone, radio, and television are largely analog systems.

Transmission channels and the signals they carry are typically classified as being either analog or digital. See Figure 3-1. An *analog signal* is a continuous function of time and, at any instant, may possess any value between the limits set by the maximum power that can be transmitted. Analog signals vary continuously and smoothly in amplitude and frequency. Cable television systems and radio relay systems equipped with amplifiers are common examples of analog channels.

Digital technology has had a significant influence on the development of newer communication lines and computer networks. *Digital signals* have only discrete values. The most common digital signal is a binary signal having only two values (for example, 1 and 0). The numbers *1* and *0* are called bits from the words *binary digits*. An older example of a digital-type channel is a telegraph circuit, in which the operation of a relay or sounder provides the coded output signal.

Digital communication is a comprehensive, umbrella-like term including such topics as digital transmission, digital radio, and digital recording. Digital transmission is the movement of digital electronic pulses between two or more points in a communication system. This form of transmitting a signal requires a physical or electromagnetic

[1]Computer Science and Telecommunications Board, National Research Council (1994). *Realizing the information future: The Internet and beyond.* Washington, D.C.: National Academy Press.

Analog

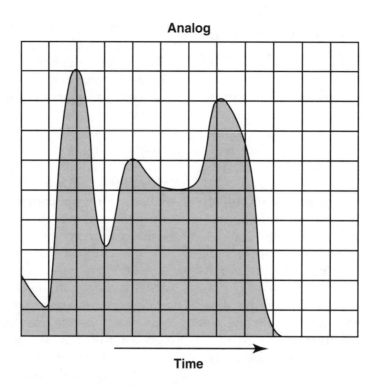

Time

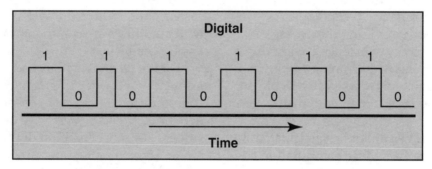

Figure 3-1.
Analog signals are smooth, continuous, and constantly varying, as shown in the top graph. They are often compared to voices, since voices change over a wide range, can be soft or loud, and can vary a great deal in amplitude and frequency. Analog signals can be difficult to receive, since they may vary in strength. Thus, weak signals, high points, or low points can distort them. Digital signals can represent only discrete numbers, as shown in the bottom graph. The most common examples are 0 and 1. Discrete means the signal change is abrupt. Digital signals are used in computers, compact discs, laser discs, and digital audiotape.

connection between the transmitter and the receiver such as a metallic wire pair, a coaxial cable, a microwave transmitter, or an optical fiber cable. In digital radio, Earth's atmosphere is the medium for transmitting the signal. Data communication and voice frequency telegraphy over telephone lines illustrate the process of transmitting digital signals over analog channels. Conversely, analog signals may be altered for transmission over digital channels by means of analog-to-digital converters. Today, most of the telephone network is digital except for what is called the "local loop." The local loop is the "last mile" or so between your home and the telephone company's local office. Most telephone lines in the local loop are still copper wire, and copper wires still carry analog signals best.

When you review the array of available technologies, there is little doubt that the computer has utterly transformed our society. Personal computers (PCs) have been around for thirty years and, more than any other office tool, have had a significant impact on the way American business works. Numerous industrial firms, educational institutions, and residential households have made huge investments in PCs, expecting increased productivity and access to unlimited amounts of information. These expectations can only be met to the extent a network exists, enabling multiple users to share and exchange information often stored at a remote database. The ability to enhance the computational power of PCs is therefore dependent on linking them together in a *local area network* or *LAN.* See Figure 3-2.

A LAN is an array of coaxial or optical fiber cables or wireless connectors and switching boxes, providing data communication links between computers and other shared devices, such as printers, in the same work area, building, campus, or business park. *Wide area networks (WANs)* connect computers or LANs together over a larger geographic area and are commonly referred to as backbone networks. LANs allow computers, modems, printers, and other peripherals to communicate with each other directly at very high speeds. Until a LAN is installed, only computers and peripherals

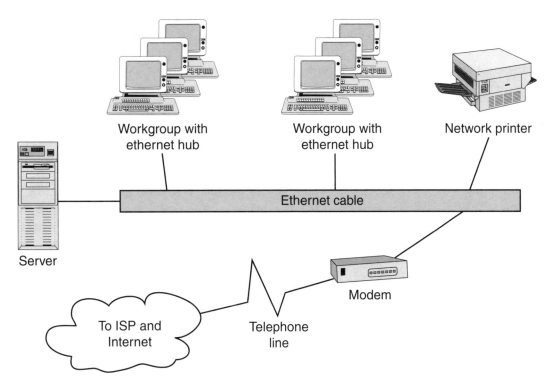

Figure 3-2.
This is a diagram of a simple LAN.

hardwired to each other can interface and interact.

The speed at which communications traffic travels or flows through a circuit is referred to as the bandwidth of the circuit, expressed in terms of bits per second. Bandwidth is the difference between the highest and lowest frequencies, measured in hertz, or cycles per second, needed for transmitting a modulated signal. The rate of travel is a crucial determining factor for which kind of information can be transmitted and who can afford to send or receive it.

Managing the flow of data from senders to receivers requires control software that generates a set of signals used by the network components to monitor the status of data transfer. As the number of options and types of intelligence in data communications terminals have increased over the years, the technology of internetworking has evolved. The network architecture outlines the way a data communications network is organized and generally includes the idea of levels or layers within the layout. Each layer consists of specific protocols or rules for communicating, which carry out a given set of functions. A set of rules governing the orderly exchange of information is commonly referred to as a data communications network protocol. Protocols, formulated by equipment manufacturers and vendors, may regulate lines, types of service, modes of operation and data transfer, circuit compatibility, or entire networks.

Today's Internet conforms to a protocol suite known as TCP/IP (Transmission Control Protocol/Internet Protocol). Since TCP/IP is an open or nonproprietary protocol, it has been incorporated into the Unix® operating system. One very functional feature of Unix operating system is a tool called Unix-to-Unix Copy Program (UUCP) that enables any computer running Unix operating system to dial and connect via modem with any other Unix computer. The successful union between TCP/IP and Unix operating system makes it a dominant communications protocol/operating system in the Internet. In reality, an open network is one capable of carrying information services of all kinds (not limited to one application such as TV distribution) to customers of all kinds (libraries, government labs, hospitals, educational institutions, research centers, industrial firms, or private households), across network service providers of all kinds—all in a seamless, accessible fashion.

TCP/IP treats information in a different way than traditional telephone service does. It divides information into packets. Information (for example, an e-mail message) is assembled into a packet. The packet is then sent to an address. The computer sends the packet to the nearest router on the network (analogous to your local post office). The packet is sent from router to router until the last router delivers it to the correct address—your e-mail mailbox.

Perhaps at the heart of information technologies research is a desire to refine the communication infrastructure to enable the realization of an expanded Internet in the near future. This requires continual improvements in the nation's voice-oriented telecommunications system in order to transmit all forms and large quantities of data between computers. Research being conducted in several areas of enabling technologies must evolve with a focus on flexibility, broad utility, and social benefit. Full computer connectivity and compatibility mean that the different technologies for providing video, audio, telephone, and data communications services must be interoperable, linked by recognized and accepted protocols and capable of sharing a common infrastructure. What all this means, in simple terms, is that all homes, places of employment, and public institutions will be inextricably linked together.

New strategies are being brainstormed to reconstruct the "last mile" in the communications industry. This is the connection or link between your home and the telecommunications industry (telephone line), cable TV industry, and electric power utility industry. Telephone and cable TV companies are positioning themselves to make substantial upgrades to their equipment. New systems including interactive television and video on demand are springing up in several states across the country. Justification for spending billions of dollars on updating the existing communications infrastructure, especially the very expensive last mile, should most certainly be based on the projected demand for the resulting services. Stated differently, which of the products or services already mentioned in this chapter would you be interested in having, and at what monthly cost? The following sections present a glimpse of several areas of study that continually redefine our images of what it means to communicate and exchange information in today's global society.

Computer Mediated Communication (CMC)

The presence of computers in our lives offers a provocative array of new possibilities for personal growth and development. Perhaps still less common than the traditional modes of communication, *computer mediated communication (CMC)* is an information-related phenomenon that cannot be ignored. CMC enables people around the world to form new kinds of relationships and master new skills. *Electronic mail,* or *e-mail,* correspondence among business and academic professionals has become more the rule than the exception. Parents are finding it easier to keep in touch with their college-aged sons and daughters—e-mail messages can be composed and sent anytime, regardless of time zone differences and complicated schedules. Radio stations no longer encourage their listeners to simply call, but announce their Internet and e-mail addresses on the air. The White House went on-line with an Internet-accessible e-mail address (president@whitehouse.gov), reported to have received more than two hundred thousand messages in a year!

Despite the highly sophisticated nature of the information technologies that constitute the technical architecture of CMC, many people who participate in these sessions are oblivious to how it all works. The ways in which bits of computer data or portions of video information are transmitted over wires and reassembled as recognizable computer files are invisible and seemingly unimportant to many users. It is only when these technologies break down and access to CMC services is restricted that users recognize how relevant the network architecture really is. It is this opaque nature (sometimes called "user-friendly") of telecommunications and computer equipment that has led to a great deal of "double-clicking" among all sectors of our society. The ability to "point and click" to manipulate images on a screen is non-threatening. It is a lot less daunting to point and click than to devise, memorize, and type long lines of computer codes just to get a program to load.

Undoubtedly, the most familiar CMC environment is e-mail. People who use e-mail type their messages into the computer and transmit them to recipients through internal network links or external telephone services. Generally speaking, e-mail software allows the user to save, delete, or respond to the message once it is read. Although e-mail has been featured in computer time-sharing environments since the 1960s, it has only been in the last several years that people have actually begun to include e-mail addresses on their business cards. Issues regarding the privacy of e-mail correspondence are discussed later in this chapter.

A second CMC application, computer conferencing systems, emerged in the 1970s, enabling people around the globe to carry on public conversations and exchange private e-mail. In computer conferencing, a person can send a message to a mainframe computer acting as a central repository for the ongoing conference. Other conference participants are able to retrieve these public messages at their convenience and save, discard, or respond to them. Two examples of computer conferencing systems are the WELL (Whole Earth 'Lectronic Link) and Usenet (Unix operating system users' network). Individuals with access to Usenet can send a specific, signed, electronic message to the entire community. The address of the message is not to another person, but to a newsgroup labeled by a particular topic of discussion, such as alt.games.marathon!, rec.autos.antique!, soc.college.gradinfo!, or talk.politics.medicine!

A third form of CMC is an innovative approach to computer conferencing called a multiuser dimension, or MUD (sometimes referred to as MOO [MUD, Object-Oriented] or MUSE [MultiUser Simulated Environments]). People who belong to these on-line social groups assume alternative identities to engage in all sorts of role-playing games. A MUD is more of a cyberspace hangout where "actors" can experience life in a medieval castle or engage in a battle onboard a starship. In a MUD, users communicate with others in the network who have also assumed alternate personas. One salient difference between MUDs and the computer conferencing environments mentioned above is that MUDs allow and promote real-time, synchronous interaction among users, while computer conferences are typically asynchronous, in that users do not need to be present at the same time to engage in the conference exchange.

CMC technological breakthroughs have totally redefined the notion of what it means for us to communicate with each other. Empirical research in the field of CMC is far-reaching and has produced numerous theories regarding the reasons people interact the way they do, as long as they can hide behind the opacity of computer technology. A visionary named Howard Rheingold used the term *virtual community* to describe this evolving social phenomenon in which people form webs of personal relationships in cyberspace. Clinical psychologist Sherry Turkle, in her book *Life on the Screen: Identity in the Age of the Internet*, provides views about the impact of CMC on the way people think about their identities and contributing roles in society. Turkle found that people construct different relationships in their digital world than in their real-life world. Also, the anonymous nature of CMC allows people to change their personas. A man can be a woman, or a woman can be a man. Over time, the cyber world can become more real and more attractive to a person than the actual world. Some people choose to live most of their lives in the digital world. As people increasingly depend on their digital interrelationships, they are less concerned about and involved in the place where they physically live.

The Internet

The number of people connected to the Internet continues to grow at an exponential rate. The United States, with its over 111 million users at the start of the twenty-first century, is the world's leading Internet nation. There are another 100 million users across the world—approximately 46 million in Europe and 34 million in Asia. Similar to the nation's interstate highway system, the Internet is an immense digital system, linking millions of computer terminals connected to thousands of networks around the world.

The Internet emerged out of the R & D labs operated by the U.S. Defense Department's *Advanced Research Projects Agency (ARPA).* ARPANET was created in the 1970s and became the electronic research conduit used by the government, universities, and corporations dedicated to defense-related projects. As the advantages of this type of network became apparent and valuable to computer enthusiasts and experts at academic institutions who did not possess defense contracts, the National Science Foundation (NSF) stepped in to fund what has become one of the best known backbone networks, NSFNET. The Internet is therefore the U.S. government-sponsored successor to the ARPANET. Today, we use domain names to identify companies (such as Microsoft.com). Before the mid-1980s, all computers were identified by their Internet Protocol (IP) addresses (a series of numbers).

The Internet has more or less exploded into public view. A significant portion of its recently acquired constituency exists in the commercial sector. The fact that the Internet and its predecessors served federally funded research projects rather inconspicuously for nearly two decades is difficult to imagine today. The connectivity and diversity of the resources available through the Internet make it an unprecedented marketing opportunity for American businesses. Most regular Internet users in the United States and abroad think of it first as the WWW or don't think of it at all, as they use the Internet to send e-mail to family and friends.

The Internet is not static, however. The United States, as well as other countries, is exploring ways to increase the capacity of the Internet so more and better applications can be delivered. Internet2® software is a nonprofit consortium of U.S. universities, industries, and government agencies that is developing new ways of using the Internet. Universities in this consortium, over 180 by the latest count, are upgrading their computing facilities and linking to the high-speed networks that are the basis of Internet2 software. Internet2 software is complemented by a new federal initiative, Large Scale Networking (LSN). This time, instead of by the Department of Defense (DOD), development in networking is coordinated and funded by interagency groups under the National Science and Technology Council (NSTC).

Through this process, the United States is funding $2 billion of new research in networking technologies and services. The focus of this research is in seven areas:

❑ Development of new technologies and services that assist in the widespread development of wireless, optical, mobile, and wired communications.

❑ New networking software that will facilitate information transfer to individuals, groups, or the entire network.

❑ Research to address the growth and increasing scale of the Internet.

❑ Improvement of the performance of the Internet and research into new ways to measure and maximize this performance.

❑ New software for developing and using distributed applications—applications or programs used by many users at different locations.

❑ Development of new and improved distributed applications, for example, e-commerce or digital libraries.

❑ Research into infrastructure support and testbeds. *Testbeds* are limited access test networks that can be used to try new technologies and applications before they are released to the entire Internet.

At the same time, the U.S. government is exploring ways to set up its own private

Internet, GovNet® network, because of the open access and burgeoning use of the regular Internet. This new development brings us back full circle to the reason the Internet was first funded by the government—to develop a stable communication system for the government. GovNet network, if funded and realized, would be a private voice and data network open to government employees only (just like the goal of the original Internet of the 1970s).

As the Internet becomes more pervasive in our lives, we are exposed to its negative aspects. In the early days of networks, hackers were the most serious threat, but hackers only went after large targets. Today, many computer users are affected by digital hackers—viruses, worms, and Trojan Horses. The term *virus* is the most widely used because, in many cases, it has become the way people refer to any program glitch in their computer. A true computer *virus* is a small program designed to alter the way a computer operates and function without the knowledge of the user. Viruses can affect either a user's computer or an entire network. When you get a virus on your computer, it starts itself and starts reproducing and infecting clean files.

Trojan Horses are files masquerading as something else; Trojan Horses do not reproduce once they get to your computer. Generally, Trojan Horses are spread by e-mail, and they can only be started if you open them. When you open them, however, they can cause tremendous damage. A recent Trojan Horse, called Eurosol, targeted consumers who used a certain electronic payment system, the Russian-based WebMoney® system. The users were told they could earn money for viewing advertisements while they were on the Internet. After the user clicked on the option to begin, the Trojan Horse began showing advertisements and, at the same time,

looked into the user's computer to find bank data. Eurosol then transferred the stolen financial data to another computer.

Worms are programs that move from one computer to another automatically. The worm program places a new document containing the worm on a user's computer. Often, a worm uses an e-mail program to send itself to another computer. The Nimda worm used a security hole in Microsoft's Outlook® e-mail program to have users' computers automatically send infected e-mail messages to anyone on their mailing lists. The Nimda worm infected over 1.3 million computers in its first release in the early part of the twenty-first century.

Whether a virus, worm, or Trojan Horse, these intruders decrease the security involved in going on-line. These programs, at the same time, lead to increased use of antivirus software, certainly a booming industry in the digital world. No matter how good the antivirus, however, some programmer will find a way to circumvent it. Eventually, one or more of these unwelcome visitors may infect home users. So far, the spread of viruses and worms has not stopped the digital revolution. For most people, it appears to be a small price to pay for the benefits of information technology.

The World Wide Web (WWW)

One of the more novel information services available on the Internet is the World Wide Web (WWW). The WWW consists of pages of information sent over the Internet using a protocol called hypertext transfer protocol (http). The codes used to make web pages, hypertext markup language (html), and http were developed by Timothy Berners-Lee, a physicist who worked at CERN, a particle physics lab in Geneva. Berners-Lee wanted to develop an easier way for physicists to share information and research. His new method was eventually adopted and expanded and has become the WWW we know today.

The WWW became available on the Internet in the early 1990s. The WWW is the fastest growing Internet tool for users and businesses desiring to establish their cyberspace presences on the Internet. It is based on a hypertext paradigm, in which a document is organized as a graphical collection of multimedia objects, each of which has pointers to other relevant objects. The WWW provides the necessary tools to let anyone attach a website, and it exemplifies an open marketplace exchange of ideas. A website is accessed using its web address, or Uniform Resource Locator (URL). For example, http://www.oswego.edu is the web address for Oswego State University of New York, and http://www.apple.com is the web address for Apple Computer, Inc. URLs can be broken into the following components:

❑ http—hypertext transfer protocol.

❑ www—World Wide Web.

❑ oswego or apple—registered name in a domain.

❑ edu or com—name of the domain and type of service or organization supporting the account.

The top level domains on the Internet include *com* for commercial; *edu* for education; *gov* for nonmilitary government agencies; *mil* for military agencies; *org* for other types of organizations, such as nonprofit or research associations; and *net* for network.[2] Web addresses, like those above, generally indicate that the website is located in the United States. The highest level internationally would indicate a country. The web address for Oxford University in the United Kingdom(UK) is http://www.ox.ac.uk/; the *uk* in the URL indicates this site is in the UK.

The number of WWW pages has exponentially expanded since the early 1990s, and currently there are more than 1 billion web pages. These web pages are becoming more interactive and complex. It is common to watch videos, download music, view photos, and play games on the WWW. In addition, more business is being conducted via the Internet. Both consumers and businesses are finding it easier to compare prices on-line to get the best deals. This new use of the Internet, e-commerce, has the potential to change the way people interact with their world.

E-Commerce

E-commerce is sales or purchase of goods and services using the Internet. E-commerce is a fairly recent phenomena, but one that is exploding. There are many different types of e-commerce activities today, and the future possibilities seem endless. There are several categories of e-commerce today:

❑ Business-to-business e-commerce includes companies obtaining supplies and contacting suppliers through the Internet.

❑ Business-to-consumer e-commerce includes on-line versions of traditional stores, as well as digital catalogs such as catalogcity.com.

❑ Consumer-to-consumer e-commerce allows individuals to become their own companies by utilizing on-line auction sites, such as eBay® trading services.

❑ Information retrieval and services from banks, government agencies, libraries, or other sites allow individuals to run errands on-line.

Although, at first, we might think of e-commerce as a consumer purchase of a product through the Internet, most e-commerce

---

[2]Two very good references regarding the establishment of a business or commercial venture on the Internet are on the list of suggested readings for this chapter. See *Doing Business on the Internet* by Mary J. Cronin, and *The Internet Business Companion* by David Angell & Brent Heslop.

is actually done through business-to-business interactions. In fact, business-to-business e-commerce generates ten times the revenue as does business-to-consumer e-commerce. The automobile industry, for example, uses e-commerce to manage its parts suppliers and to reduce the costs involved with purchasing parts. The Spring Hill, Tennessee, Saturn plant, for example, has an on-line manufacturing database including the production schedule for its cars. Preapproved suppliers do not get purchase orders from Saturn; instead, they look up the required parts on the schedule and send them when they are needed. When they ship the parts to Saturn, the vendor sends an e-mail indicating what was sent. When the parts arrive, a clerk scans their bar codes. This scanning routes the parts to the place in the plant where they are needed. In addition, this information electronically tells the accounting department to pay the vendor.

Although e-commerce is increasing, it still constitutes only a small percentage of total sales in the United States (less than 5 percent). At the beginning of the twenty-first century, consumers averaged $999 million in e-commerce each week. Over the last few years, the amount of business-to-consumer e-commerce has increased significantly each year. Also, more people are using on-line auction sites. Consumer spending on auction sites increased from $238 million in August 2000 to $389 million in August 2001 in the United States alone. The U.S. Department of Commerce estimates that all types of e-commerce will continue to grow at an ever-increasing rate over the next few years.

In many ways, e-commerce allows a company to market its products more effectively. Most sites use imbedded web browser features to collect information about where a customer goes and what he or she clicks on when web pages are visited. *Cookies* are files placed on the user's computer when that user visits a specific website. Cookies generally store personal information about the user, such as passwords and screen names. Cookies are stored in a folder called "Cookies" in the Windows® computer services directory of your PC. If you look into this folder, you will generally see a long list of places you have been on the web. Another type of web tool is a web bug. *Web bugs* are little software programs on a website that can record the actions users take when they are on a website. Amazon.com keeps all information about what individual customers purchase on its site. When you return to Amazon.com, you will be greeted with a customized web page recommending books or products based on those you have already bought. The most interesting thing about Cookies and web bugs are that most users have no idea that this information is being collected. Since most people cruise the web alone, they assume they alone know where they go on the WWW. Instead, today, the WWW is more like a market research laboratory. Whenever you go on the WWW, someone is keeping track of you. This unseen surveillance does not bother some people, but to what extent it will be used in the future is an open question. A user is protected by an Internet provider's privacy policy, but how many people read their privacy policies? The biggest U.S. Internet provider, AOL communication service, recently made changes to its privacy policy to allow it to use tracking tools, including Cookies and web bugs, to compile data about AOL communication service subscribers and to measure the effectiveness of its advertising. This may make AOL communication service more effective, but this change does reduce the amount of privacy for AOL communication service's users.

The growth in e-commerce has affected states in their tax collection efforts. Since the late 1990s, there has been a ban prohibiting taxes on Internet access, as well as new taxes on Internet transactions, but this is not the main issue for states. Currently, individual states cannot force out-of-state businesses, including e-commerce sites, to collect state sales tax. For example, if a consumer lives in California and buys a book from an e-commerce site in Georgia, then the consumer may not pay the California sales tax. So, for example, if you live in California and buy a book from Amazon.com, you would not be charged any tax because Amazon.com does not have a physical presence in California. If, however, a California resident bought a baby gift from Amazon.com, then sales tax would be charged because the baby gift supplier is Babies 'R' Us, which has stores in California. Confusing? Just imagine yourself running an Internet business.

Another aspect of e-commerce falls under the umbrella of *e-finance*. Many banks, brokerage companies, and other institutions allow customers to manage their money and investments on-line. Another aspect of e-finance is on-line lending. Today, consumers can use on-line lending companies such as E-Loan or mortgage.com to get a mortgage, car loan, or credit card. Since the late 1990s launch of E-Loan, this on-line lending company has originated over $6.4 billion in consumer loans. See Figure 3-3. On-line lending companies increase the options for consumers, but at the same time, they increase the competitive pressure on local banks.

One prevalent concern with e-commerce lies with the security of the data being transmitted. Many consumers are fearful that their personal and financial data can be

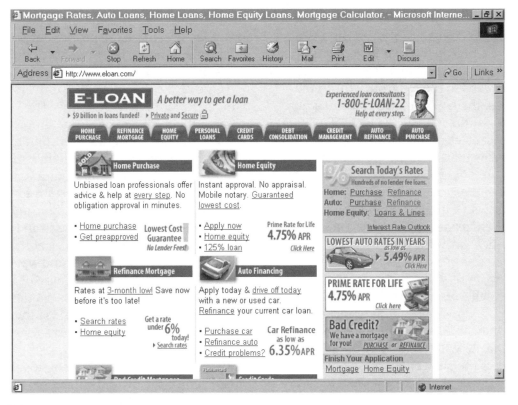

Figure 3-3.
E-Loan's web page shows there is no need to visit a bank because all loans can now be handled on-line.

stolen as they buy a book or gift on the Internet. These concerns are justified. In the past few years, various unscrupulous individuals have used numerous techniques including electronic eavesdropping, spoofing—creating phony sites that appear to sell legitimate goods, and altering data to change the dollar amounts to intercept this data. Most e-commerce companies have responded to these threats by increasing security and creating a secured network for all financial interactions. These secured networks begin with a URL of *https:* instead of the familiar *http:.* Most of the financial transactions are encrypted so only your computer can see the financial information.

E-commerce has transformed the U.S. economy as more businesses create on-line versions of themselves. In addition, the government is taking more advantage of e-commerce to assist its customers. Today, 94 percent of all customs declarations are electronic, and 60 percent of all duty is paid electronically. The Internal Revenue Service (IRS) is in the middle of a strong push to get all Americans to file electronically. In the late 1990s, the U.S. Congress passed a new law that requires the IRS to have 80 percent of Americans electronically filing by 2007.

It seems that, whether we want to or not, we are going to be required to use the Internet and e-commerce to complete our everyday work in the future. After 2007, will U.S. citizens have to pay extra if they mail in their tax returns rather than use electronic filing? Will department stores build any more stores if it is cheaper and easier to sell everything electronically? Will banks close down their local branches to save money and require customers to communicate with them electronically? No one knows the answers to these questions, but it is certain that both e-commerce and e-finance will transform the way we interact with businesses and each other.

On-Line Education

The Internet and the information technology revolution began in the educational world. Despite this, universities have lagged behind businesses in exploiting the WWW. In the past few years, universities have begun to adopt Internet business models as a means of increasing their revenue. As a medium, the Internet holds several attractions for education. Universities have extensive physical plant and, with reduced funding from state governments and research, it is more difficult to expand by building new buildings. Here, the use of the Internet seems ideal.

As compared to the rest of the world, the United States has a highly diversified higher education system, including research universities, private liberal arts colleges, public community colleges, and public universities. Today, a new entity is joining the education world—the for-profit provider. It is these for-profit ventures that are bringing new pressure to bear on educational institutions.

One of these new on-line universities is University of Phoenix On-line. The University of Phoenix On-line is an offshoot of the University of Phoenix program. The University of Phoenix began in the mid-1970s and offered accelerated undergraduate and graduate programs to working professionals. Currently, it has over eighty thousand students at thirty-two campuses and seventy-one learning centers established in eighteen states; Puerto Rico; and Vancouver, British Columbia. The University of Phoenix On-line, begun in the late 1980s, offers 100 percent of its coursework via the Internet. This makes this program particularly appealing to students who are working or having problems commuting to a traditional campus. Currently, this university has over twenty-five thousand students taking courses toward a bachelor's, master's or doctoral distance learning degree.

The growth of the University of Phoenix and other on-line universities has caused a tremor in the halls of the traditional university, and the result has been that many traditional universities have jumped on the on-line bandwagon. A growing number of schools—including Duke University, Stanford University, and New York University—hope to profit from on-line learning like the University of Phoenix did. The Massachusetts Institute of Technology (MIT) is taking another approach. MIT recently announced it would post lecture notes, course outlines, and teaching materials for all of its classes on-line—free.

Will universities join the e-commerce trend and begin selling courses on the WWW? Or will they follow MIT's lead and provide information for free and for the betterment of society? There is no simple answer to these questions. As universities feel increasing financial pressure, they will undoubtedly use the WWW and other networking technologies to alleviate this stress. The effect of on-line learning on the quality of education will be an ongoing concern.

Telecommunications Networks

The delivery of personalized communication services to all people, regardless of where they work or reside, is one of the underlying ambitions of the national information infrastructure. The United States maintains a competitive edge in the development of networking technologies. Today, many of us are so accustomed to surfing the Internet that we ignore the technology underlying it. The Internet has become as ubiquitous as the telephone system—we only think about it when it does not work.

The Internet is a vast computer network comprising many smaller networks and links. Networking, however, is no longer the sole province of large businesses and universities. Small businesses, even homes, can now utilize emerging technologies to create networks. As many small businesses and home users learn, the problem with connecting to the Internet via a modem is the slow speed. A network using a wireless connection or a cable modem can decrease the time it takes to access the Internet.

How does a network work? It begins with the individual computers. Each of the computers has to be able to connect to the network, either through a modem or a network interface card (NIC). The most common type of NIC allows the computer to use an Ethernet type of network. See Figure 3-4 for a graphic example. This network can use wires, or it can be wireless. If there are enough computers, one might add a hub to connect these users. Since a hub requires that users share the network, if there are too many users, either a switch or a router would replace a hub. A switch sends information only to the recipient, not to the entire network like a hub. A telephone analogy would be the difference between a private line and a party line. In a party line (hub), when one person is using the line, no one else can use it. In a private line (switch), each party uses a dedicated line to talk to another party.

Routers are even more complex. Routers use the complete address to decide where the information should go. Routers are used to connect one network to another in the same building and as an interface to connect a network to a wide area network (WAN). As shown in Figure 3-4, servers are added to the network when there is more demand for shared files. It is faster to place shared files on a central server than to e-mail them back and forth.

In general, local area networks (LANs) are used for people who are in the same building or location, and WANs connect users and LANs from different locations.

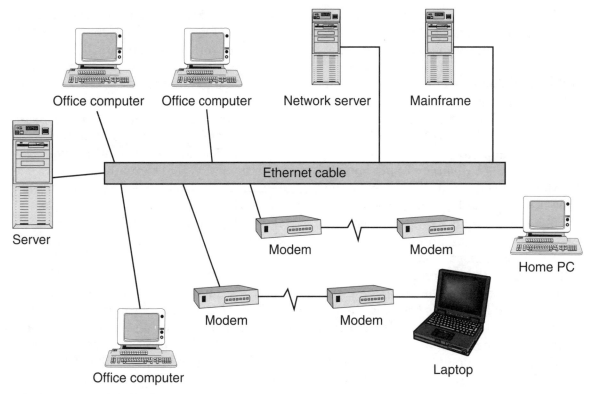

Figure 3-4.
This network for a growing business includes computers connected to a LAN, as well as the ability to access the network through a modem.

Connection speed of LANs tends to be much higher than that of WANs. A basic Ethernet LAN has a capacity of 10 million bits per second (bps). A fast modem, in contrast, can only transmit information at fifty-six thousand bps—this is about 1 percent the speed of an Ethernet LAN. A home user or a small business can get faster connections. Integrated Services Digital Network (ISDN) is a network technology that allows traditional voice or telephone service to share the network with data or digital traffic. ISDN uses regular telephone lines. An ISDN connection is about twice as fast as a standard modem, while a cable modem connection can be almost as fast as an Ethernet LAN.

Faster ways of accessing the Internet, coupled with new networking hardware,

have opened a new market—home networking. The number of home networks will continue to increase in the next few years, as dual computer users realize the value of sharing fast Internet links among different computers in the home. Several computer companies are jumping on this new market. Compaq is now selling home networking products that will allow easy networking using phone lines, Ethernet, or wireless communication. The key difference is ease-of-use. Eventually, these companies promise that home networking will be as easy as plugging a new telephone into an existing telephone jack.

In the coming decade, breakthrough technologies to support both *optical networks* and *wireless networks* are ones to watch. These R & D pursuits will guide our political representatives in their formulation of the next version of the Internet. As these networks are interlinked, it is essential to not only address issues pertinent to data communications, but also to be attentive to the manner in which information is represented, transferred, and protected.

Fiber Optics

Contemporary fiber-optic networks emerged on the scene in the 1970s, when Corning Glass researchers fabricated a useful light wave conduit. These glass fibers in use today transmit voice, video, and data at speeds ten to two hundred times faster than the standard copper wire used for more than one hundred years in telecommunications networks. There are two types of optical fibers. Single-mode fibers, used in telephones and cable TV, are used to send one signal in each fiber. The more common type used in telecommunication is multi-mode fibers that can transmit thousands of signals in each fiber. It seems that almost every year, the capacity of fiber-optic cables increase. In the late 1980s, the first transatlantic optical cable entered service; within three years, another optical cable could carry double its calls each second. AT&T's Synchronous Optical Network (SONET) fiber-optic cable has 288 glass strands; each of these strands can carry forty-eight thousand telephone calls. See Figure 3-5. The WaveStar® 40G Express computer network product, developed and being tested by Lucent Technologies, makes the SONET appear like a snail. The WaveStar computer network product can carry half a million phone calls on one single glass fiber. Overall, transmission speeds for fiber-optic cables quadrupled in the late 1990s.

For most fiber-optic systems to transmit a conversation, voice signals (electrical impulses that copy a caller's voice) are

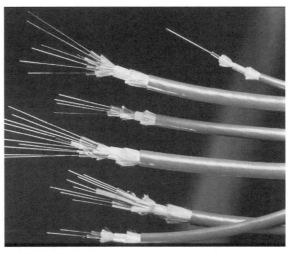

Figure 3-5.
The fiber in this cable provides the transmission medium for optical communication systems. A typical fiber-optic cable made up of one hundred or more such fibers can carry more than forty thousand voice channels. (Corning Incorporated)

converted into digital form. When computers are the senders and receivers in the link, the conversation is already in digital form. The digitized information is carried on light pulses produced by laser diodes through a glass fiber at a very fast rate (faster than forty megabytes per second). When this information reaches its final destination, it is translated into its original analog form. *Fiber optics* are being installed to replace old wiring systems at an increasingly rapid rate. Unlike electrons, using light waves makes transmission less susceptible to electromagnetic interference caused by electrical storms or faulty hair dryers. Since these conduits do not create electromagnetic fields, inductive pickup devices cannot tap fiber-optic cables. As a result, the messages are ultimately transmitted without distortion and with almost no loss of information. Satellites have lost some of their telecommunications business to optic fibers because these networks are more secure. It is significantly more difficult to steal signals from enclosed glass filaments than to pirate signals sent through the atmosphere.

In contemporary fiber-optic networks, each time a light pulse is amplified, switched, sent into, or removed from the network, it must be reconfigured into a stream of electrons for processing. Light wave signals are diminished when the fiber-optic cabling is interrupted to connect new hardware. It becomes more difficult for the network to process the smaller pulses of light required to transmit gigabits of digital information each second. In today's ultrahigh-speed networks, once the transmission rate exceeds fifty gigabits per second, it is very hard for electronic equipment to maintain the constant back and forth conversions between light waves and electrons.

Wireless Networks

In another camp of internetworking technologies, communication technologists are confronting the challenge of providing computer users access to ubiquitous wireless networks. Wireless communication is indeed a significant growth segment of the global telecommunications industry. Mobile workers and residents of remote areas anywhere may soon be the beneficiaries of personal communication services that were heretofore unimagined.

People whose employment requires them to connect to their offices from remote locations are finding it easier to do so with each passing year. Mobile workers have become accustomed to the liberation afforded by cellular telephones and now expect the same range of freedom for their digital links. The use of cellular phones exploded in the late 1990s and is considered by many to be a necessity, not a luxury. As personal digital assistants (PDAs) like the Palm Pilot™ information management software, Sony Clie® PDA, or Compaq iPaq®

PDA become more common, they will increasingly use more wireless functions.

A typical demand for a wireless link between the traveling worker and the central office is to exchange e-mail. More rigorous expectations of wireless networks entail connecting remote employees with real-time corporate data, applications running on network servers, and mainframe computers in the home office. Wireless networks might also be used to provide service technicians who come to your home or office, access to technical documentation, and up-to-date warranty information. Anyone who has called the United Parcel Service or Federal Express (FedEx), requesting information about a shipment, knows both companies currently use unique wireless software and thousands of mobile computers to track all their packages in transit.

As with any new area of development relying on concurrent advancements in several science and technology suites, many obstacles stand in the way of wireless networks. One key difference between wire and wireless service is the transmission medium for radio communications that is regulated and requires frequency administration. Unlike examining the potential for success in an optical fiber or wire environment, a test of a new concept using radio technology calls for a ruling by the FCC (Federal Communications Commission). Employees who connect to the office from a field site want high speed. Even if they can settle for low throughput rates, there are still areas of the country devoid of any form of wireless services. Finally, a lack of standards makes it necessary for people to use different wireless services such as the existing cellular telephone network.

Portable telephones, PDAs, pagers, and other wireless devices are being purchased at an increasing rate by customers across the nation. The mid-1990s saw an explosion in wireless use, with the number of customers subscribing to wireless services in the U.S. increasing 584 percent, from 16 million in the early 1990s to 109.5 million in the late 1990s. The growth recently has even accelerated; over the last two years, subscribers grew at an astounding rate of 24 percent a year—over 20 million a year. Today, there are more wireless phones sold in the U.S. each year than personal computers.

Despite the tremendous growth in wireless use in the United States, amazingly, the United States lags behind other developed countries in wireless use. The amount of cell phone use in Japan is significantly higher than in the United States, and Japanese users tend to use their cell phones for more than telephone calls. Seventy-two percent of Japanese cell phone users access the Internet using their cell phones—in comparison to only 6 percent of U.S. cell phone users. The United States tends to lag behind in cell phone use because it has a better existing land telephone infrastructure and more affordable telephone prices than most other countries. Even with the United States bringing up the rear in the use of wireless technologies, this market will still expand. In a few years, experts predict that the number of cell phones could surpass traditional telephones. By 2006, there is expected to be 1.6 billion cell phones, as compared to 963 million traditional telephones.

Cell phones are only the first level of wireless technology. As the options of cell phones expand and they become minicomputers, the demand for additional wireless technology will grow. By 2005, all cell phones sold in the United States are expected to contain another wireless feature, a Global Positioning System (GPS) chip. (GPS is discussed in more detail in Chapter 4.) In the mid-1990s, the Federal Communication Commission (FCC) passed a new rule for cell phones requiring all cellular phones to include a way to pinpoint the caller's location for 9-1-1 calls to within fifty to three hundred meters. Because of this enhanced 9-1-1 rule, all U.S. cell phones will eventually contain a GPS chip, but one wonders what other uses this chip will have. With a GPS chip, the cell phone could now notify you when you drive by a particular store or give you directions to a client's office. It is unclear how this technology will be used and how much unsolicited information the consumer will accept.

In the future, the applications in handheld computers will also expand. Today, it is possible to browse the Internet, get directions from a GPS, and link to a home computer with a PDA, all using wireless technology, but this is only the beginning of the expansion of wireless technology. See Figure 3-6. Wireless services have begun to appear in noncomputer devices. You can buy a new car equipped with an imbedded wireless system. Drivers will be able to get e-mail, driving directions, news, and stock quotes— the sky is the limit. This type of application is called *telematics*—the use of wireless voice and data technologies between a car and somewhere else. See Figure 3-7.

Telematics first was used on Ford Lincoln® cars in the mid-1990s, but it was mainly for roadside assistance. General Motors (GM) expanded the telematics to include driving directions and now offers its OnStar® system in thirty-two of fifty-four U.S. brands. As more buyers lined up to buy these systems in cars, more enhanced systems have been developed. Siemens exploits the functions of a Palm PDA into its telematics system, Quick Scout™ navigation system. Car owners can slide their PDAs into special cradles in the dashboards of their cars. Quick Scout navigation system exploits the display of the PDA and allows the driver to see or hear the driving directions, e-mail, or traffic information. Other new telematics applications

Figure 3-6.
The Palm i705 includes a built-in antenna that lets its owner access the Internet and e-mail through a wireless network.

are directed at commercial users. Daimler-Chrysler, in affiliation with AT&T Wireless, has developed a new wireless system that will be available in all of its makes and models. The system will work through the user's cell phone and will activate when the user enters the car. The system will allow all current and future services for AT&T Wireless customers to be accessible inside the car.

Universities have also adopted wireless technologies. Universities are discovering it is more efficient and affordable to use wireless technology than to provide computer laboratories for their students. Carnegie Mellon University has begun replacing its existing wire network with a wireless network, Wireless Andrew, which will cover the entire 103 acre campus. Wireless Andrew will include a scattered array of antennas that will each supply coverage to a small geographic area of the campus.

The future of wireless technology seems limited only by the imagination and resourcefulness of industry. Already, companies

are demonstrating wireless technology that can be worn as clothing. California-based Charmed Technologies displayed a wide range of these wearable, wireless devices at a recent trade show. Imagine earrings lighting up when you get an e-mail message, a belt attachment acting as a voice input device for your handheld computer, or a bracelet containing an electronic organizer. Sony is already marketing a wristband digital camera in Japan that sends its pictures to a desktop computer using wireless technology. In addition, in the aviation industry, wireless communication is being expanded through digital cockpits that include computer displays allowing the pilot to merge information gathered from GPS and other sources, including the route and terrain on the same display.

People will not be the only ones using wireless technology in the future. Experts predict that wireless technology will be used for tracking and security purposes. A cellular company in Japan, NTT DoCoMo, projects that by 2010, 20 million wireless

Figure 3-7.
Digital cockpits can already be found in many of today's aircraft. Dassault's EASy (Enhanced Avionics System) cockpit is based on Honeywell's Primus Epic® next generation avionics suite. The Primus Epic system has been selected by original equipment manufacturers for every new business aircraft, regional jet, and helicopter program since its introduction in the mid-1990s. (Honeywell)

devices will be implanted in pets and 60 million will be in bicycles. No more worrying about your dog getting lost, because you can track it through an imbedded GPS chip. Whether these projections hold true, there is no doubt that more uses will be found for wireless technology.

Intranets and Extranets

A business can utilize networking and web technologies to create a private corporate network called an ***intranet.*** Today, the computers in most businesses are connected to the Internet through one means or another. Intranets can be kept private and protected through a ***firewall.*** A firewall consists of hardware and software placed between the private intranet and the outside network (the Internet). The firewall examines all information entering and leaving the intranet to make sure there are no unauthorized messages. In this way, the firewall stops hackers from breaking into a company's computer system. The actual intranet uses web pages to run the corporate network. The difference is that the intranet web pages are only accessible to people inside the company.

An intranet becomes an *extranet* when a company allows customers and other users to have limited access to the organization's intranet. GM, for example, uses a corporate extranet, along with wireless technology, to distribute service manuals to its dealers. A few years ago, GM converted from a paper manual system to a PC-based system in which technicians accessed information from CD-ROMs. However, the company discovered this was ineffective because most repair shops only had one computer. The manual information was loaded onto a corporate extranet, and now technicians can access this information through a secured Internet site open only to authorized users. Technicians can even access these manuals at their individual workstations using special handheld computers designed by Fujitsu.

Another new type of network is a *virtual private network (VPN)*. A VPN is a low cost alternative to a private WAN. Before the Internet explosion, the only way a company could create a network linking remote offices was by creating its own WAN. A WAN generally requires that a business either lease lines between its various offices or provide dial-up access. In any case, they require that the business spend a lot more time and money monitoring and administrating their own private network. With the growth of the Internet, many companies are accomplishing the goals of a WAN without having to lease telephone lines themselves by using a VPN. A VPN is a secure connection between two points on the Internet, and in contrast to a WAN, VPNs are available through Internet providers. Remote users just use their

modems or Internet connections to access the Internet provider. This leads them to the secure VPN. Companies have quickly realized the cost benefits of VPNs. About half of mid- to large-sized companies already use some type of VPN in their business. Experts predict that spending on VPNs will surpass $32 billion in just a few years.

All these types of networks are exploding, as companies realize the cost savings involved in networking. When a FedEx client uses the company's extranet to track the status of his package instead of calling the service center, FedEx saves eight dollars. Cisco Systems has a large intranet it uses to link the corporate headquarters in San Jose, California with offices in seventy-five countries. Tupelo Grill, a restaurant located across the street from New York's Madison Square Garden, has an interesting twist on this type of network using the AT&T network and software from cEverything® transmission. The owners can watch their restaurant, using streaming video and the Internet, from anywhere in the world. This allows them to monitor the service their customers receive without actually being there. The growth of *telecommuting* (see section below) will only increase the creation and use of VPNs by companies throughout the world. Instead of costly travel to a company seminar, a company can utilize its VPN to deliver training to employees located throughout the world.

When we think of a traditional business, we imagine a quaint storefront located either in a mall or downtown business area. Now, these storefronts are digital, and the goods displayed are located somewhere else in the country. As more companies take advantage of networking over the next decade, there will be profound changes as to how we use and view the Internet. What these changes will be, only time will tell.

Intelligent Software and Virtual Reality

In the not-too-distant future, our already information-saturated, sensory-overloaded society will make use of **intelligent software agents** that know users' interests and needs and act autonomously on their behalf. In the true sense of technology designed to expand or enhance a person's physical and intellectual potential, these software agents may assume the role of personal secretary to the user. Agent programs differ from conventional software in their capacities to learn from experiences and respond to unexpected situations. An ideal software agent program strives to achieve a known goal and can sense the current state of its environment in order to make continual progress toward the goal. For example, a college professor may employ an intelligent agent program to update his or her lecture notes at the end of each semester. Specifically, the agent's knowledge that the goal is to locate projected international energy resource consumption rates for the next twenty years will direct it to patrol the Internet and beyond, to seek information for its owner.

One glitch to using intelligent software agents is they look like invasive viruses to the security systems on other users' computers. This mistaken identity could prompt the network to delete agents or refuse them entry to various sites. Software engineers perceive a clear and present need to devise standards to ease the travels of these agents across the Internet. Currently, there is no clear agreement about what an agent is or what it should be. A recent development called the Unified Agent Architecture (UAA) proposes a uniform look for mobile agents. In essence, UAA attaches special codes to software agents that both identify their owners and allow them to be destroyed if they run awry or if a virus contaminates them. The technologies to monitor as this field is further commercialized include expert systems and neural network simulators.

A few short years ago, **virtual reality** was regarded more as science fiction than as a functional tool. Television reports presenting virtual worlds as computer generated alternatives to reality have convinced us that virtual reality has the overarching potential to transform computers into extensions of our whole bodies. While we are still a long way from the world of having a virtual reality chamber in every household, virtual environments figure prominently in several industrial application areas, including marketing, prototyping, and training.

Contemporary applications for virtual reality range from firefighting to medicine to the design of automobiles. Virtual reality is a promising training medium for people who aspire to be firefighters and others who must learn to resolve dangerous tasks without risking their own lives. Virtual reality takes simulation another generation forward. It transforms (and even distorts) a person's perceptions, by appealing to several senses at once (vision, hearing, touch) and presenting graphic images that respond immediately to her or his movements. Virtual reality permits people to behave as if they were somewhere they really are not. Virtual reality equipment has been used to help people overcome their fear of heights. Pilots and astronauts train in virtual reality cockpits that combine three-dimensional graphics with the view out the window while sound systems provide environmental cues. Automotive companies are looking at virtual reality technologies as they apply to car design, construction, assembly, and repair.

With the advent of Apple QuickTime® digital video, a new type of virtual reality has sprung up using QuickTime VR. Students can visit the Louvre and take a virtual tour of its galleries by accessing its website. Instead of spending thousands of dollars on equipment and travel, one can climb Mount Everest or explore the pyramids through a virtual expedition. Another current use is in real estate. QuickTime VR allows realtors to provide virtual web tours of their properties. Instead of wasting time driving to a house, a potential buyer can view the house on the WWW. Beyond its entertainment profile, virtual reality is an aspect of information manipulation you can expect to be around for a long time.

Telecommuting to Work— Virtual Transportation

> **TELECOMMUTING**
> The concept of using telecommunications technology to transport work to employees who therefore avoid the daily commute to a distant job site.

Just as virtual reality technology allows people to behave as if they were somewhere they are not, a convergence of powerful technologies enables telecommuters to work as though they were in their corporate offices, even though they are at their home computer terminals. Alvin Toffler first coined the term *electronic cottage* in his best selling book, titled *The Third Wave*, in the early 1980s. This social phenomenon has given way to yet another new concept labeled "virtual transportation," in which work is transported to the employees instead of the employees physically transporting themselves to the job site.

History of Telecommuting

Jack Nilles at the University of Southern California's Center initially coined the word *telecommuting* for Futures Research in the early 1970s. Oddly enough, it coincided with a major international oil embargo that prompted researchers to seriously evaluate the use of telecommunications networks to allow workers to avoid gasoline-dependent automobile trips to and from work. Another term frequently used, particularly in Europe, to describe remote working is *teleworking*.

The past two decades have introduced more computers, LANs, e-mail messages, fax machines, videoconferences, and other services into the workplace. These converging technologies have greatly expanded the number and types of jobs that can be performed wholly or in part without travel. Recent estimates of the number of people who already telecommute at least one day a week range from 25 to 32 million. About 10 to 20 percent of the U.S. workforce teleworks, either at home, on the road, or in a satellite office. Very few people telework full-time, however. Most still spend about two to three days per week at their central offices. For most teleworkers, work tends to be allocated differently during office days and telework days. Office hours are used for interpersonal work activities, such as face-to-face meetings and team meetings. Telework hours are generally reserved for projects, writing reports, and research. VPNs have made it easier for workers to telecommute, since they allow the home worker to access the office computer easily and more securely.

While it is difficult to forecast future telecommuting scenarios, experts estimate that some 25 million people could be efficiently telecommuting the majority of the workweek by the decade's end. Writers, consultants, stockbrokers, realtors, software engineers, music teachers, accountants, architects, and government employees who work for countless companies and state and federal offices are already avoiding the hassles of the daily driving commute.

Public interest in telecommuting in the coming years will more than likely rest on the social goals to alleviate problems of traffic congestion and the related energy and air quality issues. A survey conducted by the U.S. Department of Transportation in the early 1990s revealed the American commuter drives an average of four thousand miles back and forth to work each year, burning just under two hundred gallons of gasoline, and the EPA found that employee commutes to work comprise 28 percent of all vehicle miles yearly on American roads. Emissions of greenhouse gases range in the millions of tons each year, which causes severe environmental problems. If just 10 percent of the nation's workforce would telecommute one day a week, there would be thirteen thousand tons less air pollution, and the country would conserve more than 1.2 million gallons of fuel each year.

Based on productivity surveys, there is a positive benefit to employees telecommuting. If one employee telecommutes, others in the firm must be more organized, which may partially explain gains in productivity. Also, telecommuting employees tend to use less sick time and work a longer workday. If telecommuting programs are critically conceived and implemented, reported productivity gains of 5 to 19 percent are typical.

Benefits of Telecommuting

There are some proponents of telecommuting who suggest consumers will be the ultimate beneficiaries as the trend toward telework continues to emerge. Companies' direct expenses may be lower. Established firms will face greater competition, as entrepreneurs take advantage of technical developments and hire workers who had not previously been part of the traditional workforce. This may ultimately lead to lower consumer costs for established goods and services.

One of the most significant economic benefits for companies considering launching a telecommuting program is the potential for saving on current or additional office space, building maintenance, and computer time. Unlike costs associated with the computers and networking, construction prices continue to climb. When a company doubles its workforce in a short period of time, the cost of building expansion may be prohibitive.

Telecommuting programs are also a mechanism through which organizations can recruit and retain valuable employees. These programs can also cut relocation costs for both new and veteran employees. Companies can eliminate the fees associated with moving their personnel to housing within commuting distance by allowing these individuals to work from their own homes. Telecommuting workers may then be asked to commute only occasionally to either the main office or to a nearby satellite office.

Telecommuters who sincerely enjoy working at home tend to work harder and longer than their office-restricted counterparts. One of the most prevalent productivity nightmares is the winter blizzard. If the office does not have to close down completely, it is probable that the few employees who do manage to show up to work will spend most of the day worrying about the trip home instead of getting any useful work done. The employees who never left home are easily going to be more productive under these circumstances.

In short, telecommuting promises benefits for all concerned. It may yield lower costs for businesses, more freedom for workers, and a fuller family life for parents. Telecommuting also appears to increase the range of services and products available to consumers and lowers prices along the way.

Potential Drawbacks

Before you close the book to start dreaming about the luxurious life of the telecommuter, it is necessary to review some of the not-so-wonderful aspects of telecommuting. From the corporate perspective, one of the most frequent excuses used in opposition to telecommuting programs references the loss of managerial control. Some executives believe it is impossible to manage people whom they never see. Overcoming this out-of-sight prejudice may take more time for some than others. Telecommuters must be measured strictly by their output and have fewer opportunities to "snow the boss," so to speak.

From the employees' viewpoint, the out-of-sight, out-of-mind syndrome may be more serious. They may be forced to deal with the possibility they will be overlooked when it comes to such things as promotions, new projects, and raises. Prime decision-makers within organizations who give promotions tend to work during conventional daytime hours. Those workers who work outside of the nine-to-five routine have always been at a disadvantage. Since it is frequently the casual interactions taking place on the job site that open doors for career advancement, the invisible employee may remain looking in from the outside.

Many people look to their work places as respites from domestic chores and responsibilities. If the home becomes the workplace, these people may experience a sense of imprisonment and isolation. Adult workers often need a chance to interact with colleagues for socio-psychological reasons. Face-to-face meetings also encourage workers to remain involved with the organization and keep abreast of technological changes that may be pending.

More importantly, working at home requires a considerable amount of self-discipline, coupled with tremendous concentration.

Although the office environment has its share of interruptions, a person's home is full of distractions. Unsuccessful telecommuters report they have gained weight, learned how to procrastinate in the garden, and become hooked on a daytime soap opera. On the opposing side of these stories is the tale of the man who insisted on dressing in a three-piece business suit each morning before walking down the hall to his den to begin the workday with a tape of the sounds of his former office. Family members may not be able to adapt to the ever-changing roles the telecommuter is forced to play when the home is where the job is, which may be viewed as the "Mommy or Daddy doesn't work anymore" syndrome.

Finally, the issue of security eventually comes to the mind of corporate executive officers who are responsible for managing large amounts of data. Is it wise to allow sensitive corporate data out of the office? Information becomes vulnerable as it is conveyed electronically via a modem, diskettes, or e-mail. Papers shipped to and from home workers can easily be lost or stolen. Who is liable for the security of the information once it leaves the corporate office—the chief executive officer (CEO) or the telecommuter?

Employee Selection Criteria

Some telecommuting advocates say the question is not who should be allowed to work at home, but who should be *forced* to work in the office. This statement may be somewhat premature in light of the above concerns. Other proponents indicate the key to initiating a successful telecommuting program is the selection process itself. The company, rather than the employee, should set the selection criteria. Once these criteria have been established and potential candidates have been identified, the process should then become one of voluntary self-selection.

In other words, approved employees should have the option of either accepting or rejecting the offer.

Some of the numerous personnel selection guidelines used by corporations who have been satisfied with their telecommuting programs are summarized below. If you ever find yourself in the position to make this type of high-level corporate decision, perhaps these criteria will be insightful:

❏ Employees whose work is dependent on direct interaction with customers who visit corporate offices cannot work effectively at home.

❏ Employees who are regularly involved in team projects and those whose work is regularly dependent on interaction with their colleagues are usually eliminated from consideration. Some companies, however, allow team employees to meet at specified times at the office. The trick is to avoid wasting time playing "telephone tag."

❏ Some companies have had success in hiring employees specifically to work at home. Most recommend that telecommuters be selected from personnel who have been with the company for some time and understand how it operates.

❏ The psychological makeup of the employee is also crucial. Not everyone is suited for telework. Telecommuters must be intrinsically motivated individuals who can work well without immediate supervision. It is also prudent to eliminate those people who obviously place a high priority on the social relationships fostered in the office.

❏ Successful companies do not offer the telecommuting option as an employee benefit. It is not a reward. It is simply another way of getting the job done.

❏ Lastly, successful telecommuting requires that telecommuters return to the office periodically. As a rule, most effective programs tend to require their telecommuters to return to the main or area office at least one day per week to "plug back into" the daily office routine.

Videoconferencing and Teleconferencing

Videoconferencing was first introduced in the mid-1960s, at the New York World's Fair when AT&T unveiled its videophone that delivered voice and video over standard telephone lines. It took at least twenty years for corporations to make serious investments in videoconferencing technologies. The current pace at which U.S. companies are installing videoconferencing capabilities continues to accelerate.

VIDEOCONFERENCING
The ability to communicate visually using computer technology, television monitors, cameras, microphones, and special modems. An interactive group communication takes place through an electronic medium.

The basic philosophy of videoconferencing has focused on moving electrons rather than people. These information and communication technologies provide an alternative to face-to-face meetings and the frustrations of playing telephone tag. Advocates of this type of electronic meeting credit videoconferencing for reduced business travel costs and much improved communications. It is estimated that a significant percentage of all business meetings requiring travel are intercompany conferences between people who already know each other. Electronic meetings can be scheduled and might eliminate the necessity of being physically together.

Businesses who want to establish offshore subsidiaries might be able to enter foreign markets sooner by using videoconferencing to reduce the number of required trips to destinations around the globe. For instance, traveling to Argentina once a year and then maintaining contact via bimonthly videoconferences is considerably more manageable and less expensive than making multiple intercontinental trips.

Concurrent engineering (CE) is another area that can benefit greatly from videoconferencing applications. CE (discussed in Chapter 6) calls for a refinement of the product design cycle. Basic stages, such as conceptualization, preliminary design, final design, and manufacturing engineering and assembly, are no longer completed independently of each other. They are executed concurrently among various work teams. Videoconferencing allows these often geographically dispersed groups to be virtually colocated. They are able to refine engineering drawings and argue interactively about key design aspects of a prototype without leaving their work sites. Technical issues can be worked out a lot quicker using videoconferencing than if a series of face-to-face meetings had to be arranged.

The potential applications of videoconferencing using today's technology alone are limitless. Universities and public schools are using videoconferencing technology to teach courses to students at distant locations. The capacity for even more people to telecommute to work will undoubtedly be influenced by videoconferencing technologies that are becoming more affordable.

Teleconferencing is the newest videoconferencing option found in LAN and desktop systems. Instead of telephone lines, these systems use personal computers connected through a LAN or the Internet. Vendors, including Intel, InSoft, and PictureTel, have released these systems costing between fifty and two thousand dollars. CU-SeeMe®

TELECONFERENCING

A videoconferencing option found in desktop sytems connected to a network utilizing digital video transmission.

---

software was developed a few years ago at Cornell University. Cornell researchers have distributed their software over the Internet, free of charge. To become a CU-SeeMe software videoconference participant, a user needs an inexpensive video camera, a computer with either a MacIntosh or Microsoft Windows operating system, and access to the Internet through a modem or other connection with a minimum 28.8 kilobits/second capacity.

Another simple and inexpensive avenue to experience teleconferencing is found in the Microsoft NetMeeting® computer program. The NetMeeting computer program is software allowing remote users to hold face-to-face meetings; share programs; and communicate with anyone in the world by video, sound, text, or an electronic whiteboard. The NetMeeting computer program works with a wide range of digital cameras and Internet technologies. Since this software works on any Windows computer services–based computer, it is convenient for teleconferencing. Toys 'R' Us uses the Microsoft NetMeeting computer program to train employees located in its nineteen regional distribution centers, while Boeing uses the NetMeeting computer program to allow electronic weekly meetings of the project team building two new demonstrator jet fighters for the DOD.

Teleconferencing has expanded the vision of educational institutions as well. See Figure 3-8. Most large universities have offered traditional distance education programs using videoconferencing, or similar technologies, for many years, but teleconferencing has increased the possibilities for education at a distance. A professor in San Diego, California taught an on-line geography class using the NetMeeting computer program for students at Western Michigan University. Macquarie University in Australia uses the NetMeeting computer program as a way for students both on and off campus to talk to a reference librarian.

Figure 3-8.
Education Service Center, Region VI in Huntsville, Texas hosts a virtual field trip with the National Aeronautics and Space Administration's Johnson Space Center as part of a staff development training initiative for teachers. Elementary, intermediate, and middle school science teachers participated in an hour-long two-way interactive videoconference with NASA Learning Outpost staff and research assistants, featuring a customized tour of the International Space Station mockup. The left monitor shows Sherri Jurls, the tour guide, at the National Aeronautics and Space Administration remote site. The right monitor shows an audience of Region VI science teachers at the local site of Huntsville. (Sandy Goss, Science Education Specialist; Lynda Degeyter, Distance Learning Specialist of Education Service Center, Region VI in Huntsville, Texas)

Both videoconferencing and teleconferencing are expanding at an ever-increasing rate. As they do, the line between these two technologies seems to fade. If one uses a dedicated videoconferencing facility, but also uses the NetMeeting computer program, is this videoconferencing or teleconferencing? Whatever one calls it, it has the potential to dramatically change both education and the workplace.

Potential Effects of Remote Conferencing

Both videoconferencing and teleconferencing can be considered as forms of remote conferencing. The primary reason for utilizing these in the business environment is reduced cost through travel substitution. Higher-level executives are generally attracted by this claim. It is no secret that the cost of moving people from place to place is escalating, as is the discomfort of the people being moved. Although some employees genuinely enjoy traveling, most find it very draining. Interestingly enough, remote conferencing has not yet proven to be a direct substitute for travel.

If remote conferencing is to be effective, it must be reviewed critically for what it is—a branch in the family of office systems technologies. Remote conferencing should be viewed as a business communications support system rather than as a rigid replacement for face-to-face meetings. Not all organizations have the same potential for success using remote conferencing. This is not necessarily due to a lack of technical competency or even insufficient investment capital. It may very well be this technology does not address a specific task or area of need within every corporate setting.

In order for the decision to implement remote conferencing into one's operational strategies to be made intelligently, several aspects of the technology should be evaluated. In other words, do the potential cost advantages outweigh the possible negative side effects that may develop over time? Take a moment to place yourself in the position of the CEO who is attempting to make this type of commitment to a new form of business communication. Consider the following potential cost advantages of remote conferencing:

❑ Reduce travel expenses.
❑ Reduce unproductive time for people doing the traveling, as well as their inevitable travel fatigue.
❑ Reduce mistakes made due to the fact the critical person was not at the meeting as scheduled.
❑ Shorten the normal business cycle and facilitate more efficient decision-making.
❑ Reduce equipment downtime when the most skillful repair person is at another site.
❑ Reduce the effects of unforeseen interruptions due to weather, fuel shortages, or political upheavals.

On the other hand, consider the following potential negative effects of remote conferencing:

❑ Increase the amount of unproductive time spent in meetings that are easily arranged, but unnecessary.

❑ Decrease freedom of operation for the remote field sites due to an excessive amount of managerial control.
❑ Lower employee morale because of decreased personal contact.
❑ Encourage people to become overspecialized and narrow-minded.
❑ Foster an overdependence on technology that could easily fail as a result of equipment limitations or sabotage.

No decision regarding technology comes with a menu of guaranteed outcomes. Remote conferencing is a set of technologies that can bring many benefits to the economy as a whole and could potentially serve the political sector as well.

Electronic Privacy

The concept of privacy is quickly changing with the increasing prevalence of electronic communication in today's society. While personal interactions once were only between the individuals directly involved, technology has made it much easier for information to travel to outsiders. Personal rights to privacy have been questioned, and privacy in the workplace has almost become extinct. Considering the ease of gathering "private" information on the Internet, it is not surprising that many legal and ethical issues have come along with the rise of telecommunications.

Personal Privacy on the Internet

Years ago, in small towns across the nation, community residents made it their business to be in tune with what everyone else was up to. Although this meddling came to be expected as a way of life, it existed long before we entered the so-called "information age." People did not concern themselves too much with their individual *rights* to privacy. More often than not, they simply accepted the fact that sooner or later their neighbors would be wise to their daily activities, as well as their deepest family secrets. With our current electronic record keeping systems and numerous communications networks in operation, these people have a lot more to worry about than busybodies in their neighborhoods.

Many of the technological developments reviewed in this chapter have a direct effect on the extent to which personal information is considered public or private. Just consider that an innocent trip to the grocery store allows bar code scanners to create a permanent record of your product preferences. Advancements in e-mail, digital telephony, remote conferencing, biometric ID databases, and other forms of computer-to-computer communications have raised serious questions about the protection of privacy. The exponential growth in the collection, storage, and communication of personal data seems to have been accompanied by an increased potential for abuse and misuse. Satellite technology has introduced an international dimension into the dilemma. See Figure 3-9. Personal information can be indiscriminately exchanged back and forth around the globe.

Figure 3-9.
This photo illustrates a communication satellite being placed into orbit from a space shuttle. The use of orbiting satellites allows electronic communication to circle the globe with potentially negative impacts on personal privacy. (NASA)

In the mid-1970s, the U.S. Congress passed the Privacy Act to address the tension between individuals' interest in their personal information and the federal government's collection and use of that information. Privacy is a value that has always been regarded as fundamental in America, but its meaning has remained obscure. The definition serving as the basis for this legislation read something like this: Privacy is the claim of individuals, groups, or institutions to determine for themselves when, how, and to what extent information about them is communicated to others.

Numerous case studies and popular editorials have been published on the topic of personal privacy. Others address the issue of legislation, or the lack of it, designed to protect our individual rights to privacy. For the most part, these writings are generously sprinkled with horror stories of people who have been refused credit, insurance, employment, or welfare assistance. What makes these personal accounts horrifying is the fact that the actions taken were usually predicated on incorrect, outdated, or deliberately manipulated data. In addition to inaccurate data, there is more risk about data being stolen. Hackers have become increasingly bold and destructive. Recently, several banks in the Washington, D.C. area had to cancel thousands of Visa® debit cards after a hacker stole a data file from an on-line merchant.

Privacy continues to be a significant and enduring value held by the American public. It is apparent through a review of several opinion polls and numerous casual interviews that many people believe computer electronics and communications technologies exemplify pervasive threats to their individual privacy. General concern regarding the lack of adequate safeguards to protect personal records has grown over the past several years. It seems a large percentage of citizens perceive the need for new and revised laws to govern how personal information records can be used by organizations who possess the computer files. The legal system is presently lagging too far behind the technological capabilities related to this issue. Simply stated, there is increasing public support for additional government action to protect privacy in this country.

Personal information in the private sector can be collected, stored, and transmitted with very few governmental restrictions. Key areas of information acquired include consumer credit, banking, insurance, employment, medical care, and education. Now, because most grocery stores and gasoline stations allow their customers to use their bank cards to make purchases, this list will undoubtedly expand to include our eating habits and automobile usage as well.

Business e-commerce sites are collecting more data from visitors. Researchers have estimated that more than 90 percent of dot.com sites collect at least one piece of identifying information (for example, name or e-mail address), and at least 50 percent collect demographic information (such as gender, preferences, or zip code). All this data being collected can be used to identify and target WWW users.

Computer blacklists have also become a burgeoning growth industry in the field of information technology. They represent another electronic assault on individual privacy. For example, TRW® information services and Equifax Inc. have long relied on huge banks of mainframe computers to conduct their businesses. These credit bureaus provide consumer credit records for banks, department stores, finance companies, and employers. On an average working day, TRW information services' machines process hundreds of thousands of requests for information culling through a massive database containing detailed records of the payment habits of hundreds of millions of people.

Contemporary ethical and legal controversies on the Internet are causing users around the world to ponder such things as the definition of appropriate content, freedom of electronic information, software piracy, thefts of trade secrets, and policing or censorship of communications. The social norms and etiquette of CMC are still being defined and refined.

Electronic Monitoring of the Workplace

Along a similar vein, organizations whose members or employees have access to e-mail accounts are in the throes of establishing rules and norms to guide the use of this technology. Today's massive population of e-mail advocates who expect privacy may risk embarrassment, loss of employment, lawsuits, or worse. For many, the physical act of typing in a personal password seems to reinforce the notion that e-mail missives are private. Further, the interface design of the e-mail systems often gives the illusion no one is watching. Three forms of policy seem to prevail in contemporary organizations depending on the use of e-mail:

❏ Right to monitor policies. The organization publishes a statement saying all e-mail messages are monitored since e-mail is a company resource and provided as a business communications link.

❏ Hands-off policies. The organization publishes a statement saying e-mail messages are not accessed—although these policies *support* privacy, they cannot legally *guarantee* it.

❏ No policy. The organization does not publish a statement, which is the situation in most organizations today. If there are no formal or explicit policies regarding management's access to e-mail, employees generally behave as if it were a private communications medium. In the future, having no e-mail policy is likely to create more legal and ethical problems than even having a bad one might. Whatever employees believe, without a set policy, companies have the right to monitor and open e-mail.

Beyond e-mail, there is the issue of the WWW. As more employees are linked via the Internet, they can use the Internet for non-job-related activities. This is a growing concern among businesses. Surveys from various monitoring agencies have revealed that approximately 25 percent of WWW use at work is not job-related. For companies, this represents a dramatic reduction in the working hours of their employees. Nonwork sites most commonly visited were general news, sex, investments, entertainment, and sports. Companies have responded to this trend by increasing the monitoring of their employees. According to a major survey by the American Management Association, over 74 percent of large and midsized companies admitted they monitored their employees electronically.

The Federal Bureau of Investigation (FBI) is undertaking a more in-depth monitoring of certain U.S. residents. It is using a new program that can monitor the e-mail of an individual person through his Internet Service Provider (ISP). This program, commonly referred to as the Carnivore system, has to be installed at an ISP data center. Although the FBI has only used this technology under court order, if the law is changed, it could be used to monitor anyone's e-mail in the country.

Electronic monitoring is not only the province of the FBI. Any user can add software to their computer to monitor WWW use. This type of software is generally referred to as "snoopware." There are many different types of snoopware available, some designed for businesses, some for home users. There is snoopware that takes screen shots of a user's activities, logs every keystroke, and can intercept "private" passwords. One such program is WinWhatWhere Investigator® software. This program collects detailed records of all activity on an employee's computer—the programs used, the e-mail sent, the websites visited. The Investigator software then puts the data into a database. The manager can assess this data to see what the employees were doing on their computers. Snoopware has been used in the past two years to discipline employees. Dow Chemical and Xerox, among other large employers, have fired employees for surfing the Internet on company time.

One technique to limit electronic surveillance is encryption. A user can electronically alter his e-mail so only the designated user can read it. Encryption works by dividing the e-mail message into sections and using a private key to convert the e-mail into gibberish. In order to reach the e-mail, it must be decrypted with the same key. The most widely used encryption standard today is Pretty Good Privacy® (PGP) computer programs. Phil Zimmermann originally wrote PGP computer programs in the early 1990s. PGP computer programs are the standard for e-mail encryption today and have millions of users across the world.

Despite the millions of users, most people who e-mail do not use encryption. Many do not even know about it. Until there is more awareness of snoopware among users, encryption will remain a small part of the e-mail process.

Summary

Twenty-five years ago, it was unusual to own a home computer. Today, most households in the United States have their own home computers, as well as an array of other digital devices. As information technologies invade more everyday appliances, one can see a future where all home appliances are linked in a vast network. The purely digital home is still in the future, but portions are here today.

Information technologies are changing the faces of business and government, and even our personal outward appearances. In many places, we are known by our electronic personas rather than by our own faces. As governments increasingly depend on electronic communication to govern, will our electronic personas begin to dominate us and define us to the world?

Technology has changed the face of communications in our society. Telecommunications breakthroughs have transformed our views of each other and the international community as a whole. Direct broadcast satellites enable us to be on the scenes of entertainment activities, as well as terrorist uprisings, instantly. All-optical and wireless networks hold many promises yet to be realized, with reference to the quantity and types of information we can transmit anywhere through the telephone system, computer networks, and the airwaves.

If we truly live in an age of information saturation, perhaps the technologies associated with information transfer have converged to help us. As the amount of information available to the general public has expanded, advancements in electronic record storage and retrieval have followed. Rather than be fearful of our inability to keep up with the rampant pace of change in the telecommunications industry, we should systematically accept these developments for what they are—a series of mechanisms and devices allowing us to interact with one another within a vast array of CMC environments.

There are several social issues related to the technological hardware associated with communications networks. Videoconferencing is quickly rising as a corporate business communications strategy. Thousands more workers may soon be given a telecommuting option to complete their job-related assignments. Finally, individual citizens have begun to express their concerns regarding the extent to which new legislation may be essential to mediate the impact of technology on the privacy of personal information. The degree to which these issues will be resolved or widely adopted in the near future remains tentative.

Discussion Questions

1. What is the primary purpose of installing a local area network (LAN) in a campus community? Give at least two examples of LANs in your hometown with which you either have had experience or are familiar.

2. Identify and describe at least two forms of computer mediated communication (CMC). Explain how each of these might enhance your capacity to perform your academic or employment duties more efficiently.

3. Briefly explain the concept of virtual reality and conjecture as to how this technology might one day figure prominently in your daily routine.

4. Besides the examples listed in this chapter, can you think of any other advantages or drawbacks to the practice of telecommuting in our society?

5. What are some of the immediate applications and advantages of remote conferencing in the corporate world? Is there a downside to this information technology?

6. Describe some of the problems you might expect to encounter if you suddenly found out that your "private" e-mail correspondence to a friend in Australia had been pilfered by your employer.

Suggested Readings for Further Study

Aldersey-Williams, Hugh. 1996. Interactivity with a human face. *Technology Review* 99 (February/March): 34–39.

Angell, David and Brent Heslop. 1995. *The Internet business companion: Growing your business in the electronic age.* Addison-Wesley.

Arthur, Charles. 1994. The future of work: It's all in the mind. *New Scientist* 142 (16 April): 28–31.

———. 1995. Identity crisis on the Internet. *New Scientist* 145 (11 March): 14–15.

———. 1995. Spamming could be more than your job's worth. *New Scientist* 145 (25 March): 22.

Berleant, Daniel and Byron Liu. 1995. Robert's rules of order for e-mail meetings. *Computer* 28 (November): 84–85.

Berners-Lee, Tim, Mark Fischetti, and Michael L. Dertouzos. 1999. *Weaving the Web: The original design and ultimate destiny of the World Wide Web by its inventor.* San Francisco: Harper.

Binder, John. 1995. Videoconferencing: Yesterday's science fiction, today's telephone. *Aerospace America* 33 (February): 17–19.

Brody, Herb. 1996. The web maestro: An interview with Tim Berners-Lee. *Technology Review* 99 (July): 32–40.

Brown, Kathryn. 2001. Online, on campus: Proceed with caution. *Science* 293 (31 August): 1617–1619.

Buderi, Robert. 2001. The commuter computer. *Technology Review* (June): 76–81.

Caldwell, Barrett S. and Shiaw-Tsyr Uang. Technology usability issues in a state government voice mail evaluation survey. *Human Factors* 37 (June): 306–320.

Callaghan, Robert. and Thomas. W. Mastaglio. 1995. A large scale complex virtual environment for team training. *Computer* 28 (July): 49–56.

Chan, Vincent W. 1995. All-optical networks: Fiber optics will become more efficient as light waves replace electrons for processing signals in communications networks. *Scientific American* 273 (September): 72–76.

Computer Science and Telecommunications Board, National Research Council. 1994. *Realizing the information future: The Internet and beyond.* Washington, D.C.: National Academy.

Cook, Terry. 1995. It's 10 o'clock: Do you know where your data are? *Technology Review* 98 (January): 48–53.

Cosentino, Victor J. 1994. Virtual legality: Once on-line, some people totally disregard legally and socially acceptable behavior. *Byte* 19 (March): 278.

Cronin, Mary J. 1995. *Doing more business on the Internet: How the electronic highway is transforming American companies.* New York: John Wiley & Sons.

Davenport, R. John. 2001. Are we having fun yet? Joys and sorrow of learning online. *Science* 293 (31 August): 1619–1620.

Davis, Andrew W. 1995. Face to face. *Byte* 20 (October): 69–72.

Dodson, Weldon. 1995. Hitched to everything. *Sierra* 80 (September/October): 26–27.

Galkil, William. 1996. Names on the Net, e-mail in court, and trade secrets for hire. *Computer* 29 (January): 110–111.

Gobel, Martin. 1996. Industrial applications of VEs. *IEEE Computer Graphics and Applications* 16 (January): 10–13.

Goldstein, Harry. 1995. Is virtual reality for real? *Civil Engineering* 65 (June): 45–48.

Green, Paul E. Jr. 1994. Toward customer-useable all-optical networks. *IEEE Communications Magazine* 32 (December): 44-49.

Gross, Lynne. S. 1995 *Telecommunications: An introduction to electronic media.* 5th ed. Madison: Brown & Benchmark.

Harley, Diane. 2001. Higher education in the digital age: Planning for an uncertain future. *Syllabus* (September): 10–12.

Hodges, Mark. 1996. Videoconferencing for the rest of us. *Technology Review* 99 (February/March): 17–19.

Holmes, Hannah. 1993. Telecommuting. *Garbage* 5 (April/May): 32–37.

Horrocks, Nigel. (2001). Busted at work. *Netguide.* [Internet]. 49 (April). <http://www.netguide.co.nz/magazine/pulp/49/busted/>.

Hughes, David. 1995. Desktop video conferencing may offset some air travel. *Aviation Week & Space Technology* 143 (17 July): 37–38.

[Internet]. *Electronic Frontier Foundation.* <http://www.eff.org/>.

[Internet]. Khufu's pyramid. *Virtual Worlds on the WWW.*
    <http://www.pbs.org/wgbh/nova/pyramid/explore/khufuall.html>.

[Internet]. Mount Everest Project. *Virtual Worlds on the WWW.*
    <http://www.pbs.org/wgbh/nova/everest/>.

[Internet]. *The Well On-Line Network.* <http://www.well.com/>.

[Internet]. *The World Lecture Hall.* <http://www.utexas.edu/world/lecture/>.

Jeffcoate, Judith. 1995. *Multimedia in practice: Technology and applications.* New York: Prentice Hall.

Knorr, Eric and Stuart Bradford. 2001. Mobile web vs. reality. *Technology Review* (June): 56–61.

Kovach, Kenneth A., Sandra J. Conner, Tamar Livneh, Kevin M. Scallan, Roy L. Schwartz. 2000. Electronic communication in the workplace—Something's got to give. *Business Horizons* 43 (July): 59–64.

Laudon, Kenneth, Jane Price Laudon and Kenneth C. Laudon. 2001. *Essentials of Management Information Systems.* 4th ed. Upper Saddle River, NJ: Prentice Hall.

Levine, Ron. 1995. Surprise! Fax servers smarten up. *Datamation* 41 (15 May): 63–64.

Maes, Pattie. 1995. Intelligent software. *Scientific American* 273 (September): 84–86.

Marx, Gary T. 1994. Taming rude technologies. *Technology Review* 97 (January): 66–67.

McCarthy, Michael J. 2000. "Erase" won't wipe out a rant—Workplace keystroke loggers save every typed word—When else can you peer into someone's raw thought process? *The Wall Street Journal Europe* (9 March).

McLellan, Mark R. 1995. Electronic information management—A model for industry, academia and IFT. *Food Technology* 49 (March): 74.

Morgenstern, Barbara L. and Michael M. A. Mirabito. 2000. *New communications technologies.* 4th ed. Boston: Focal.

Owen, Jean V. 1994. Making virtual manufacturing real. *Manufacturing Engineering,* 113 (November): 33–37.

Pahlavan, Kaveh, Thomas H. Probert, and Mitchell E. Chase. 1995. Trends in local wireless networks. *IEEE Communications Magazine* 33 (March): 88–95.

Patterson, David A. 1995. Microprocessors in 2020. *Scientific American* 273 (September): 62–67.

Rheingold, Howard. 2000. *The virtual community.* Cambridge, MA: MIT Press.

Ritter, Gary and Stan Thompson. 1994. The rise of telecommuting and virtual transportation. *Transportation Quarterly,* 48 (summer): 235–248.

Salamone, Salvatore. 1995. Radio days. *Byte* 20 (June): 107–108.

Savage, Neil. 2001. Building a better backbone. *Technology Review* 104 (June): 40–46.

Schraeder, Terry. 1995. How private are e-mail messages? *PIMA Magazine* 77 (October): 12.

Schneider, Gary P. 2002. *Electronic commerce.* 3d ed. Boston: Course Technology.

Seymour, Jane. 1996. Virtually real, really sick. *New Scientist* 149 (27 January): 34–37.

Tomasi, Wayne. 2000. *Advanced electronic communications systems.*5th ed. Englewood Cliffs, NJ: Prentice Hall.

Turkle, Sherry. 1995. *Life on the screen: Identity in the age of the Internet.* New York: Simon and Schuster.

Wallich, Paul. 1995. The chilling wind of copyright law? *Scientific American* 272 (February): 30.

Ward, Mark. 1995. Uniform look for Internet agents. *New Scientist* 147 (26 August): 18.

Wayner, Peter and Alan Joch. (1995). Agents of change. *Byte.* [Internet]. 20 (March): 94–96. <http://www.byte.com/art/9503/sec10/art1.htm> [30 May 2002].

Weisband, Suzanne P. and Bruce A. Reinig. 1995. Managing user perceptions of e-mail privacy. *Communications of the ACM* 38 (December): 40–47.

Whittaker, Steve. 1995. Rethinking video as a technology for interpersonal communications: Theory and design implications. *International Journal of Human-Computer Studies* 42 (May): 501–529.

Chapter 4

Space Exploration

DATELINE

Baikonur, Kazakhstan

A Russian government commission approved the inclusion of the first extraterrestrial tourist today. American millionaire Dennis Tito became the first tourist to visit the International Space Station, arriving as a guest of the Russian Space Agency on a Soyuz spacecraft. Mr. Tito paid $20 million to the cash strapped Russian space program. Also, he signed an agreement that he will pay for anything he breaks and, if he is hurt, neither he nor his heirs can sue for compensation.

The other partners to the International Space Station, particularly the National Aeronautics and Space Administration (NASA), were very unhappy about this turn of events, but they realized there was little they could do to stop their Russian partners from bringing along a guest. NASA decided to make the most of it and have the resident astronauts at the Space Station spend their time baby-sitting the visiting California millionaire. Tito's first words as he arrived at the International Space Station were "I love space!"

The Russian Space Agency is planning more tourist flights in the future....

Introduction

When it comes to discussions about space exploration, many historians and political scientists seem to share a common sentiment. They tend to concur that no scientific venture has captured the fancy and emotional support of the American people as has the space program.

In the mid-1980s, it quickly became apparent that the twenty-fifth mission of this nation's reusable spacecraft, the shuttle (technically called America's Space Transportation System), had attracted international attention. While its previous missions were devoted to sophisticated scientific research projects, this spacecraft was supposed to carry the first private citizen, a schoolteacher, into orbit. When the shuttle *Challenger* burst into flames less than two minutes after its liftoff, people around the world were forced to grasp the intensity of an unexpected tragedy. Stunned into

silence, it is likely that many lost all faith in technology—even if just momentarily.

Following the crash, NASA was called on to rethink its long-range goals with regard to national space policy. In the wake of a disaster of this magnitude, there are never any easy choices or decisions to make. Some observers initially felt the shuttle accident would bring an end to manned space flights for many years to come. In comparison with previous setbacks in the U.S. space race, such as *Apollo 13*, the shuttle in which one of the oxygen tanks caught on fire during flight, the *Challenger* accident caused a shift in the public's perception of spaceflight, particularly in the safety of these flights. NASA itself was faced by a more critical analysis of its missions and procedures, by both the U.S. Congress and the general public. Until then, this country's space community had repeatedly become obsessed with expensive programs that had not been dedicated to meeting any clearly defined long-term goals. As you will see in this chapter, NASA was forced by the public and Congress to revise its procedures and approach its work in a different way, including finding international partners to meet the challenges of new demands.

Four decades ago, it appeared that humankind had finally begun its conquest of outer space. The events of the early years brought to life scenarios that had previously been the material of science fiction writers. The 1950s and 1960s introduced many feats in space travel: the first artificial satellite, the first human in space, the first planetary probe, the first space walk, and the first human on the Moon. It seemed almost certain that the remaining decades of the twentieth century would unveil the development of space stations and space colonies, trips to the planets, and perhaps even vacations in space. At least one forward-thinking individual envisioned building the first Hilton® hotel in space. Unfortunately, as we proceed further into the first decade of the twenty-first century, few, if any, of these dreams appear to be forthcoming.

Added to the space challenges for the United States is the increased internationalization of space, with over forty countries already having satellites orbiting Earth. The biggest international project, the International Space Station, will include the scientific expertise and participation of sixteen nations: the United States, Canada, Japan, Russia, eleven members of the European Space Agency (ESA), and Brazil. Space exploration has grown beyond the Cold War scenario of the United States versus the Soviet Union. How different countries will balance their own interests against international concerns has not yet been determined. In this chapter, we will examine a variety of technical, socioeconomic, and political concepts that presently characterize space ventures being envisioned around the world.

The National Aeronautics and Space Act of the late 1950s did not establish an aerospace trucking company; primarily, it did set up a research and development agency. If NASA continues to place numerous scientific projects on hold, it may be forced to redefine its primary responsibility to the American public. Is it plausible for NASA to continue promoting its politically favored flagship programs, such as the shuttle and the International Space Station, while it also conducts a vigorous unmanned program in applied technologies, exploration, and science? Before anyone can honestly attempt to respond to this inquiry, it is essential to determine first what we wish to accomplish in space and then identify the means through which these goals can be best attained. This chapter provides a foundation for a continuing dialogue that can ultimately respond to the following questions:

❏ In light of the history of the space shuttle, what should contemporary space transportation systems realistically be designed to accomplish?

❏ The persistent challenges to colonize and industrialize space have become international in scope. What interest, if any, do national defense departments have in space development projects?

❏ What are the primary technological objectives of the American satellite industry? What contributions has this industry made toward the social and scientific advancements of society?

❏ What concerns have been expressed about prolonged periods of time spent living and working in space?

❏ Are there any specific long-range goals that should define international space policies? Would these goals ensure future support?

Key Concepts

As you review this chapter, you should familiarize yourself with the following terms and phrases as they relate to space exploration:

aerobraking
artificial satellite
bioastronautics
Deep Space Network (DSN)
expendable launcher
faster, better, cheaper (FBC)
geostationary orbit
geosynchronous orbit
glass cockpit
global positioning system (GPS)
gyroscope
International Space Station
Skylab
smart cockpit
space adaptation syndrome
space junk
spectrograph
spectrometer

The Conquest of Outer Space

Outer space for decades has been emotionally regarded and applauded as the final frontier. Its romance seemed almost blinding at times. Although the risks were always there, and the failures of man-made spacecraft were statistically inevitable, we continued to press on with an ever-increasing ration of the "pioneer spirit." The ability of mankind to conquer and industrialize outer space within our lifetime seemed inevitable. Why? One of the more prominent features of the earliest days of the "space age" during the late 1950s was the intense competition between the Soviet Union and the United States. People around the world were stunned in the mid-1950s by the Soviets, who successfully launched the first *artificial satellite, Sputnik 1,* into orbit. American scholars hung their heads and questioned the credibility of our educational system, the scientific community, and our technological leadership. The United States subsequently embarked on an accelerated scientific journey, containing a series of technological obstacles, in order to overtake the Soviets. Along the way, public support of the emergent space program was borne out of patriotism, coupled with blind faith in new and exciting forms of technology.

In the early 1960s, the Soviets were once again victorious after they sent the first human, Yuri Gagarin, into space. The United States followed a few months later, when Alan Shepard became the first American to be launched into outer space in the Mercury Project. Project Mercury arose directly as a U.S. response to the launching of the *Sputnik* by the Soviet Union. Project Mercury made six unmanned flights around Earth in the early 1960s and provided NASA with valuable experience in how to recover both man and spacecraft safely. The Mercury Project led to the Gemini Project, which was seen as a necessary intermediate step before the Apollo flights to the Moon. The goals of the Gemini Project, which included twelve flights, were to support two men in space for long duration flights up to two weeks and to practice rendezvous and docking with other vehicles in space.

Despite the successes of the Mercury and Gemini projects, it became evident that the Soviets were still ahead of the United States in most areas of space technology. Feeling frustrated and perhaps embarrassed, the Kennedy administration challenged NASA to land a man on the Moon before the close of the decade. The Apollo Moon Program was inaugurated and ultimately captured the hearts and minds of American men, women, and children throughout the 1960s. Apollo clearly represented a bid for national prestige and a demonstration of technological savvy.

The race between the United States and the Soviet Union wore on in the early 1960s, as NASA forged through its Mercury, Gemini, and Apollo programs. Throughout this series of missions, astronauts were commonly viewed as our heroes. The pressure of the space race forced us to underwrite the tremendous expense of each spacecraft launch. Congress approved each request NASA submitted, culminating with a tremendous victory just five months before the close of the decade. Millions of viewers around the world were glued to their televisions screens, as the *Apollo 11* mission created another significant entry for our historical journals. Astronauts Neil Armstrong and Buzz Aldrin became the first men to take a walk on the Moon—the United States had won the race.

The thrill of victory, however, was short-lived. Congress grew tired of financing adventures that had, at least on the surface, the same plot each time. Government officials called for more scientific fieldwork and fewer flashy stunts performed by our nation's finest men. The problems of *Apollo 13* added to the public's uncertainty with the program, although before the mission was aborted because of an explosion onboard, *Apollo 13* was barely noticed by the public. Also, NASA tended to downplay the problems it had with *Apollo 13*, so the public was never really fully informed about the dangers that existed for the crew. One example of the type of behavior that, unfortunately, did receive national media coverage, much to NASA's chagrin, involved the 1971 lunar mission of *Apollo 14*. The major "scientific experiments" included Alan Shepard driving a golf ball and Edgar Mitchell attempting to send telepathic messages to fellow parapsychologists on Earth. Although Shepard did make the longest drive in the history of golf (thanks to the Moon's gravity), congressional expectations were far from satisfied.

The United States finished its abbreviated series of Apollo lunar landings one year later with a geologist onboard *Apollo 17*. By this time, the American public had become somewhat disinterested and was trying to envision what its next space triumph would look like. On the flip side, the nation's leaders were not in any sort of agreement on what the next step should be for the space program. The magnificent success of the Apollo Program was partially related to the fact that its goal was precise and short-term: Send a man to the Moon and return him safely to Earth. We developed and refined the technology to accomplish that goal, and then we stopped. Less than fifteen years after NASA had been formally inaugurated in the United States, its very existence was in a state of jeopardy. The follow-up Apollo applications program was cancelled due to insufficient funding, and NASA did not have a long-range goal that could sustain the country's space program. NASA was in a bind—it needed to find new projects that would allow it to survive into the next millennium.

Space Technology of the Present

Although most public and press attention centers on the shuttle, it is not the only space technology today. The scientific value of space exploration is expanding more through other lesser-known missions, including the Hubble Telescope and the *Chandra Observatory*, an orbiting X-ray telescope. Also, there have been major space missions funded by the ESA during the last ten years, including *Ulysses,* a spacecraft studying the solar wind blowing from the Sun, and the *XMM-Newton Space Observatory,* a space telescope using a series of nested, curved mirrors to collect X-ray information about the galaxy, although these non-American missions are barely noticed by the American public. It is clear from the work of the ESA, as well as of other entities, including the Russian and Japanese space agencies, that the pursuit of space in the future will have a more international flavor.

A unique exploration being sponsored by the ESA is the Cluster Program. Today, there are four Cluster spacecraft, carrying the same eleven instruments, zipping around Earth, collecting data. The goal of the Cluster Program is to study space weather inside and outside Earth's magnetic field. The four Cluster spacecraft, named *Rumba, Tango, Salsa,* and *Samba,* are taking simultaneous measurements of the space around Earth. The orbits of these four are designed so each one is located at the vertex or endpoint of a tetrahedron. Using the data from these four spacecraft, scientists will create the first detailed three-dimensional model of Earth's magnetic shield.

Despite the Cluster and the many other space missions, the shuttle is the one space technology that captures the imagination of the American public today. Perhaps it is the image of traveling to space as a passenger on a spaceship. Also, since the shuttle has a mixed crew (civilian and military), the public can better identify with the astronauts. As NASA has changed the gender and racial makeup of its crews, these new astronauts are more appealing to the American public. In the early 1980s, the first American woman in space, Sally Ride, even had her own slogan, "Ride, Sally Ride."

The Saga of the Space Shuttle

The idea of the shuttle evolved at the end of the Apollo era, when NASA was looking for another project to continue the exploration of space. Under a certain amount of political and economic pressure, NASA officials decided to replay the Apollo technique. They contemplated the need for a costly, long-term project that would appeal to the hearts of the American people and, therefore, their congressional decision makers. NASA had previously initiated the idea of developing a reusable spacecraft, after its first lunar landing, when the nation was extremely concerned about numerous environmental issues. The central premise was they could save millions of wasted dollars by reusing the launcher's components (the boosters, the fuel tanks, and the orbiter itself). NASA believed a vehicle of this configuration would assure its survival, since it would take many years and a large sum of money to develop, construct, and deploy. Out of this proposal emerged the Shuttle Project, which was formally referred to as America's Space Transportation System (STS).

From its inception, the Shuttle Program was forced to operate under a fair amount of pressure. NASA's initial proposal to develop a reusable space shuttle for routine transportation into space was predicated on two main opinions:

❑ Humanity belongs in space.

❑ A permanent space station was the next logical step to putting the species out there on a long-term basis.

NASA further promoted the concept by saying the space station would be serviced and restaffed by the shuttle. Payload costs onboard the shuttle had to be kept low enough to make this type of activity affordable.

NASA placed itself squarely behind the Shuttle Project and proceeded to gain the support of numerous federal agencies. Specifically, it fought hard to attain the support of the Department of Defense (DOD) in order to assure White House approval. In doing so, NASA agreed to incorporate a number of military specifications into the shuttle design. For instance, the Air Force requested a huge payload bay measuring fifteen by sixty feet, sixty thousand pounds of lift capacity, and the ability to land on either the East or West coast. Early in the project, military officials were mainly interested in launching large reconnaissance satellites. Even still, they also stipulated that the spacecraft should be able to return quickly from orbit and maneuver easily under threatening conditions. As you can well imagine, NASA immediately said farewell to any hopes it may have had for low developmental costs. NASA placed all of its space eggs in one basket and promised Congress that a fleet of five orbiters would be operational by the end of the 1970s. Its original cost estimate ranged between $5 and $6 billion.

Following its approval by the Nixon administration, the Shuttle Program got off the ground when Rockwell International won the design contract in the early 1970s. Construction began two years later. Delays were encountered repeatedly along the way, due to inadequate funding and continual problems with the main engines and heat shield tiles. Realizing the Shuttle Project had severely overrun its budget, NASA made several decisions to scrap a variety of valuable research programs already in operation. One of the most memorable of these abandoned projects was Skylab. See Figure 4-1.

Skylab was a preliminary version of a U.S. space station that had evolved during the Apollo era. NASA launched this huge orbiting space laboratory in the early 1970s.

In less than one year, three crews of astronauts visited Skylab and respectively spent one month, two months, and eighty-four days living and working inside the space station. During these visits, western scientists were finally given an opportunity to make an exhaustive study of the physiological effects of prolonged space flight. NASA's decision to abort the Skylab Project resulted in its eventual demise. One of the first proposed tasks of the new shuttle vehicle was to visit Skylab and keep it functioning, but the shuttle was not ready in time, so Skylab burned in the atmosphere in the late 1970s. Noting the irony of the fact that we no longer had a space station of some sort in orbit, some observers began to question the validity of the Shuttle Program.

Plagued by a series of engine fires, the shuttle schedule was modified and set back on a regular basis during the late 1970s. No one had built throttled liquid rockets of this size before. The problems of pressurizing, heating, and containing the explosive fuel were (and still are) quite formidable. Leaks, fires, and turbo pump failures occurred over and over again. From the original target launch date in the late 1970s, the shuttle slipped almost three years behind schedule, and the *Columbia* was finally launched for the first time in the early 1980s. Many of the technological bugs had not been worked out, and the heat shield tiles had not been flight-tested to ensure the safe return of the astronauts. The cost of the project had approached the $11 billion mark by that point—double NASA's initial projections.

The space shuttle has three main parts: the orbiter, which holds the crew and supplies; a large external fuel tank, which holds the fuel for the main engines; and two solid rocket boosters, which provide the fuel for the shuttle's liftoff and first two minutes of flight. See Figure 4-2. The external fuel tank is burned in the atmosphere after

Figure 4-1.
Skylab flies over the Amazon River delta during Skylab 3 mission. (NASA)

launch, but the other parts are reused. The solid rocket boosters detach after the first two minutes of flight and fall to Earth, while the orbiter (or as it is usually called, the shuttle) returns to ground and lands like an airplane. Currently, there are four U.S. space shuttles in operation: *Columbia, Discovery, Atlantis,* and *Endeavor.* Each of them is designed to fly at least one hundred missions.

Prior to the *Challenger* disaster in the mid-1980s, NASA successfully executed

Figure 4-2.
The liftoff of Space Shuttle *Atlantis* began an eleven-day mission to the International Space Station in 2000. While onboard, the seven-member crew performed support tasks, transferred supplies, and prepared the living quarters in the newly arrived Russian Zvezda Service Module. (NASA)

twenty-four shuttle flights over a four and a half year period. 1985 was by far the best year for the program, with three shuttle orbiters in operation and nine successful flights. As impressive as this might sound, NASA's original projections during congressional hearings in the early 1970s indicated that each orbiter would be launched once each month after the program had gotten underway. In other words, they optimistically envisioned some sixty flights per year! An accurate history of the Shuttle Program is given in Figure 4-3, which shows the number of shuttle flights per year since the early 1980s. It is interesting to compare this history with NASA's original projection. Using NASA's original plan, the number of 101 total shuttle missions should have been reached in only two years.

Even before the twenty-fifth mission of the Space Shuttle Program ended in a fatal explosion, experts had begun to question the economic viability of the star player in this country's space program. Costs per pound of payload were twenty to fifty times greater than NASA's initial projections.

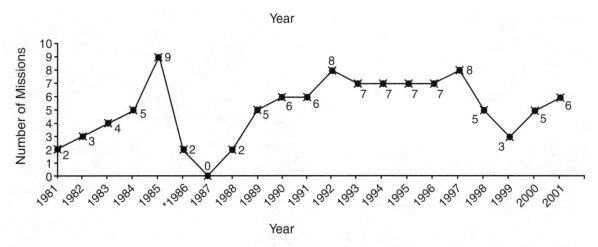

Figure 4-3.
This graph shows the number of shuttle flights each year from 1981 to 2001.
*1986 was the year of the *Challenger* disaster. There were no shuttle flights in 1987.

NASA's budget for the Space Shuttle Program alone in 2000 was almost $3 billion. This budget, since there were only five missions, results in an average cost of $596 million for each mission. One of the main reasons for these very high operational costs for the shuttle stems from its most significant design feature—it must be operated with human beings onboard. The life support systems and safety features necessary to make a spacecraft safe for human passengers weigh a lot. Also, there are other costs, including safety and performance upgrades; ground operations before, during, and after the flight; and flight hardware (such as engines and external fuel tanks). It cost more than $93 million to lift one astronaut aboard the shuttle during 2000. It is possible these costs could be justified if scientific experiments or manufacturing procedures requiring human input were the primary purpose of the shuttle. Through the 1980s and 1990s, however, the shuttle had been used mostly to deploy unmanned satellites. This type of task can be done more cost efficiently from expendable unmanned boosters, or *expendable launchers,* such as the Ariane rockets being used by the ESA.

The shuttle is the most complex machine ever built, having more than 2.5 million parts and weighing 4.5 million pounds at launch. It is this complexity that keeps the shuttle expensive despite its reuse. To add to this complexity, the design of the shuttle has constantly changed over the last twenty years to increase both its efficiency and safety. Since the early 1980s, the shuttle's engines have been overhauled three times. Imagine redesigning and replacing the engine of your car three times during your ownership! The cockpit has dramatically changed in the past twenty years, with a new ***glass cockpit*** that was first used in the Shuttle *Atlantis* in 2000. The design of the glass cockpit replaced thirty-two gauges and electromechanical displays and four cathode-ray tube (CRT) displays with eleven new full-color, flat panel display screens. See Figure 4-4. These changes allowed the space shuttle to use cockpit displays that have become common in many new commercial aircraft, and this new cockpit has an added benefit. It is seventy-five pounds lighter and less power hungry than before. The glass cockpit, however, is not the last change—it is projected that by mid-decade,

A

B

Figure 4-4.
These photos compare the old cockpit and the new "glass cockpit." A—The crew of the first space shuttle, Commander John Young (left) and pilot Bob Crippen (right) aboard *Columbia* before the first shuttle launch. The old cockpit featured green screens and gauges. B—This "fish-eye" view illustrates NASA's "glass cockpit." This is the fixed base Space Shuttle mission simulator in the Johnson Space Center's Mission Simulation and Training Facility. (NASA)

a *smart cockpit* will be installed in the shuttle that will reduce the shuttle pilot's workload during critical periods by simplifying the pilot's job.

The shuttle has found a new purpose and justification with the building of the *International Space Station.* Construction and installation of the various components occupy most of the shuttle's time, with six out of eight missions scheduled for 2001 related to the construction projects for the International Space Station. After the space station's completion, will the shuttle become merely an expensive commuter plane, or will its role evolve into yet another new area? For the near future at least, the shuttle will remain an active part of the U.S. Space Program, as the construction work on the International

Space Station continues. Through 2004, at least, the U.S. government budget provides for six shuttle flights each year.

The Mir Space Station

The Mir is the last in a long history of Russian (formerly, the Soviet Union) space stations. While the U.S. Space Program focused its energies on the Shuttle Program and held its sights firmly on the construction of a permanently manned space station, Russia made steady strides toward extending its civilization into space. There were three different generations of Russian stations. The first space station launched in the early 1970s, the Salyut I—about the size of a mobile home—had a checkered history as a space station. The first crew, Soyuz 10, sent to occupy the Salyut I was unable to enter the station because of a problem with the docking mechanism. The second crew, Soyuz 11, lived on the station for three weeks, but died during their return to Earth because of a malfunction in their spacecraft. Three other first generation Salyut space stations were more successful and supported a total of five crews successfully.

One of the most important achievements of the Salyut Program was that it proved astronauts could successfully live in space longer than expected, and it was long-term space exploration that was the goal of the second generation of Russian space stations. Seventy-five percent of all manned experience in space through the mid-1980s belonged to the Soviet Union. Its cosmonauts worked onboard their Salyuts for more than eight months at a time, with the longest stay being 237 days. Compare this to the longest American space mission onboard the Skylab of less than three months. Salyut 6 and Salyut 7 were active from the mid-1970s to the mid-1980s and together were visited by twenty-six different crews, including twelve long-duration crews. These cosmonauts routinely performed in orbit repairs and tested several furnaces for producing specialized materials. The station was resupplied on a regular basis by smaller docking craft called progress service modules.

The Soviet Union boosted its successor to the Salyuts, known as Mir (peace), into a low Earth orbit (LEO) in the mid-1980s. The Mir was actually many spacecrafts in one, and it was expanded over the years into a complex of many parts. It was more than 108 feet long, when both the *Soyuz* spacecraft

and progress service module were attached, and about ninety feet wide at its largest point. The central Mir module was designed as the living quarters and included six docking ports. Over time, six more modules were added to the Mir space station, giving it a total weight of 143 tons. The Mir served as host to twenty-eight crews representing sixty-three other countries, including seven Americans. One crewmember on Mir, Dr. Valeri Polyakov, holds the record for the most consecutive time in space—439 days. Anatoly Solovyov, another Russian astronaut, spent 651 days on Mir during five missions.

After the fall of the Soviet Union, the United States and Russia started a four-year series of joint missions using the Mir space station. Since the Mir had, at that time, been used for ten years, NASA believed it could use Mir as a learning experience before beginning the construction of the International Space Station. At the same time, the Russian Space Agency was under increasing financial pressure because of the changes in that country after the end of the Cold War. Cooperation with NASA allowed the launch of the last two modules: *Spektr,* which contained more than sixteen hundred pounds of U.S. scientific equipment, and *Priroda,* a science module that added Earth–remote sensing capability to Mir. The Mir, however, was showing its age and had a series of problems that overshadowed the scientific efforts of the crewmembers, including fires, leaks, and a collision of a progress service module that punctured the *Spektr* module and resulted in its closure. Much of the efforts of the Mir crews had to be diverted into repairing the space station, and it became clear that the Mir would not survive to become part of the International Space Station.

In 2001, after fifteen years circling Earth, the Mir finished its voyage. Russian space controllers turned on its engines for the last time and sent it into the atmosphere, breaking it apart. Mir crashed in the Pacific Ocean between New Zealand and Chile, but the importance of the Mir has outlived the space station itself—its contributions as the first permanent space station will never be forgotten. In its fifteen years, it orbited Earth 86,331 times and hosted a total of 104 astronauts. The lessons learned from the Mir about living in space will aid all space exploration in the future.

Telescopes in Space

Over the last ten years, NASA and the ESA have pursued a scientific study of the universe using a series of orbiting telescopes or observatories. Each of these was designed with a particular technology to measure specific subjects or events. The Hubble is an optical and ultraviolet telescope, the *Compton* observes the emission of gamma rays, the *Chandra* and *XMM-Newton* operate using X rays, and the Infrared Solar Observatory (ISO) studied the star system using the infrared range of the light spectrum. All together, they have provided a new view of the universe for scientists.

Telescopes and observatories in space are superior to their counterparts on Earth because these space-based instruments are free of the distorting effects of Earth's atmosphere. Because they are in space, they can see farther and clearer than any telescope ever could on Earth. All of these space observatories made significant scientific contributions, but unfortunately, they rarely are known or understood by the general public. As unmanned vehicles, they are inherently cheaper to both build and operate than NASA's flagship programs, the shuttle and the International Space Station.

Infrared Solar Observatory (ISO)

Funded by the ESA, the Infrared Solar Observatory (ISO) was active for a short time in the late 1990s. Infrared radiation can be considered as heat, and every space object, no matter how cold, will emit some heat (or infrared radiation). Because the ISO looked for heat in space, it was able to see space objects not viewable using conventional optical telescopes. Particularly, the ISO was effective in studying comets because comets are cold objects, consisting of dust and ice. Throughout its three years in space, ISO made over twenty-six thousand scientific observations of the solar system.

ISO was truly an international mission. Multinational teams, with team leaders in France, Germany, the Netherlands, and the United Kingdom, developed the four instruments on the observatory. What was unique about the ISO was the design of the instruments. Since these instruments were used to observe cool objects in space, the instruments themselves had to be kept very cold. In fact, the instruments were kept at temperatures close to absolute zero—about minus 273 degrees Centigrade. The engineers of the ISO kept the spacecraft cold with liquid helium. The mission ended when the supply of liquid helium ran out.

Hubble Space Telescope

The Hubble Space Telescope is the first large orbital optical observatory. The Space Telescope is named after American astronomer, Edwin P. Hubble. It consists of an Optical Telescope Assembly, a Support Systems Module, and a series of sophisticated scientific instruments. The Hubble is very big, about the size of a large school bus, and very heavy. On Earth, it would weigh over twenty-five thousand pounds.

The $2.2 billion telescope was launched in 1990 from the Space Shuttle *Atlantis*. However, soon after its launch, NASA discovered that the primary mirror was defective. It had a flaw called spherical aberration, caused by a manufacturing mistake in the grinding of the outer edge of the mirror. Images, when seen through the Hubble, appeared fuzzy. The Hubble had to be repaired in the early 1990s by astronauts from the Space Shuttle *Endeavor,* who added corrective mirrors to the Hubble.

NASA sent two additional servicing missions to the Hubble in the late 1990s. The second servicing mission added two new scientific instruments to the Hubble. The Space Telescope Imaging Spectrograph (STIS) expands the spectrographic ability of the Hubble by thirty times for spectral data and five hundred times for spatial data. A *spectrograph* separates the light from the telescope into different components so, for example, the composition and temperature of a planet or star can be analyzed. The second instrument, The Near Infrared Camera and Multi-Object Spectrometer (NICMOS), allows the Hubble to see objects farther away than before.

The third servicing mission, originally scheduled for preventative maintenance of the Hubble, was made more urgent when the fourth of six *gyroscopes* failed in 1999, causing the Hubble to go dark. A gyroscope is used to keep a spacecraft in balance. Usually, there are gyroscopes to monitor motion on a spacecraft in one or more of the three spatial movements: vertical (called yaw), longitudinal (called roll), and lateral (called pitch). When the spacecraft shifts its position in space, the gyroscopes are measured electronically. The information is applied to the spacecraft's navigation computer that then can adjust the ship to keep it in balance. For most of 1999, the Hubble had been operating with only three of its six gyroscopes working; when the fourth failed, the Hubble went to sleep. The third successful mission added or replaced old equipment on the Hubble, including all six gyroscopes, a faster computer, and a new data recorder, among other improvements.

Since its repair by the shuttle in the early 1990s, not counting the one-month downtime in the late 1990s, the Hubble has fulfilled its goal and provided scientists with extensive data and photos of space. The Hubble has a 7.9-foot mirror that can make observations twenty-four hours a day. Also, since it orbits about 350 miles above Earth, it has a clear view of space. The Hubble can detect objects as faint as thirty-first magnitude. In contrast, the human eye can see celestial objects as faint as the sixth magnitude. Each magnitude is approximately 2.5 times dimmer than the one before; therefore, the Hubble can see objects 8 billion times dimmer in space than we can see with the human eye. This would be like if you could look out of your window in Washington, D.C., and you were able to see two fireflies six feet apart in Tokyo.

NASA scientists declare that the difference between the Hubble Telescope and contemporary optical telescopes viewing the sky from ground observatories is analogous to the difference between Galileo's first telescope and the human eye. More specifically, consider these assertions:

❑ Man's unaided eye limit—six hundred thousand light years distant.

❑ Ground observatory limit—2 billion light years distant.

❑ Hubble Space Telescope expected limit—14 billion light years distant.

Fourteen billion light years equates to a volume of space 350 times greater than can presently be seen from Earth.

Each year, astronomers compete for observation time on the Hubble. They must submit a proposal to NASA; these proposals are then submitted to peer review committees. The Hubble Institute director makes the final decision as to which proposals are funded. Because of the large number of astronomers and their divergent interests, the Hubble has been used to observe a wide range of celestial phenomena. In a ten-year period spanning most of the 1990s, the Hubble has probed over fourteen thousand celestial objects and made over 330,000 exposures. Each day, it generates three to five gigabits of data—enough to fill a typical home computer.

In its short lifetime, the Hubble has provided astronomers with hundreds of dazzling images and has affected almost every area of astronomy. As shown in Figure 4-5, the Hubble has helped astronomers in their understanding of the life of stars, the properties of galaxies, the atmospheres of the planets, and the properties and existence of black holes.

Compton Gamma Ray Observatory (GRO)

Compton Gamma Ray Observatory (GRO) was launched from the Space Shuttle *Atlantis* in the early 1990s. *Compton* orbited Earth, collecting data about the gamma ray emissions of the solar system until it was forced to reenter Earth's orbit in 2000. *Compton,* named for Nobel prize winner Dr. Arthur Holly Compton, carried four different instruments that could detect a broad range of high-energy gamma rays. *Compton* was unique because of the large size and increased sensitivity of its instruments. Although the number of gamma rays is small in the universe, these rays are the most powerful form of radiation. Also,

because the number of gamma rays is small compared to light rays, instruments need to be large in order to detect them.

The $670 million spacecraft had a broad mission including the study of different bodies in space: solar flares, gamma ray bursts, pulsars, novas, and supernovas. During its brief nine-year life span, it collected volumes of gamma ray data and discovered new and unique types of objects in our galaxy. The Compton mission was cut short when one of its gyroscopes used to maintain its orbits failed. NASA decided to force *Compton* into the atmosphere rather than wait for it to enter on its own. Similar to the Mir, but with a much shorter active life, *Compton* ended its life scattered across the Pacific Ocean.

X-Ray Telescopes

At the start of the twenty-first century, there are two X-ray telescopes circling Earth: the *Chandra X-Ray Observatory,* funded by NASA, and the *XMM-Newton,* funded by the ESA. NASA's newest telescope, the *Chandra X-Ray Observatory,* had the longest development cycle of any recent mission. The *Chandra* was first proposed in the mid 1970s, and funding for this project began in 1977. As time and NASA's budget (and priorities) shifted, the project was downsized in the early 1990s. Overall, it took over twenty years to design, build, and launch this spacecraft, and it was not until the late 1990s that the *Chandra* was launched in the cargo bay of the Space Shuttle *Columbia.*

The *Chandra* has three major parts: the X-ray telescope, the scientific instruments recording the X rays, and the spacecraft itself. The X-ray telescope system of the *Chandra* is unique. Since X rays would penetrate a normal mirror, the *Chandra* was designed with cylindrical mirrors resembling tubes within tubes. See Figure 4-6. The

Area	Pre-Hubble Knowledge (1990)	Hubble Contribution (2000)
Distant galaxies and galaxy evolution	Very little known about galaxies beyond a few billion light-years.	Deep imaging traces the evolution of galaxies and rate of star formation. Galaxies seen within a billion years of the Big Bang.
Remote supernovae as distance indicators	Not possible to discriminate supernova light from light of host galaxy.	HST detects supernovae all the way back to half of universe's age. The results show universe is accelerating.
Universe's rate of expansion	Two research groups disagree by a factor of two; this yields estimates for age of universe as between 10 to 20 billion years.	Value converges toward 10 percent accuracy, suggesting an age of the universe of 12–14 billion years.
Super-massive black holes	Data collected from Earth suggests the existence of black holes.	HST precisely measures gas velocity around black hole, providing definitive proof. HST surveys reveal that black holes are common to the cores of galaxies.
Quasars	Quasars originated early in the universe. Some are surrounded by a "fuzz," which is interpreted as the host galaxy. Quasars are likely powered by black holes.	HST clearly resolves a variety of galaxies hosting quasars. Some are involved in mergers with other galaxies. These collisions fuel the central black hole.
Gravitational lenses	A few examples are known.	Many small lenses uncovered in medium deep survey. Lenses have potential to contribute to cosmological tests of the curvature of space and age of universe.
Pluto	Transits and eclipses of Pluto's moon Charon yield a brightness map of surface.	HST confirms earlier map by showing a variegated surface. Images clearly separate planet from its moon Charon.
Supernova 1987A	The nearest supernova in four hundred years, an armada of telescopes watches its changes, following the February 1987 explosion.	Only HST has the resolution to trace yearly change happening at the sub-light-year scale, including changes in the fireball debris and circumstellar ring of enriched gas.
Galactic bulge structure	Only the Milky Way's central bulge and those of nearby galaxies can be viewed in detail.	Large bulges formed early in the universe, along with elliptical galaxies. Smaller bulges can be "inflated" by ongoing star birth, fueled by disk instabilities or galaxy mergers.

Figure 4-5.
This chart summarizes ten years of scientific achievement from the Hubble. (Adapted from NASA and the Space Telescope Science Institute)

mirrors of the *Chandra* are the smoothest ever created, and the largest of the mirrors is almost four feet in diameter and three feet long. Overall, the entire spacecraft is over forty-five feet long and sixty-four feet wide when the solar panels are deployed. At

12,930 pounds, the *Chandra* was the largest and heaviest payload ever launched by the space shuttle.

The *Chandra* has an unusual elliptical orbit taking this spacecraft from sixty-two hundred miles above Earth to a third of the

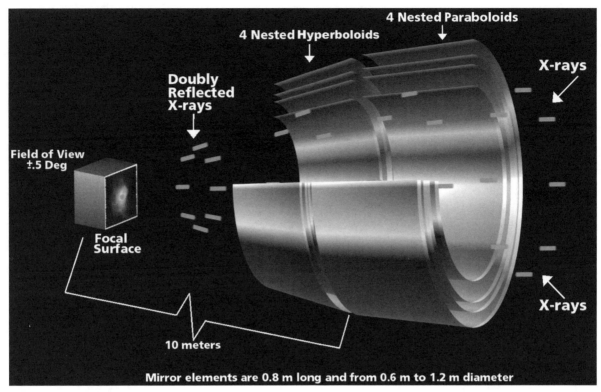

Figure 4-6.
This cutaway illustrates the design and functioning of the High Resolution Mirror Assembly (HRMA) system on the *Chandra*. (NASA/CXC/SAO)

way to the Moon. It takes the *Chandra* sixty-four hours and eighteen minutes to complete one orbit. Named after the late Indian-American Nobel prizewinner, Subrahmanyan Chandrasekhar, the *Chandra* studies X rays rather than conventional photos to give scientists information about the structure and evolution of the universe. Since the *Chandra* orbits in a higher orbit than the Hubble, it is able to get clearer readings while away from the belts of radiation surrounding Earth. In fact, the *Chandra's* orbit takes it two hundred times higher than the Hubble Telescope.

This telescope is designed to observe X rays from high-energy regions of the universe, including the remnants of exploded stars. Its higher resolution can allow scientists to see more features of these cosmic events than ever before. The *Chandra* can produce images with fifty times better resolution than was obtained before. Astronomers use the *Chandra* to study black holes and use the *Chandra* data to confirm the existence of the event horizon, the outer edge of the black hole, which was predicted by Einstein's theory of relativity.

The *Chandra* is complemented by the *XMM-Newton*, launched in the late 1990s by the ESA. Like the *Chandra*, the *XMM-Newton* uses a series of nested, curved mirrors to collect X-ray information about the galaxy. The mirrors on the *XMM-Newton* are the most powerful ever developed in the world. That allows the *XMM-Newton Space Observatory* to see even further back in time than the *Chandra.* It is an interesting fact about space that the farther the distance a telescope sees in space, the further it sees back in time. This is because what a telescope is actually seeing is light. Since light travels at a set rate, the farther away one can see light, the older it is. Even though the *XMM-Newton* can see farther back in time than the *Chandra,* the *Chandra's* telescope has a crisper focus than the *XMM-Newton* does.

In its one year of operation, the *XMM-Newton* has detected hundreds of new X-ray sources in the universe. One of its tasks is to explore X-ray background emissions; these emissions occur widely throughout the universe, but the cause of them is not known. The *XMM-Newton* was able to begin to solve this celestial mystery with its ten hour-long observation of Andromeda. The information gathered by this probe showed the presence of various chemical elements in this diffusion. This discovery gave astronomers more clues about the behavior of the galaxy.

Exploring Earth's Neighbors

In addition to the manned missions the various space agencies undertake, there are many unmanned spacecraft that have scientific exploration as their primary missions. These smaller missions receive less press acclaim from the public, but their long-range contributions to science are more significant than the more popular programs. Each of these missions has a more narrow focus; therefore, they tend to be significantly cheaper than the manned missions.

A landmark, unmanned mission is *Pioneer 10,* first designed to study the planet Jupiter. After almost thirty years, it is still making its way across our galaxy and is the first Earth spacecraft to leave our solar system. *Pioneer 10* carries a plaque, designed by Carl Sagan, which shows an illustration of a man and a woman and a diagram showing Earth's location in the galaxy. See Figure 4-7. *Pioneer 10* is now over 10 billion miles away from Earth, and its signal continues to be tracked by scientists on Earth. When *Pioneer 10's* formal mission ended in the mid-1990s, it took over nine hours for its radio signal to reach Earth.

Voyager is another long lasting mission. The two *Voyager* probes have been collecting data about the outer planets of our solar system since the mid-1970s. In the late 1990s, *Voyager 1* became the most distant Earth spacecraft, as it passed *Pioneer 10* (although they are traveling in different paths out of the solar system). Because of their sturdiness, the *Voyager* spacecraft have surveyed farther than ever expected. NASA estimates that these two spacecraft will continue to operate well into the first two decades of the twenty-first century—over forty years in space.

Not every mission launched by NASA or another country can expect the longevity or output of *Voyager* or *Pioneer.* During the 1990s, NASA and the ESA launched many different missions that had as their goal the exploration of our neighbors in space. In this section, we will highlight five successful missions: *Ulysses, the Solar and Heliospheric Observatory (SOHO), Galileo, Near Earth Asteroid Rendezvous (NEAR),* and *Cassini.*

Exploring the Sun

Two spacecraft are currently studying the nature and behavior of the Sun. Interestingly, both of these projects are joint missions of the ESA and NASA, although the ESA is the senior partner. *Ulysses,* launched in 1990, is studying the solar wind blowing from the Sun, while *SOHO's* main goal is to observe the Sun from its interior to its surface and atmosphere. These two spacecraft complement each other; together, they give scientists a deeper understanding of how the Sun works to sustain our galaxy.

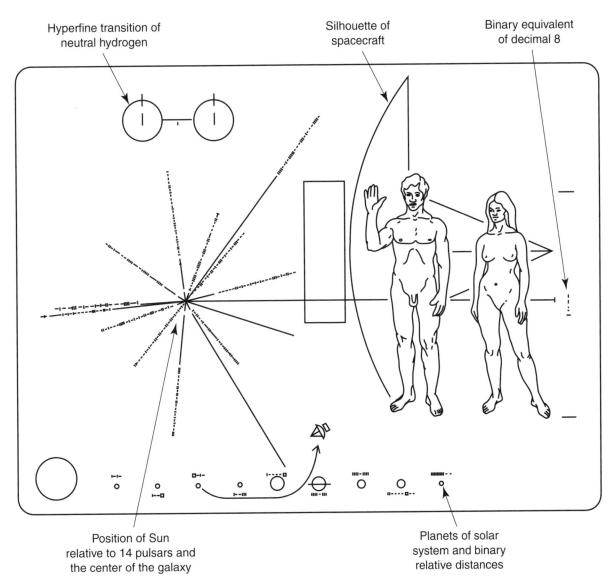

Hyperfine transition of neutral hydrogen

Silhouette of spacecraft

Binary equivalent of decimal 8

Position of Sun relative to 14 pulsars and the center of the galaxy

Planets of solar system and binary relative distances

Figure 4-7.
Pioneer 10 carried a graphic message in the form of a six- by nine-inch gold anodized plaque bolted to the spacecraft's main structure. (NASA)

Ulysses follows a polar route around the Sun, and it took over six years for this satellite to complete one rotation. During its journey around the Sun, *Ulysses* was able to detect and measure magnetic fields, cosmic rays and dust, solar X rays, electrons, and wind ions. Before *Ulysses* began its observation of the Sun, physicists knew that the solar wind left the Sun at two speeds, fast and slow. The data collected by *Ulysses* showed there is a boundary between these two solar wind areas, and the slow wind is pushed toward the Sun's equator by the fast wind.

SOHO was built by companies in fourteen European countries and had a total weight of about thirty-five hundred pounds at its launch in the mid-1990s. The spacecraft has twelve sets of instruments, and more than five hundred scientists in fifteen countries are involved in conducting research using data collected by *SOHO*. NASA monitors *SOHO* at the Goddard Space Flight Center in Maryland and uses its **Deep Space Network (DSN)** antennas to collect data from this Sun probe. Many international missions like *SOHO* use DSN and its antennas. DSN has three different sites on Earth: at Goldstone, California; Madrid, Spain; and Canberra, Australia. See Figure 4-8.

This spacecraft holds a position between the Sun and Earth, about nine hundred thousand miles from the Sun's surface. The original plan was to shut *SOHO* down after three years, but it has been so successful in its mission that NASA and ESA extended its life for an additional five years. *SOHO* has operated continuously since its launch, with the exception of two time periods, about four months in the late 1990s. During the

first period, NASA lost contact with *SOHO*, although scientists were able to reconnect with the spacecraft after a few months. The second operational lapse was due to gyroscope failure in the spacecraft.

SOHO has proven to have unexpected benefits for scientists on Earth. It has discovered more comets than any other machine (or person) in the history of astronomy. Within the first six years of its deployment, it had found more than three hundred comets. Also, *SOHO* can detect sunspots on the far side of the Sun. Since sunspots and flares can have harmful effects on satellites and power lines, this early detection is invaluable for minimizing the harmful effects of solar flares on Earth. Also, since space agencies continue to send astronauts into space, reducing astronaut exposure to dangerous particles from solar flares is very useful.

Galileo

After a six-year journey through space, the *Galileo* spacecraft arrived at Jupiter in the mid-1990s. Since that time, this sturdy craft has been studying the planet. *Galileo* has been an unqualified success for NASA, as it has more than doubled its original life span of two years. Today, it still is surveying Jupiter and its moons for NASA.

Galileo has two parts: the spacecraft and the probe, which together weighed almost six thousand pounds at launch. When it reached Jupiter, *Galileo* released the seven hundred-pound probe, which entered Jupiter's atmosphere and descended ninety-five miles through the planet's atmosphere. The probe relayed data to the *Galileo* spacecraft, which then transmitted this data to Earth. After fifty-eight minutes, the intense heat of Jupiter's atmosphere vaporized the probe.

Galileo has an unusual design. One section, rotating at three rotations per minutes, contains six instruments that gather Jupiter

Figure 4-8.
Deep Space Network's largest antenna is 230 feet (70 meters) in diameter, nearly the size of a football field. (NASA/JPL/Caltech)

data while the section is rotating. The other section of *Galileo* has a fixed position and holds four instruments.

When the original mission ended in 1997, *Galileo*'s goals were expanded, and it began a two-year study of the Jupiter moons, Io and Europa. *Galileo* sent back data showing an ocean under the surface of Europa, and it has studied the behavior and effect of the many volcanoes on Io. In fact, images from Io show so many volcanoes that the entire surface seems to be lava in various stages of cooling. Also, data from *Galileo* shows that some of the volcanoes on Jupiter resemble those on Earth. The Prometheus volcano on Io, for example, seems similar to the Kilauea volcano in Hawaii.

Galileo's mission was extended two more times. Its last mission extension will involve five more orbits of the moons of Jupiter. *Galileo* will take pictures and measure magnetic forces as it completes it final orbits. This $1.4 billion spacecraft is scheduled to take its final voyage in 2003, when it will be crashed into Jupiter. You might ask, Why crash *Galileo*? The short answer can be found in a document from the beginnings of space travel. In the mid-1960s, the United States signed the United Nations' Outer Space Treaty—this treaty requires that space missions minimize the transfer of living organisms from one body to another. Since *Galileo* has been traveling through space for over eleven years and might have picked up some microscopic living organisms, it must be crashed into Jupiter to avoid contaminating that planet or its moons and possibly transferring any living organisms to that planet.

Near Earth Asteroid Rendezvous (NEAR) Shoemaker

The *Near Earth Asteroid Rendezvous (NEAR) Shoemaker* spacecraft, named in honor of the late astronaut Gene Shoemaker, was a project under the NASA Discovery Program, which is a series of small-scale, lower cost spacecraft missions. *NEAR Shoemaker,* launched in the mid-1990s, is mechanically simple in its design. It has a five-foot antenna and four solar panels mounted on the outside of the spacecraft. See Figure 4-9. The scientific

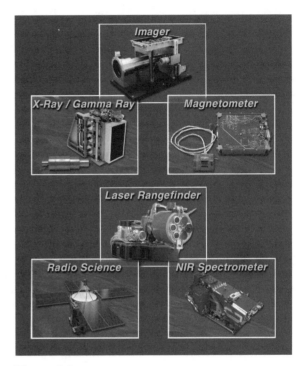

Figure 4-9.
The *NEAR Shoemaker* spacecraft carried six scientific instruments. The Magnetometer searched for a magnetic field on asteroid Eros (and didn't find one). The X-Ray/Gamma-Ray Spectrometer (consisting of two sensors) measured key chemical elements like silicon, magnesium, iron, uranium, thorium, and potassium. The Multispectral Imager—*NEAR Shoemaker*'s electronic camera—mapped the asteroid's shape, landforms, and colors. The Laser Rangefinder scanned the precise shape of the asteroid. The Near-Infrared Spectrometer mapped the mineral composition of the surface by measuring the spectrum of sunlight reflected by Eros. Also, a radio science experiment determined the asteroid's mass by tracking tiny changes in *NEAR Shoemaker*'s radio frequency. (The Johns Hopkins University Applied Physics Laboratory)

instruments are mounted on the side opposite to the antenna. The spacecraft also has a thruster that allows it to be moved. As spacecraft go, *NEAR Shoemaker* is light and inexpensive—it was about 1,774 pounds at launch and cost $223 million to build.

The mission for *NEAR Shoemaker* was to study Eros, a near Earth asteroid having the potential to collide with Earth within the next 1.5 million years. It took *NEAR Shoemaker* four years to travel the 2 billion miles to Eros. *NEAR Shoemaker* had six instruments for its analysis of this asteroid. The Multi-Spectral Imager (MSI) took both standard and near-infrared pictures of Eros to determine its shape and surface features. Also the MSI was able to map mineral deposits on the asteroid. Further surface information was gathered by the NEAR Laser Rangefinder (NLR), a laser altimeter that allowed scientists to develop a high-resolution topographic profile of Eros.

NEAR's Infrared Spectrometer (NIS) uses information gathered from infrared measurements to determine the distribution and amounts of surface minerals. The X ray/Gamma ray Spectrometer (XGRS) is two instruments in one. Together, these instruments detect X rays and gamma rays from elements on the surface. The last two instruments are the Magnetometer (MAG), which maps any magnetic fields, and a radio transponder, which enables *NEAR* to establish its position and velocity in space. The design of *NEAR Shoemaker* highlights the new design methodology of NASA—reuse technology whenever possible. The transponders (the core of the communication system) on *NEAR Shoemaker* were modified from those used on the *Cassini* spacecraft (discussed in the next section).

The spacecraft *NEAR Shoemaker* spent one year orbiting and analyzing the asteroid Eros before it landed on the asteroid in 2001. During its time in orbit, it sent back over 160,000 images to Earth, making it another successful project for NASA. The information that *NEAR Shoemaker* gathered gave scientists a tremendous amount of information about the composition and history of this asteroid.

Cassini-Huygens Mission to Saturn

The *Cassini* spacecraft, launched in the mid-1990s, is designed to orbit Saturn and gather information about the planet; its rings; its principal moon, Titan; and its other satellites. *Cassini* is one of the most complex, largest, and heaviest interplanetary spacecraft ever designed. See Figure 4-10. The total weight of the orbiter, Huygens probe, launch vehicle adapter, and fuel was 12,346 pounds at launch. *Cassini* is about twenty-two feet high and thirteen feet wide. *Cassini* is so large and complex because it will conduct twenty-seven scientific investigations once it reaches Saturn. This mission has a large international flavor, with lead scientists representing thirteen countries, including the United States, and over two hundred scientists from around the world working together on this project. By comparison, eleven countries are working together on the International Space Station. The *Cassini* mission has a NASA orbiter, an ESA probe, and an Italian communications system.

Cassini will take almost seven years to reach Saturn; after that, it will spend four years orbiting the planet and collecting data. In its data collection, *Cassini* will follow the model used during the *Magellan* mission to Venus. *Magellan* used one large antenna both for communication with Earth and for mapping the planet. Like *Magellan*, during its orbit of Saturn, *Cassini* will spend about twelve to fifteen hours each day observing Saturn. Then, *Cassini* will turn its antenna toward Earth and spend nine to twelve hours transmitting the science data collected to DSN. *Cassini* will use DSN's 112 feet and 230 feet antennas at each of the

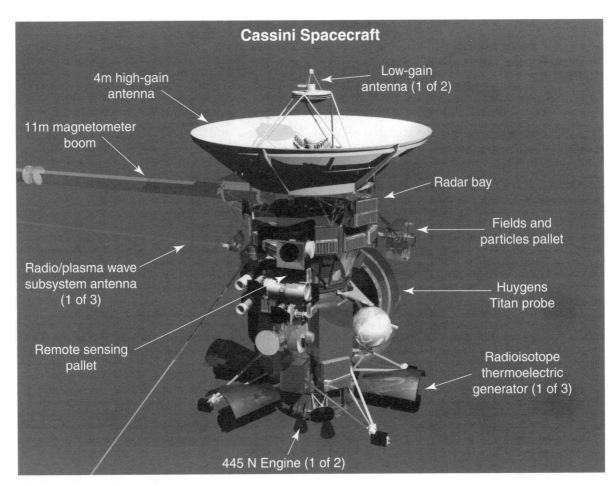

Cassini Spacecraft

4m high-gain antenna

Low-gain antenna (1 of 2)

11m magnetometer boom

Radar bay

Fields and particles pallet

Radio/plasma wave subsystem antenna (1 of 3)

Huygens Titan probe

Remote sensing pallet

Radioisotope thermoelectric generator (1 of 3)

445 N Engine (1 of 2)

Figure 4-10.
This is a diagram of the *Cassini* spacecraft. (NASA)

three sites. Over its four-year planned observation mission, *Cassini* should collect and send about two trillion bits of data to Earth—the equivalent of eight hundred sets of Encyclopedia Britannica. On its way to Saturn, *Cassini* made observations of Jupiter in conjunction with *Galileo,* which had been orbiting Jupiter for more than five years. Together, the data gathered from these two spacecraft provides more information about Jupiter's weather patterns and volcanic activity on its moon, Io.

Missions to Mars

Throughout history, we have looked to Mars as our nearest neighbor. Mars holds a fascination for us, as we consider whether there really is life beyond Earth. Each of us can probably think of a movie or book featuring little green men from Mars. Also, it appears Mars is the only other planet in our solar system that could possibly sustain life. So, it is only natural that Mars is the first planet we want to explore.

The first Earth spacecraft to visit Mars was a Russian craft that did a flyby of Mars in 1960. The first successful U.S. mission, *Mariner 4*, returned twenty-one photos of Mars to Earth in the mid-1960s. Since these early beginnings, Mars has been the focus of twenty-nine missions. Some U.S. successes include *Mariner 9*, launched in the early 1970s, which returned over seven thousand photos of Mars, and *Viking I* and *Viking II*, combination orbiters and landers that returned over fifty thousand photos.

In the 1990s alone, there were eight different missions to study the planet Mars. NASA ran six of these space missions; the other two were funded by Russia and Japan. Unfortunately, the success of these missions has varied. A summary of these missions is shown in Figure 4-11. Why were so many of these missions unsuccessful?

Despite a few failures, including the *Challenger* and *Apollo 13*, NASA had enjoyed an enviable record in space until the late 1990s, but changes were already underway that would lead to the problems that occurred in the exploration of Mars. In the early 1990s, because of shrinking budgets and the large amount of money earmarked for the shuttle, NASA adopted a policy to reduce costs called *faster, better, cheaper (FBC)*.

International Mars Missions in the 1990s				
Year	Mission	Sponsoring Agencies	Outcome of Mission	Cost
1992	Mars Observer	NASA; Centre National d'Etudes Spatiales/France; Russian Space Agency	Lost on August 21, 1993, three days before it was scheduled to go into orbit around Mars.	$980 M
1996	Mars 96	Russian Space Agency	Crashed soon after launch.	Unknown
1996	Mars Pathfinder	NASA	Successfully landed on Mars; sent back signals until March 10, 1998.	$440 M
1996	Mars Global Surveyor	NASA	Reached Mars orbit in 1997 and is collecting data on Mars.	$219 M to build and launch. Costs $20 M/year.
1998	Nozomi (Planet-B)	Institute of Space and Aeronautical Science, Japan	Launched successfully, but the spacecraft used more propellant than planned in a course correction maneuver and is scheduled to arrive at Mars in December of 2003.	$848 M
1998	Mars Climate Orbiter	NASA	Destroyed when a navigation error caused it to miss its target altitude by 80 to 90 km.	**
1999	Deep Space 2	NASA	Presumed lost on December 3, 1999.	$29.2 M
1999	Mars Polar Lander	NASA	Presumed lost, or crashed during landing.	**

** The Mars Surveyor '98 program is comprised of two spacecraft that were launched separately, the Mars Climate Orbiter and the Mars Polar Lander. NASA spent $327 M total on both of these projects. M = million.

Figure 4-11.
This chart summarizes international Mars missions in the 1990s.

The goal of FBC was to shorten development time, reduce costs, and increase the scientific output of NASA by having more missions in less time.

The first project using the rules of FBC was the *Mars Global Surveyor,* which was an unqualified success and is still collecting information about the planet today. The next three major NASA Mars projects were, however, failures—very spectacular failures.

Two of these failed projects, the *Mars Climate Orbiter* and the *Mars Polar Lander,* were part of the Mars Surveyor '98 Program designed to study the Martian weather and climate. According to a follow-up report by NASA, a software flaw probably caused the *Mars Polar Lander* to shut off its engines too early in its descent. The *Mars Climate Orbiter* was destroyed by an even simpler human error, a mix-up over metric and English measurements. The prime contractor, Lockheed Martin, gave data for the navigation program to the Jet Propulsion Laboratory (JPL) team writing the navigation software program in English units rather than in metric units. This error caused the *Climate Orbiter* to pass too close to the planet and burn in Mars's atmosphere.

The third project was *Deep Space 2,* a mission with two probes designed to send back data on the properties of Mars below the surface, but NASA never received any signal from *Deep Space 2,* leading it to conclude it was lost or crashed. In any case, the loss of three Mars missions over a year's time brought NASA into greater scrutiny.

The future of FBC is in doubt. From a review held after the loss of all three missions, an independent assessment team determined the Mars Surveyor '98 Program was inadequately funded, and the pressure of maintaining the budget led to too many corners being cut during engineering design and manufacture. As a result of trying to do more with less, NASA's FBC strategy went too far and failed. As long as budgets are tight, there is an increased possibility missions will fail, but at the same time, both Congress and the U.S. public are unwilling to return to the days when NASA had a blank check for its work.

In 2000, NASA announced a new plan for Mars exploration including six new missions during the following decade. In contrast with most previous missions, these Mars missions will include international partners, like the French and Italian space agencies. The first mission was the *Mars Odyssey,* launched in 2001.

Despite the three recent setbacks in Mars exploration, there have been successful missions. A few of the more important missions are discussed in the next sections.

Mars Pathfinder

The *Mars Pathfinder* had two different parts: a lander and a surface rover. The *Pathfinder* reached Mars in the late 1990s, entered the planet's atmosphere, and bounced on inflated airbags when it landed. The surface rover, named Sojourner, then exited the lander and started exploring the surface of Mars. The main objective of the *Pathfinder* was to demonstrate the feasibility of low cost landings on Mars.

Sojourner, the rover, was tiny—about eleven inches high, twenty-five inches long, nineteen inches wide, and powered by solar cells. The rover was operated by remote control from Earth, although the rover did have a hazard avoidance system. However, since there is a ten to fifteen minute time delay between Earth and Mars, it was a challenge to control the rover remotely, certainly nothing like the immediate response we get from our experiences using a remote controlled vehicle on Earth.

The mission was designed so the rover would communicate with the lander, and the lander would then communicate with Earth. Sojourner could take both color and black and white pictures as it moved along

Mars's surface. Both the lander and the rover lasted longer than expected. The lander functioned for almost ninety days, while the rover lasted for eighty-four days. More importantly, a tremendous amount of data was sent to Earth, including 16,500 images from the lander, another 550 images from the rover, sixteen chemical analyses of rocks and soils, and 8.5 million measurements of pressure, temperature, and wind. Overall, the *Mars Pathfinder* mission cost approximately $265 million, including the launch and operations.

Mars Global Surveyor (MGS)

The *Mars Global Surveyor (MGS),* currently in orbit around Mars, includes six different parts. *MGS* has a camera taking both high- and low-resolution pictures of the planet's surface, as well as five different instruments measuring Mars's gravity, weather, climate, surface, and atmospheric composition. In addition, *MGS* has an altimeter that is building a topographic map of Mars.

MGS is small and relatively cheap for a space probe. It consists of a rectangular box about four feet on each side, weighed only about twenty-three hundred pounds at launch, and cost $154 million to build and $65 million to launch. Actually, the center section is really two smaller boxes on top of each other—the equipment module contains all the instruments and is on one side of the spacecraft, and the propulsion module holds *MGS*'s engine and fuel tanks and is located on the other side. *MGS* has two solar panels extending about thirty-nine feet from tip to tip.

MGS began orbiting Mars in 1997 after a journey of ten months from Earth, but it did not begin collecting data until 1999. Why the delay? When *MGS* arrived near our neighbor planet, it was in an elliptical orbit that took forty-eight hours to complete. NASA scientists had to slow *MGS* down and also change its orbit to a circular orbit.

It took sixteen months to move *MGS* into a circular seven-day orbit around Mars. NASA scientists took advantage of the atmosphere of Mars to slow down the spacecraft. They used a technique called *aerobraking.* First, *MGS*'s orbit was moved into the upper portion of the Martian atmosphere. Then, each time *MGS* orbited the planet in the atmosphere, it slowed down slightly because of air resistance and lost altitude. Over the sixteen months, *MGS* was moved from an orbit of 34,900 to 250 miles above the Martian surface. Originally, this aerobraking was scheduled for four months, but a malfunction in one of the solar panels delayed the total process for one year.

Despite the problems with its solar panel, *MGS* completed its primary mission of mapping the surface of Mars. Its mission was extended, and it was given the job of collecting information on potential landing sites for future robotic landers. Since *MGS* can spot details as small as ten feet, there is a lot of information it can collect for future Mars missions.

Mars Odyssey

Odyssey is the newest spacecraft launched to study the surface of Mars. Its success is critical for NASA, given that the two previous Mars missions, the *Mars Climate Orbiter* and the *Mars Polar Lander,* were both spectacular failures back in the late 1990s. For an agency somewhat accustomed to success, this first mission after a string of failures increased the stress and put NASA's design of the *Odyssey* under a magnifying glass.

After *Odyssey* reaches Mars, it plans to use the technique of aerobraking successfully used by *MGS. Odyssey* will orbit Mars in conjunction with *MGS,* but it has a different mission and carries different instruments. The focus of *Odyssey* is geology; in fact, its three primary instruments will allow scientists to get a complete geological picture of the red planet. All of these three instruments involve *spectrometers,* which are instruments separating light into its different wavelengths to create spectra. Spectrometers allow scientists to determine what elements are present in an object. The Thermal Emission Imaging System (THEMIS) will analyze the distribution of minerals on the Martian surface, while the Gamma Ray Spectrometer (GRS) will use the gamma ray part of the light spectrum to study how Mars has changed over time by measuring the presence of twenty primary elements (including carbon, oxygen, and iron). The last instrument, Mars Radiation Experiment (MARIE), will monitor radiation levels on the surface.

Overall, the *Odyssey* has four main goals. The first goal is to determine whether life ever existed on Mars. To achieve this goal, it will look for water under the Martian surface and search for mineral deposits from past water on the surface. The second goal is to study the climate of Mars. The third goal is to analyze the geology of Mars using the THEMIS. The last, and ultimately most important, goal is to prepare for human exploration of Mars. MARIE will collect data about radiation levels that will be used for future manned missions to Mars.

The International Space Station

The International Space Station had its roots in the later years of the Cold War. At that time, Russia had a permanent space station, the Mir, and the United States did not. Then-President Ronald Reagan, in the mid-1980s, approved the development of a U.S. space station called Freedom, a name chosen to show the contrast to its Russian counterpart. Over the next few years, the design underwent several changes. The

---

INTERNATIONAL SPACE STATION
An orbiting space station that will serve as an international laboratory for research.

---

space station has a unique phased approach. It is built in modular segments; the segments are then connected in space like a giant LEGO® toy. Because of the tremendous costs involved, international cooperation was a necessity from the start. The prime international partner in the original space station was the ESA. The ESA, tired of being treated by NASA as an afterthought, demanded concessions from NASA. Because of the continued conflict between the United States and Russia, the ESA wanted to guarantee free access to the space station. Also, it wanted technology transfer from the United States in return for its financial contributions to the space station. Because of these pressures, the United States, ESA, Japan, and Canada agreed to limit space station use to peaceful purposes, and the final agreement for the Freedom space station was signed in the late 1980s.

The Freedom space station, however, was to undergo another transformation in the 1990s. As the Cold War period drew to a close in the late 1980s and early 1990s, the station was redesigned once again to reduce costs and increase the level of international cooperation. It also received a new name, the International Space Station, when the Russian government agreed to supply major hardware for the space station that had originally been planned for Mir II. The new design uses 75 percent of the original Freedom design.

The International Space Station is a very complex feat of engineering. The completed station will have a mass of about 1 million pounds, and it will be about 260 feet wide and 290 feet long. At the same time, it is being built with the scientific expertise and participation of sixteen nations: the United States, Canada, Japan, Russia, eleven members of the ESA, and Brazil. It is no wonder it is so complex.

The International Space Station is being built in stages. See Figure 4-12. It is an interesting point that the historic competitors in

Current Construction Sequence of Major Modules in the International Space Station		
Module	**Country of Manufacture**	**Date**
Zarya control module. Zarya means "sunrise" in English.	Russia	November 1998
Unity connecting module	United States	December 1998
Spacehab Double Cargo Module	United States	May 1999
Zvezda Service Module	Russia	July 2000
Destiny Laboratory Module	United States	February 2001
Raffaello Multi-Purpose Logistics Modules (MPLM)	Italy	April 2001
Ultra High Frequency Antenna	United States	April 2001
Canadarm2	Canada	April 2001
Docking Compartment	Russia	August 2001
Integrated Truss Structure S0	United States	April 2002
Mobile Base System	Canada	June 2002
Science Power Platform	Russia	Future
Universal Docking Module	United States	Future
Docking Compartment 2	Russia	Future
Node 2	United States	Future
Japanese Experiment Module Experiment Logistics Module	Japan	Future
Kibo Experiment Module. Kibo, in English, means "hope."	Japan	Future

Figure 4-12.
This chart shows the current construction sequence of major modules in the International Space Station.

space, the United States and Russia, contributed the first two modules, Zarya and Unity, but these two countries are not alone in the construction program. Canada provided a fifty-seven-foot long robotic arm for assembly and maintenance on the space station, and Italy is providing a series of Multi-Purpose Logistics Modules (MPLM) that serve as moving vans for the space station. These MPLMs, named Leonardo, Rafaello, and Donnatello, carry equipment, experiments, and supplies to and from the space station using the shuttle's cargo bay. After the MPLM is unloaded in space, used equipment and trash are loaded into the module to return it to Earth.

The cornerstone of the first stage is the Unity connecting module because it has six ports, and the space station will be built by connecting onto these six ports. Connection of the Zvezda service module to Unity allowed the space station to become operational because the Zvezda module provided the first living quarters, life support, electrical power, data processing, and propulsion system. The addition of the Destiny Laboratory Module in 2001 allowed the International Space Station to begin its scientific mission. Although wide scale experimentation on the space station is still in the future, about 120 researchers have already been chosen to conduct experiments using this laboratory. See Figure 4-13.

Figure 4-13.
The International Space Station was photographed against clouds over a landmass on Earth during a fly around by the Space Shuttle *Atlantis.* (NASA)

The International Space Station welcomed its first crew in 2000. The international crew of two Russian cosmonauts and one American astronaut reached the space station using a *Soyuz* spacecraft. The Russians supplied a *Soyuz* spacecraft to the space station as a lifeboat. Over the life of the station, the *Soyuz* craft will be replaced so it is always ready to be used. The flight of the first space tourist, see *Dateline,* was one such replacement flight. The California millionaire took probably the most expensive vacation ever, with his $20 million trip to the space station, but he assisted the crew and was their guest, as they replaced the *Soyuz* escape craft. The Russians have also delivered progress supply vessels to the space station. As discussed in the section on the Mir, the unmanned progress resupply ships were successfully used to deliver food, fuel, oxygen, and supplies to the station.

Most of the construction of the station is done in space, so it is critical that all parts mesh and work well. After all, there is no local Radio Shack in space in case the astronauts need a spare component! The early construction of the space station has been hampered by computer problems on the U.S.-built sections of the station. In one day, all three command-and-control computers failed on the space station, and the station was forced to rely on a pair of computers in the Unity module. Although NASA engineers eventually fixed this computer glitch, the precarious nature of construction in space is still very evident.

In addition to the technical problems inherent in a construction project of this complexity, there are financial and political pressures as well. The Russian Space Agency is feeling financial pressure because its budget has been cut resulting from the changes in Russia since the end of the Cold War. Added to these financial woes is a $4 billion price hike in the cost for the space station that became public in 2001. Either NASA has to scale back its plans for the space station, or it must eliminate other programs in order to cover the increased costs. Currently, the plan is to stop work on the U.S. habitation module, crew return vehicle, and propulsion module. Unfortunately, these modules are needed before the space station can host its full crew of seven. Until these modules are in place, only three crewmembers can be supported at the space station at one time. Despite these problems, there is no doubt construction will continue on the International Space Station. Its final shape and size are all dependent upon the budget decisions being made over the next few years.

Satellites

The successful launch of the first man-made satellite marked the beginning of man's exploration of outer space. Sometimes referred to as handymen of the heavens, artificial Earth satellites transmit communications at the speed of light, serve as military sentries, and explore the universe. These electronic servants are used to locate mineral deposits, provide incredible records of the world's weather patterns, transmit transcontinental phone calls, and guide ships through darkness and storms. Acting as silent spies in the sky, some satellites maintain surveillance that sustains peace in our volatile world.

Just how many objects are floating around out there? As incredible as it may sound, the North American Aerospace Defense Command (NORAD) center counted nearly nine thousand objects floating in space, although close to 70 percent of those items could be classified as *space junk.* For example, when satellites are launched into space, the rocket bodies, nose cones, and spent fuel containers all remain in orbit for a certain period of time. Other pieces of space junk are just accidents. In 2001, American astronaut Jim Voss added to the space junk total when he dropped a foot restraint system during a spacewalk on the International Space Station. Wherever the junk comes from, NORAD tracks everything and watches objects as they fall from orbit in order to spot things that should not be there, such as enemy missiles. Less than four decades after the Soviet Union's surprise launch of *Sputnik 1,* NORAD tallied more than twenty-six thousand man-made objects that had been shot into space.

NORAD estimates there are 2,837 satellites in orbit that are actually operational. Of these, 1,335 belong to Russia, while 741 belong to the United States. Several other countries and the ESA launched the remaining satellites. Regardless of who owns them, man-made satellites ultimately lose momentum and, therefore, altitude. Without additional servicing, their orbits decay, and they fall into the atmosphere to disintegrate. If the disintegration is planned, as with the Mir and *Compton GRO,* the target is usually the Pacific Ocean between New Zealand and Chile, but not all reentries are as successful. Skylab crashed in Western Australia in the late 1970s, and a Russian spacecraft hit New Zealand's South Island. All the debris circling Earth makes it even more likely that unexpected landings will happen in the future.

Satellite Orbits

Just as artificial satellites come in a variety of shapes and sizes, so do their orbital paths around Earth. A satellite positioned in a *geostationary orbit* or *geosynchronous orbit* always stays in the same spot over the equator. Located some 22,300 miles away from the planet surface, the satellites in this orbit move at the speed of Earth's rotation. This type of an orbit provides a prime location for both meteorological and communications satellites. When viewed from Earth stations, these satellites appear to be standing still.

Generally speaking, satellites tend to remain in an orbital plane defined by the latitude of their launch sites. For example, a satellite launched from an equatorial site, like ESA's Ariane base in French Guiana, will remain in an equatorial orbit unless a correction is made. U.S. satellites launched from Cape Canaveral have a lesser angle of inclination with the equator than Soviet satellites launched from higher latitudes. Elliptical orbits are useful for placing satellites that make scientific measurements at differing altitudes, such as ozone levels. See Figure 4-14. Satellites launched into a polar orbit can observe the entire surface of Earth as the planet revolves beneath them. Polar orbits are commonly employed by the military to position their huge spy satellites. Finally, a Sun synchronous orbit is one

inclined a few degrees beyond Earth's pole. This type of orbit permits an imaging satellite, such as *Landsat,* to create its pictures with the sunlight always striking Earth from the same angle.

By definition, a satellite's orbit is a condition in which the pull of Earth's gravity is matched by the outward fling of the craft's speed. There are an assortment of subtle pressures that may cause a satellite's orbit to decay, including solar flares and outer atmospheric disturbances. If a satellite starts to wander off course, watchful ground crews can systematically fire its small fuel jets, steering it back into orbit. This procedure must be used sparingly, since the exhaustion of these gases spells the end of the spacecraft's useful career.

Commercial Development

In just over thirty years, our society has become accustomed to its reliance on an assortment of satellites orbiting our world. The broadcaster's phrase "live via satellite" doesn't have nearly the luster it once had. These technological contraptions have become indispensable tools to the communications industry, the scientific community, and our military personnel. It did not take long for space scientists to recognize the promise of satellites for the international communications network. Satellites were envisioned and described as great antenna towers in the sky that would be capable of transmitting messages almost instantaneously across oceans and landmasses.

The United States became quite active in the satellite launching business during the early 1960s. *Echo I* became the world's first communications satellite and one of the largest man-made objects ever sent into orbit. This satellite was actually a huge Mylar™ balloon covered with a thin coating of aluminum able to reflect radio signals

Figure 4-14.
This image of Invesco Field and Mile High Stadium was taken in 2001, by Space Imaging's IKONOS satellite from an orbit of 423 miles above Earth. IKONOS is the world's first and only commercial, one-meter resolution, Earth imaging satellite. (Space Imaging)

from coast to coast. NASA launched the world's first weather satellite, called *Tiros 1.* Equipped with twin television cameras, this spacefarer returned pictures of Earth's surface for several months.

The Bell System's *Telstar I* satellite ushered in the era of live transatlantic television broadcasting. Unlike *Echo 1*, which passively bounced back signals, *Telstar* carried transponders powered by solar cells.

These amplified signal strength thousands of times before relaying them to their destination. *Telstar* set a pattern for all following communications satellites. Since its orbit was low, it could only relay messages for short periods of time, while it was in the correct position between sender and receiver. If a newsworthy event was taking place in Paris, you may or may not have been able to receive the televised report in New York. In other words, the geostationary belt had not yet been utilized successfully.

Intelsat I was soon launched into a geostationary orbit. Nicknamed "Early Bird," it provided the initial space link for a remarkable venture known as the International Telecommunications Satellite Organization (Intelsat). The Intelsat cooperative, headquartered in Washington, D.C., provides telecommunications services to over 144 countries using a fleet of nineteen satellites. The participating agency for the United States is the Communications Satellite Corporation (Comsat). Intelsat's newest satellite, *Intelsat IX,* can juggle tens of thousands of telephone calls and several television programs simultaneously. By humble comparison, *Intelsat I* relayed up to 240 calls and carried a single television station less than twenty-five years ago.

Several more benchmark events occurred in the field of satellite technology during the 1970s. NASA launched a series of *Landsat* satellites that sent back television-like pictures and infrared images of distinct features of Earth's surface. These powerful tools have been instrumental in improving agricultural yields, locating oil and mineral deposits, monitoring volcanoes and other geologic events, and mapping optimal routes for railways and pipelines. The current version, *Landsat 7,* can capture and process 250 *Landsat* scenes each day. The Landsat Program currently is governed by a joint agreement between NASA and the U.S.

Geological Survey (USGS). In this partnership, NASA is responsible for the development and launch of the satellite, and USGS is responsible for *Landsat*'s operation and for distributing and maintaining the data *Landsat* collects.

The history of space commercialization witnessed a milestone event in the mid-1990s. According to a study by the accounting firm KPMG Peat Marwick, commercial spending in space exceeded government spending in space. Satellite technology continues to progress and improve in recent years, with the addition of new capabilities for Internet and multimedia transmission, and as the number of satellites increase and the costs of these satellites decrease, companies have begun exploiting the untapped for satellite images. One such company, Aerial Images, went into business with a Russian government spin-off, Sovinformsputnik, which promotes and markets products developed during the Russian Space Program. Aerial Images now markets high-resolution images taken from space to customers across the globe. These images can show objects as small as three feet across.

Along with the increased use of satellites in our daily lives comes increased dependence on them. Often, we are unaware of our reliance on these orbiting stations until a problem happens. The nation was reminded of the role satellites play in our everyday lives when the *Galaxy IV* satellite malfunctioned in the late 1990s and shut down 80 percent of the nation's 40 million pagers, as well as credit card authorization networks and video for cable networks. In order to restore service, thousands of ground antennas and satellites had to be moved—this process took weeks to resolve. Over the next two years, two other satellites malfunctioned. The first, in the late 1990s, disrupted the transmission of several organizations in the United States, including CNN, the Public

Broadcasting Service, and The Associated Press. Also, in 2000, a problem in the ground stations interrupted satellite transmission for three hours from a number of U.S. satellites.

As more people in the United States utilize wireless technologies, such as cell phones and pagers, and entertain themselves through cable and satellite television, they have an increased dependence on the satellites providing these services. Although most people do not think of satellites as parts of their cable television networks, for example, a disruption would lead to an unwelcome lesson in the importance of satellites to our society. An increasing number of satellites are being launched—about twenty new ones each year. As satellites play an increasingly greater role in our communication, any possible disruptions have the potential to further affect our everyday lives.

Military Applications of Satellites

With increasing frequency, military officials looked to the satellite industry for reconnaissance assistance. Subsequently, satellites are presently equipped with visible light detectors, which can see an enemy's weapons directly, and infrared detectors, which can sense the heat emitted by a rocket exhaust plume during tests. Navigational satellites are the workhorses of the military's (and civilian) global positioning system (GPS).

Global positioning system (GPS) is becoming a popular application of satellite technology for consumers, although it was first developed for military purposes. It was originally designed to solve a persistent problem in warfare. During the heat of battle, with all the confusion and distractions, it is difficult for soldiers to determine their precise locations. At the same time, in order

not to cause damage to your own tanks, trucks, and troops, military managers need to know where their troops are at all times. Enter GPS. Since GPS can give an exact position, it allows for more efficient mobilization of soldiers in battle.

GPS works using a system of twenty-four satellites, including three spares, that are positioned so a person on Earth can receive a signal from four of them at one time. A GPS receiver finds its location by getting signals from three different satellites. GPS was developed by the DOD to provide navigation assistance to its forces. The GPS satellites have two different signals—military and civilian. The civilian signal is accurate to about one hundred meters, while the military signal is accurate to twenty meters. See Figure 4-15.

U.S. forces used GPS extensively during Operation Desert Storm in the early 1990s. Since the war was fought in a desert, it was very difficult, if not impossible, to move troops, especially during the night or in sandstorms. With GPS, the soldiers could be moved quickly—this increased mobility led to a swift victory. Over nine thousand GPS receivers were used by foot soldiers; on vehicles, on helicopters, and in aircraft; and on Navy ships, during Operation Desert Storm.

Today GPS is used in a variety of civilian applications. During construction of the Chunnel (the tunnel under the English Channel), there were two different starting points at each side of the Channel. One crew started from Dover, England, and the other crew started from Calais, France. These crews used GPS to check their positions, so that the two sides of the tunnel would meet. Among the hundreds of other uses of GPS, automobile manufacturers are now offering GPS as an option in cars. Imagine an electronic map that will keep you from ever being lost!

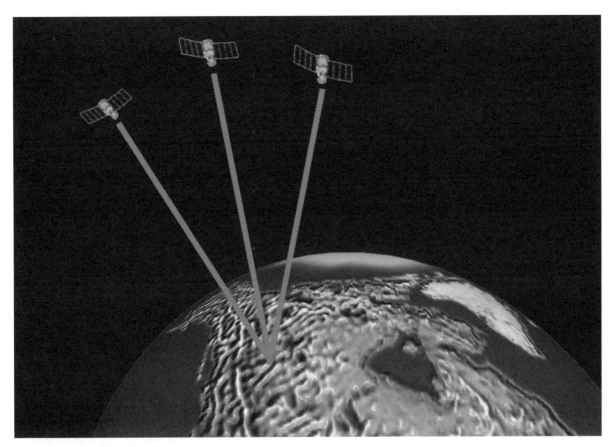

Figure 4-15.
This graphic depicts how GPS works.

The military uses satellites to provide communications, conduct reconnaissance, collect data for maps and charts, and gather information from natural and man-made disasters. The military also is involved in satellite technology that will stop the jamming of a signal. Today, for example, one can buy a Russian GPS jamming system, about the size of a cigarette pack that can jam a GPS signal up to eighty kilometers.

As commercialization of space continues, the military is becoming more dependent on commercial companies. As mentioned above, the U.S. military used commercial satellites in Desert Storm for communication. This type of dependence allows for more flexibility, but at the same time increases the security threats for the United States. In its recent study on space, the Commission to Assess U.S. National Security Space Management and Organization, sponsored by the DOD, recommended increased funding to address national security issues in space. This Commission believes future military conflict in space is a virtual certainty. Because of the perceived threat of enemy action from space, it recommended that the United States increase funding for space defense and develop superior space capabilities. Whether Congress will grant this wish in a time of budgetary constraints is unclear.

Future Developments in Space Exploration

For as long as historians have kept written records of society, they have been mindful of one simple tenet or principle: A civilization that ceases to explore begins to die. For more than four decades, government officials around the world have shown interest in space exploration, some more than others. At the same time, however, only a very small percentage of our civilization is involved in space exploration. Since the costs of space exploration are high, the potential benefits may be too small for most people to justify these costs.

It is no secret that the United States and Russia have been the primary players in this exploration game. The Russian government has made slow but steady progress toward the attainment of a series of long-term objectives characterizing its national space program. On the other side of the ocean, the U.S. program has made great technological leaps, some of which appeared on the scene in flashy, explosive spurts, toward the completion of short-term projects. It is imperative for the U.S. government to make some hard decisions and then adhere to some reasonable long-term goals in space.

The Future Goals of the National Aeronautics and Space Administration (NASA)

Sally Ride, the first woman astronaut in space, headed a National Aeronautics and Space Administration (NASA) task force in the mid-1980s that looked to develop new priorities for the space program. The task force developed four future scenarios for NASA.

❏ Conduct Earth studies to collect information and knowledge for the protection of the world's environment.

❏ Accelerate the unmanned mission program to explore the Moon and other planets in the solar system using robotic explorers. As a forerunner to human tended outposts, NASA should consider launching a robotic lunar prospector to examine the entire surface of the Moon.

❏ Resume manned visits to the Moon. These should not be for brief expeditions, but for longer, systematic explorations, culminating in permanent settlements.

❏ Conduct a manned expedition to Mars.

The immediate stages of the commission's program contemplate new technologies, including the production of three new reusable space transport vehicles. The first is a low orbit cargo vehicle. The second is a passenger vessel smaller than the shuttle for travel to and from a LEO space station. The third new orbiter is described as a workhorse transfer ship to carry both people and cargo beyond the Moon. This will also provide ferry service within the inner solar system (interspace station or colony excursions).

As we can see from the recent space expeditions discussed in this chapter, NASA has pursued this strategic plan, particularly with respect to the exploration of other planets in our solar system. Since the mid-1980s, however, there have been political changes on Earth that have dramatically impacted the space program, including the end of the Cold War and the high amount of international collaboration in space travel. Both of these events have caused shifts in the future prospects in space from an adversarial competition between the United States and the Soviet Union to cooperative relationships among the Americans, Russians, Japanese, and the European

Community. Today, almost no major space mission is conducted solely by one country. Therefore, NASA's future has to be reevaluated in light of this new political environment. Also, the public's support for the Space Program has waned since the Apollo years. Unless NASA recaptures the optimistic imagination of the American public, it will be more difficult for it to fund its ambitious projects in the future. Because of its awareness of the changing nature of space exploration, in 2000, NASA released its strategic plan for future space exploration. Imbedded in this plan are short- and long-term goals for NASA and its international partners in space.

As part of its strategic plan, NASA wants to develop the ability to send humans further into the solar system. There are three phases to this goal. The short-term focus is on LEO missions, including space shuttle voyages of up to two weeks and International Space Station missions of thirty to ninety days. NASA hopes it can develop and test new technologies during this first phase so it can move into the middle phase, missions outside of LEO lasting about one hundred days. It has already begun integrating robotics and computer controlled components into its spacecraft to allow it to complete these missions. The *Deep Space I* probe, for example, included an onboard software using artificial intelligence (AI) to control navigation. During its flight in the late 1990s, NASA allowed the spacecraft to navigate itself for thirty-five hours using this AI package.

In the future, NASA sees a series of telescopes and remote structures further away from Earth that could be used as intermediary stops for voyages farther into the solar system. The long-term plan for NASA includes complex missions of five hundred to one thousand days, including the long hoped for manned mission to Mars. A challenge of this magnitude might actually foster global unity and a healthy attitude toward living together in harmony on this spaceship Earth.

Exploration of Mars

There are two prongs to future exploration of Mars. In the long-term, NASA has the goal of sending a manned mission to Mars. The technical, economic, and social constraints of this manned voyage place it far into the future. For the short-term, NASA and the other space agencies are content to send increasingly more complex unmanned missions to study this planet. Europe is making a major contribution to the unmanned exploration of Mars with the ESA's *Mars Express* mission scheduled for 2003. NASA has a mission of its own, *Mars Odyssey*, in flight toward Mars today. Japan will add to this information gathering with its *Nozomi* Mars orbiter in 2003. These three unmanned missions will add to the data provided by the *MGS*, presently in orbit around the planet.

All of these missions lead to the next logical step, the collection of a sample of Martian soil and its return to Earth, targeted for sometime in the 2010s. Presently, the design of this mission would involve a U.S. lander and rover to collect the samples that would be returned into Mars's orbit by rocket. A French orbiter would pick up the samples from the rocket and return them to Earth's orbit, where they would be picked up by a space shuttle. Even at this early stage of planning, the cost for this sample collecting project is estimated at $1 billion.

All of these missions and probes are designed for the ultimate goal—a manned mission to Mars. Although far in the future, an international consortium of space agencies is already planning the first steps to this mission. NASA has written its first manual for the Mars manned program, titled "The Reference Mission." The first and most critical aspect is crew safety, which includes both the takeoffs from and landings on Earth and Mars, as well as the long trip and stay on the planet. Any crew would have to spend about four to six months in transit to Mars and up to twenty months on the surface. The transit portion of the trip is most dangerous for crew safety, as the astronauts would be exposed to radiation and under conditions of zero gravity. NASA is considering a split mission strategy, which would use multiple launch vehicles. The surface equipment, surface living structure, and return vehicle would be sent to Mars in a series of three flights. The crew would arrive on a different spaceship twenty-six months later. The crew would then connect with the surface equipment on their arrival at Mars.

Any manned mission to Mars would require new technologies. Obviously, a mission of this length has never been undertaken before, and along with the development of new technologies, there needs to be increased research on the pressures of living in space for a prolonged period of time. Astronauts will have to live for an extended time in a close situation and will likely feel isolated, vulnerable, and unprotected from the dangers of space. In addition to their technical expertise, these Mars explorers will have to be experts in conflict negotiation and diplomats as well. All of these factors will stress the crews as they undertake their mission.

Bioastronautics

When Soviet cosmonaut Valentina Tereshkova became the first woman in space during the early 1960s, there was more than the normal amount of concern about her well-being. Although it was reported that she experienced some space sickness, she returned home essentially unscathed following a successful mission. She later married a fellow cosmonaut, and they gave birth to a healthy daughter who

was the first child to be born of parents who had both traveled in space. This incident calmed some of the fears that spaceflight may expose human hereditary material to damaging radiation. Nevertheless, two decades passed before either the United States or the Soviet Union included a woman onboard its space vehicles.

Throughout the entire space program, NASA has studied and monitored astronauts' reactions to weightlessness and an assortment of conditions associated with prolonged periods of confinement. With the end of the Cold War, much of the research on long-term space exposure conducted by Russian scientists has become available to other space agencies, including NASA. Also, the joint U.S.-Russian Mir missions in the 1990s provided additional information about the effects of space travel on human beings.

Weightlessness is not the only threat to the safety of astronauts. Sanitation is an important issue. Studies from space show that some microbes can multiply quickly in a zero-g environment. This was a severe problem on the Mir. The space station experienced widespread fungal growth. So, all potential contamination sites in space—dining areas, toilets, and sleeping areas—must be cleaned and sanitized regularly. On the International Space Station, garbage and trash are sealed in plastic bags to reduce the levels of dangerous microbes.

When NASA selects you to ride the space shuttle, the chances are pretty high that one of your first experiences onboard will be throwing up. Don't worry, you won't be the first to encounter this reaction! In the first few days in space, about one half of astronauts suffer from nausea sometimes accompanied by vomiting. Within the astronaut corps, few emotional subjects are discussed more than space sickness. Although it would rather not discuss its astronauts' complaints in public, NASA refers to this malady as *space adaptation syndrome.* It continues to be a number one medical concern of space agencies internationally. Besides the urge to vomit, other undesirable effects of weightlessness include lethargy, lack of appetite, drowsiness, stomach awareness, malaise, cold sweat, and nausea. As you can well imagine, even the most physically fit person's productivity level is bound to suffer radically under these conditions.

Researchers have found that humans gradually adapt to the microgravity environment, and these afflictions tend to subside. This process of adaptation to an environment devoid of gravity (zero-g) comes with a complement of other side effects, however. After a few days, an astronaut's body experiences what is called deconditioning, in which muscles grow weak, and the heart and blood vessels don't operate as fast as normal. After a few months in space, an astronaut's body undergoes another change, a loss in minerals that weakens the bones. The two Soviet cosmonauts who spent close to eight months aboard the *Salyut 7* space station could barely stand when they returned home. Their back and leg muscles had atrophied, and their hearts had shrunk, due to a severe loss in blood volume. Much of the calcium had been sapped from their bones.

Physicians and scientists who are interested in this process of adaptation and re-adaptation in all life forms have created a science called bioastronautics. Quite simply, *bioastronautics* is the study of life in space. The discipline grew out of sophisticated engineering feats directed toward a reconciliation or harmony between biology and hardware. This is the study of how technology can be used to sustain life (natural biological processes) in spacefaring vessels.

In recent years, bioastronautical studies have shown that space hardware is not necessarily synonymous with high technology. One problem space doctors were asked to confront involved how to restrain an incapacitated astronaut. NASA's first solution was a cumbersome and costly system of bungee cords. Their next solution came from a recommendation made by an astronaut who had ridden aboard *Columbia* on the fifth shuttle mission. He told medical experts to use duct tape to hold them in place! Apparently, he and his colleagues had used a lot of duct tape to stick things down on that particular voyage. Duct tape has since become a standard item in the onboard shuttle first aid kits.

An indirect consequence of this incident is related to the fact that hardware development is no longer the main preoccupation of bioastronautics research. In the past few years, increasing emphasis has been directed toward the study of physiological changes in zero-g.

The absence of gravity causes acute changes in the cardiovascular system, muscles, bones, metabolism, and even anatomy. As humans adapt to life in space, they disadapt to life on Earth. A sudden return to a one-g environment could overload a weakened heart or snap demineralized bones. Based on the records of the cosmonauts on the Mir, astronauts today can minimize, although not eliminate, some of the effects of zero-g. Astronauts use either a centrifuge bicycle or a treadmill every day to minimize deconditioning of their bodies. Despite intensive exercise while in space, an astronaut's body still is weakened. American astronaut Norm Thaagard spent four months on the Mir during the joint missions with Russia cosmonauts. When he returned to Earth, he had a 20 percent muscle loss from his calf muscles and a 10 percent from his thigh muscles. Also, he had anemia caused

---

BIOASTRONAUTICS

The study of how living organisms adapt to conditions of weightlessness and then readapt to life on Earth. Bioastronauts research and develop technological innovations to make this cycle occur more smoothly.

by a 20 percent drop in his red blood cell mass. Two years after his flight, he had not completely recovered from the mineral loss.

It is possible that one of the most serious problems in bioastronautics at present has nothing to do with bioengineering, adaptation, or medicine. Regardless of their assignments, space doctors are generally crippled by a lack of useful data. Ask them a simple question, such as, "Is heart disease more common among astronauts than among the U.S. general public?" Most physicians will tell you that the current sample of astronauts is not large enough to provide a valid response to your inquiry. A fundamental rift between ground-based scientists and the astronauts themselves further aggravates the problems associated with obtaining reliable data. Since astronauts insist on limited veto rights over medical experiments, many potentially significant studies never receive sanction.

In view of all the studies still to be done, bioastronautics has firmly established itself as a scientific discipline for the present and future space programs around the world. Mounting interest in space station habitation has led to enthusiasm in the space science community. The 2001 NASA budget included funding for a new Bioastronautics Initiative to conduct research about methods that will improve the health and safety of astronauts. Some space doctors hope their space biology, psychology, and sociology will produce yet another field of study—space civilization.

Summary

It is difficult for most of us to imagine what it feels like to be lifted off the ground by huge rockets, onboard a vessel destined for an outer space journey. Unlike the early days of the astronaut program, however, a greater number of people who dream about space travel may actually earn the opportunity to pursue those dreams. Several futurist writers have prepared optimistic scenarios regarding space colonies and the development of a whole new civilization in space. Their portrayals of the future do not seem to exclude the layperson in favor of the scientist or astronaut. Other journalists paint a less optimistic view of our space program, insisting there are still too many unknowns, and a fair percentage of the knowns seem speculative.

There is no doubt the unmanned space missions of today are providing a tremendous amount of scientific information about the nature and origins of our solar system. The impact on the normal citizen, however, is not as clear. People around the world have gazed at the heavens for centuries. The technological devices that have been developed to enhance this age-old pastime are truly impressive. The National Aeronautics and Space Administration is certainly not without its shortcomings, but its list of accomplishments cannot be ignored. Additionally, the Russians have broken numerous barriers to further the expansion of humanity into outer space.

We have not only become intrigued by the mysteries of space travel, but we have also become accustomed to the technological by-products of its investigation. Outer space truly is mankind's final frontier—our destiny there will unfold for many years to come.

Discussion Questions

1. Identify and describe at least three long-term projects for the National Aeronautics and Space Administration (NASA), as it leads this nation's space program into the next century.
2. In your best judgment, what was NASA's primary motivation when it presented the Space Transportation System proposal to government officials in the early 1970s? If you had been a congressional representative at that time, would you

have been favorable toward NASA's bid for financing? Why or why not?

3. Choose one of the orbiting space telescopes. Visit the telescope's web page and write an essay describing its contributions to science.

4. What is the difference between a geostationary orbit and a polar orbit? Explain why an operator would request either orbit for the location of its satellite.

5. Prepare a list of pros and cons you would consider for critical evaluation before making a decision to spend ninety days working onboard an orbiting space station.

6. Write a short (five hundred words) newspaper article on the International Space Station.

Suggested Readings for Further Study

Calamities of technology: What is the meaning of the *Challenger?* 1986. *Science Digest* 94 (7): 58–59.

Covault, Craig. 1986. Rogers commission charges NASA with ineffective safety program. *Aviation Week and Space Technology* (16 June): 18–22.

Crowther, Richard. 2002. Space junk: Protecting space for future generations. *Science* 296: 1241–1242.

Elbert, Bruce. 1999. *Introduction to Satellite Communication.* Norwood, Mass.: Artech House.

Eshleman, Von R. 1987. Space: It's bigger than politics. *San Jose Mercury News* (4 January): P1, P4.

Flight safety critical items analyzed in wake of accident. 1986. *Aviation Week and Space Technology* (17 March): 87–88, 93, 95.

Foley, Theresa M. 1986. Private space efforts face financial hardship. *Aviation Week and Space Technology* (29 September): 14–15.

Foust, Jeff. 1999. E.T.—Don't call home. *Technology Review* (September/October): 36.

Gorton, Slade. 1986. Pioneering the space frontier. *Vital Speeches of the Day* (1 September): 674–678.

Govoni, Stephen J. 1986. The race for priority in space. *Financial World* 155 (5): 20–22, 26–28.

Grey, Jerry. 1983. *Beachheads in space.—A blueprint for the future.* New York: Macmillan.

Hanging out in space. 2001. *PRISM* (February): 8.

Harris, Philip R. 1985. Living on the moon: Will humans develop an unearthly culture? *The Futurist* 19 (2): 30–35.

Hoffman, Stephen J. and David L. Kaplan, eds. 1997. *Human exploration of Mars: The reference mission of the NASA Mars exploration study team.* Houston: Lyndon B. Johnson Space Center.

[Internet]. Cassini-Huygens. *NASA.* <http://www.jpl.nasa.gov/cassini/>.

[Internet]. Chandra X-ray Observatory. *NASA.* <http://chandra.harvard.edu>.

[Internet]. Commercial Remote Sensing Program. *NASA.* <http://www.esad.ssc.nasa.gov/>.

[Internet]. *European Space Agency.* <http://sci.esa.int/>.

[Internet]. Galileo. *NASA.* <http://Galileo.jpl.nasa.gov/>.

[Internet]. Hubble Space Telescope. *NASA.* <http://www.stsci.edu/>.

[Internet]. Landsat 7. *NASA.* <http://geo.arc.nasa.gov/landsat/landsat.html>.

[Internet]. Mars missions. *NASA.* <http://mars.jpl.nasa.gov/>.

[Internet]. *NASA.* <http://www.nasa.gov/>.

[Internet]. NASA's Discovery Program. *NASA.* <http://discovery.nasa.gov/>.

[Internet]. NASA human spaceflight web. Including the space shuttle and International Space Station. *NASA.* <http://spaceflight.nasa.gov/>.

[Internet]. *National Space Development Agency of Japan (NASDA).* <http://www.nasda.go.jp/index_e.html>.

[Internet]. *National Space Science Data Center.* <http://nssdc.gsfc.nasa.gov/>.

[Internet]. *North American Aerospace Defense Command (NORAD).* <http://www.spacecom.af.mil/>.

[Internet]. *Space.com.* <http://www.space.com/>.

[Internet]. *Space Daily.* <www.spacedaily.com>.

[Internet]. Voyager. *NASA.* <http://voyager.jpl.nasa.gov/>.

Joyce, Christopher. 1984. Space travel is no joyride. *Psychology Today* 18 (5): 30–37.

Kelly, Thomas. 2001. *Moon lander: How we developed the* Apollo *lunar module.* Smithsonian History of Aviation and Spaceflight Series. Washington, D.C.: Smithsonian Institution.

Klinger, Karen. 1986. Next stop is Neptune. *San Jose Mercury News* (28 January): A9.

Kraft, Chris. 2002. *Flight: My life in Mission Control.* New York: Plume.

Launch delays pose problems for space commercial ventures. 1986. *Aviation Week and Space Technology* (7 April): 127, 129.

Launius, Roger D. 1994. *NASA: A history of the U.S. Civil Space Program.* Melbourne, Fla.: Krieger.

Lemonick, M. D. 1989. U.S. plans talks with Soviets, allies on space. *Time* (24 July): 50–51.

Leopold, George. 1986. A bruised NASA hangs in with space-factory program. *Electronics* (5 May): 44–47.

Lewis, Ruth A. and John Lewis. 1986. Getting back on track in space. *Technology Review* (August/September): 30–40.

Mann, Paul. 1986. Commission sets goals for moon, Mars settlements in 21st century. *Aviation Week and Space Technology* (24 March):18–21.

Marshall, Eliot. 1986. The shuttle record: Risks, achievements. *Science* (14 February): 231, 664–666.

Marshall, Sam A. 1986. NASA after *Challenger. Public Relations Journal* 42 (8): 17–19, 22–24, 39.

Merrifield, John T. 1986. Lockheed researchers identify issues of human productivity in space. *Aviation Week and Space Technology* (16 June): 89–92.

Miller, Russell. 1985. America's space entrepreneurs: What's up in the satellite business? *Management Review* 74 (7): 19–23.

Parker, J. F. Jr. and V.R. West, eds. 1973. *Bioastronautics data book.* 2d ed. Washington, D.C.: U.S. Government Printing Office.

Ryan, A. J. 1989. Moon landing + 20 years: A giant leap for space data. *Computer World* 23 (17 July): 1–2.

Schefter, Jim. 1985. Sky-diving satellites. *Popular Science* 226 (l): 60–61.

Shuttle tragedy, The. 1986. *Newsweek* (10 February): 26–42.

Simpson, Ted, ed. 1986. *Pioneering the space frontier: The report of the National Commission on Space.* Report made by the National Commission on Space. New York: Bantam Books.

Space shuttle: The first 20 years—The astronauts' experiences in their own words. New York: Dorling Kindersley.

Toner, Mike. 1985. It's pay off or perish for the shuttle. *Science Digest* 93 (5): 64–67, 87–88.

Tyler, Patrick E. 2001. Space tourist returns from "paradise": Wobbly, but ecstatic, Tito savors great ending to a great adventure. *The San Diego Union—Tribune.* San Diego. (7 May): A-1.

Waff, C. B. 1989. Hubble telescope spreads its wings. *Astronomy* (September): 17, 44–53.

Waldrop, M. Mitchell. 1986. After *Challenger:* Painful choices. *Science* (13 June): 232, 1335–1337.

Wilford, John Noble. 1986. America's future in space after the *Challenger. The New York Times Magazine* (16 March): 38–39, 102, 104, 106.

Wiskerchen, Michael J. 1986. The space program must go on. *San Jose Mercury News* (2 February): P1, P4.

Chapter 5

Breakthroughs in Medicine

DATELINE

Ithaca, New York

Scientists working in the Boyce Thompson Institute for Plant Research at Cornell University have genetically engineered potatoes to deliver an edible vaccine against a common virus. The Norwalk virus is named for Norwalk, Ohio, where researchers first identified it in the late 1960s. It is an intestinal virus infecting more than 23 million people in the United States each year. It is distressing in that it causes diarrhea, nausea, and stomach cramps, but it is survivable. Cornell scientists added the capsid protein to potatoes, which, when eaten, stimulated recipients' immune systems to create a flood of antibodies (the capsid is the shell covering of a virus particle). This physiological response prevented the Norwalk virus from latching onto cells lining the walls of the intestine and causing disease.

In developing nations around the world, occurrences of the Norwalk virus can be more deadly, especially if young children acquire the disease and become dehydrated. In those countries, and in ours as well, injectable vaccines are expensive to administer, and needles can more easily transmit other diseases. Edible vaccines show promise to become a better approach to battling outbreaks of this virus.

As their work continues at Cornell, researchers hope to develop edible vaccines for papilloma virus (which can lead to cervical cancer) and the hepatitis B virus. Dried tomatoes and bananas are under study to replace potatoes as the vaccine carrier....

Introduction

Most of the sociological surveys that have been conducted to determine how people feel about the advancement of technology seem to report at least one common finding. Respondents tend to be in general agreement with regard to the positive influence technology has had on their physical well-being. Technology has made us healthier and increased the average life span for human beings throughout the world. In other words, people often look to the medical discipline for examples of significant technological breakthroughs.

Technology has made numerous and diverse contributions to medicine over the years. Some historians use the medical field to chart the chronology of technology. It has become increasingly evident over the past few decades that medicine and the health care system in general have become dominated by science and technology in our society. At the core of most contemporary diagnostic and treatment procedures, one will find a host of technological devices (hardware) and control systems (software). These devices are the machines that can extend our human capabilities and serve to separate us from the perils of the natural environment. An assortment of computerized and electronic control techniques and systems has been developed to monitor, evaluate, and operate these devices.

We are living in a new era of medical miracles that may ultimately serve to correct many of our physical and mental afflictions. Some of the more visible medical forces that have entered the awesome war against injury, disease, and distress include laser surgery, electronic pain killers, over-the-counter detection kits, in vitro fertilization (IVF), artificial organs and body parts, genetic therapy, computer-assisted robotic surgery, a variety of new vaccines, stem cell research, antibiotics, and "smart" methods of delivering these drugs into the human body. Individuals who would have had a minimal chance for survival less than twenty years ago can presently look forward to a normal, healthful existence. In some ways, these technological accomplishments have begun to serve the purposes that religious ritual once did. Technology conveys the hope and promise of vitality and longevity.

Despite the many promises afforded by medical technology, some observers believe the medical profession is experiencing overwhelming dependency on new technology. During the first half of the twentieth century, medicine focused largely on the observation of disease from a relatively passive clinical and diagnostic perspective. Today's medicine is dedicated to the aggressive treatment of disease from an applied scientific point of view. The medical profession embraces and endorses modern technology and the many contributions it has made to the diagnosis and treatment of its patients.

As with most other narratives concerning technological advancement, the blessings of modern medicine cannot be counted without also reviewing the negative aspects. It seems the rise of medical technology should be paving the way for better medicine. On the contrary, the burden of absorbing new technical information has landed squarely on the physicians' shoulders. Doctors are confronted with the almost full-time problem of keeping up with new technologies and procedures in order to provide the best patient care (and to avoid malpractice suits). The pace of new discovery in the medical field is so intense, even recent graduates are likely to be out of date in a few years. Dramatic increases

in the amount of available clinical information about treatment effectiveness make it even more difficult for doctors to acquire and retain all the information they need to make sure they order all of the appropriate services for all of their patients.

This makes matters even more stressful because it seems keeping up with the latest advances in one's own field of specialization is not enough any more. Doctors have become anxious about technology in general. In some cases, if they are not familiar with the latest discoveries in other fields, they may be deemed negligent for failing to make a referral. This seems to be a strange process to enforce quality in a discipline whose primary objective has always been expressed as the healing of the sick and injured.

Technology in the medical industry can be loosely defined as the rational selection and use of devices and procedures toward the achievement of measurable, useful, and relatively immediate human outcomes. There are both costs and benefits associated with the technological breakthroughs that have been made in health care. The aim of this chapter is to examine the current status of the nation's health care industry. Technology's influence is examined through a review of some of the major advances that have either taken place recently or are still being developed. The extent to which people are able to afford many of these high-tech miracles is presented as an issue demanding much attention. This chapter provides a foundation for discussion that can ultimately respond to the following questions:

❑ What are some of the critical issues needing to be addressed, if the profile of the medical profession and the technology associated with it are to remain intact and unscathed?

❑ Should medical specialists be instructed to become less dependent on machines in their treatments and diagnoses? In other words, is contemporary medicine addicted to a "technological fix"?

❑ Why does it seem that what *can* be done in terms of health care technology (sustain life in a brain dead patient by machine or other medical intervention) is sometimes a lot clearer and easier to understand than what *should* be done (shut down the machine or discontinue medication)?

❑ What projections are being made with regard to the future cost of high-tech (or even low-tech) medical treatment? What is driving up the cost of health care in the United States?

❑ How has the introduction of over-the-counter home health detection and diagnostic test kits affected the role of the medical profession in society?

❑ What sorts of moral and ethical concerns regarding quality of life have surfaced as a result of medical breakthroughs in areas such as surgery and procreation?

Key Concepts

In your reading of this chapter, you will be introduced to a variety of technology and science concepts relevant to the contemporary medical industry. Several of them for which you should acquire an understanding include the following:

amniocentesis
balloon angioplasty
diagnosis-related group (DRG)
DNA chip
false negative/positive
health maintenance organization (HMO)
in vitro fertilization (IVF)
lithotripter
minimally invasive surgery
pharmacogenomics
preferred provider organization (PPO)

A Profile of Health Care in the United States

More progress has been made in medicine over the past few decades than through all of history. Technological innovations including artificial joint replacements, artificial heart valves, heart-lung machines, robotic surgical assistants, cardiac pacemakers, and even brain pacemakers have been introduced. An assortment of medical procedures such as organ transplants, microsurgery, coronary bypass surgery, angioplasty, kidney dialysis, minimally invasive surgical techniques, and amniocentesis are being performed with increased regularity and success. We have access to new diagnostic systems for studying the interior features of the body, including computerized axial tomography (CAT), nuclear magnetic resonance (NMR), and the use of endoscopic instruments. See Figure 5-1. A variety of new drugs that are very "smart," along with edible vaccines, like the one mentioned in the *Dateline* for this chapter, have arrived on the scene. Advancements in genetic engineering (previously discussed in detail in Chapter 2) and in a discipline known as tissue engineering are bringing us closer to being able to treat and possibly cure various forms of cancer and actually design and grow living replacement body parts.

This brief introduction, as futuristic as it may sound, barely scratches the surface of the numerous technological and scientific breakthroughs that have occurred in recent years. The benefits, in terms of increased longevity and human productivity, associated with these accomplishments are very noteworthy. However, the United States has the absolute highest medical expenditures of any nation in the world. We presently devote more than 14 percent of our gross national product (GNP) to the medical and health care of our citizens. Figures computed

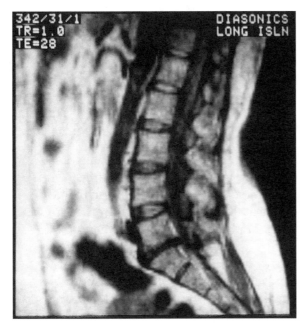

Figure 5-1.
Radiology has emerged from the ranks of fuzzy pictures of broken bones to color images that diagnose brain disorders, heart attacks, and epileptic seizures. The image shown here is a product of an imaging technique called NMR spectroscopy. This highly sophisticated technique works by creating a very powerful magnetic field around the patient and bombarding him with radio waves. When these waves are withdrawn, hydrogen atoms in the body emit a radio signal reflecting the difference between healthy and damaged tissue. A primary advantage for the patient is the absence of radiation damage. (El Camino Hospital, Mountain View, California)

in the mid-1990s revealed that approximately four thousand dollars per person were spent on health care per annum in this country; these figures compare with fifteen hundred dollars for each citizen in the mid-1980s, when approximately 10 percent of our GNP went to health care. The government's share of these health care expenditures rose from 34 percent in the early 1970s to over 50 percent in the mid-1990s. This overview of trends in our health care

system seems to indicate that a most critical issue demands effective strategies to contain these soaring costs in the years ahead.

The medical era preceding World War II was a bit simpler with regard to services rendered by practicing physicians. Most family doctors were generalists, and most patients paid what they could afford for their own care. Since patients most commonly had to pay for medical services themselves, doctors often felt obligated to adjust their fees to match individual financial capabilities. During this earlier time period, the ratio of physicians was much lower than today. Most physicians saw and treated as many patients as they could handle. There was little economic incentive to do any more for an individual patient than was clearly essential. Sophisticated medical technology was not widely available.

More often than not, primary care physicians were diagnosticians who had little more than their time, guidance, advice, and local pharmacy prescriptions to offer. All of this has changed dramatically as the technology and organization of medical practice have become more specialized, complex, and expensive. In fact, finding a local doctor such as the one described above in your hometown is becoming less possible. This may be due in part to a decreasing percentage of medical students who select residencies that might lead to careers in primary care (for example, internal medicine, pediatrics, and family medicine). Much greater numbers of medical school graduates have begun to opt for specialist practice, in which they are trained to use high-tech diagnostic and therapeutic procedures. Trends in the mid-1990s suggested that fewer than 25 percent of U.S. physicians would be working as primary care practitioners as we entered the twenty-first century. When a disproportionate number of medical doctors are licensed as surgical specialists, primary care may be adversely affected, thus contributing to escalating expenditures by allowing minor medical problems to devolve into serious and expensive procedures.

The profile of the medical profession in the early twenty-first century portrays a dynamic interplay among several factors—medical care costs, a better informed citizenry, phenomenal technological changes, and confusing health insurance and health care options. Further, demographic data reveals that we are an aging society due to the huge size of the "baby boom" generation (those people born between the mid-1940s and the mid-1960s). See Figure 5-2. This fact alone is expected to heighten the overall demand for health care significantly over the next several decades. Furthermore, the types of care elderly people are going to require will often be for serious and complex "end-of-life" illnesses. The times are not calm for medical professionals. The individuals in our country who are responsible for the practice of medicine to ensure public health and our individual rights to physical well-being are under intense public scrutiny.

There are a variety of other factors emerging to influence health care systems in the United States. A few of these include government intervention, insurance coverage and premium subsidies, changing health care delivery methods emphasizing more outpatient treatment, lack of price competition among providers, and malpractice litigation. When you add these items together, it is no wonder there is growing

The Aging of Society Makes Reform Urgent

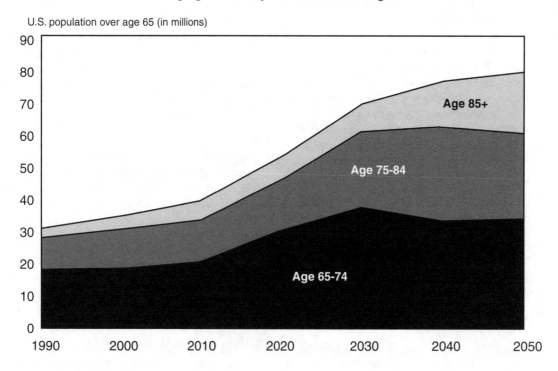

Figure 5-2.
This chart shows the aging of our society. It is adapted from the U.S. Bureau of the Census, Current Population Reports. 1995.

pressure to contain rising medical expenditures emanating from the general public, government officials, private insurers, and the business community at large.

The results of a mid-1990s Harris Poll revealed that only 16 percent of Americans believed our health care system in this country was working pretty well. More recently, in a study supported by the Robert Wood Johnson Foundation, the top three issues of concern for Americans regarding health care are costs of health care and services, lack of or inadequate health insurance or coverage, and costs of prescriptions and drugs. It is imperative to determine what sorts of trade-offs will ultimately be necessary to bring the cost of health care to a tolerable level. By comparison to the more than 14 percent of our GNP expended on health care, most developed countries with universal coverage have stabilized their health care costs at right around 8 or 9 percent of GNP. It is difficult to admit that, despite our economic sophistication, our scientific expertise and advanced research, and our high-tech medical interventions, we have much to learn from our international partners. The following sections introduce several topics pertinent to the contemporary state of affairs in our nation's health care system.

A Concern for Quality

We live in a society quite prone to seeking technological solutions to its problems. Part of this tendency is illustrated by our use of and reliance on medical technology to extend and sustain life. Medical specialists often adopt and become skillful at using technology more so than generalists due to their advanced training. Hospitals invest in more expensive equipment in order to present a higher quality and more prestigious image to the local community. Furthermore, some financing plans and government policies seem to have stimulated the use of technology at almost any expense.

Specifically, the buyers and users (patients and doctors) are often insulated from the high cost of specialized technology interventions, while the providers (hospitals and clinics) of medical technology are rewarded with greater revenue from greater use. In an increasing number of cases through the 1990s, however, hospitals found themselves in jeopardy. Many were being shortchanged because rate increases from insurers lagged behind the costs associated with providing inpatient care. Hospitals today are grappling with soaring labor, drug, and technology costs. The government sets the reimbursement rates hospitals receive from Medicare and Medicaid, our nation's two major public health insurance programs. In efforts to curtail the pace at which government expenditures on health care were accelerating (as mentioned previously), Medicare rate increases slowed down in the late 1990s. This has prompted hospitals to become more assertive and aggressive as they negotiate and "haggle" with private insurers.

Despite the fact so much money is being spent on health care, the United States encounters serious quality problems throughout its health care institutions and practices. From time to time, leading national newspapers have run feature stories about quality problems with local doctors, hospitals, clinics, and nursing homes. Undoubtedly, you could find a story of this genre in your local newspaper if you simply reviewed last week's seven issues. These types of media releases are typically followed by intense policy discussions among physicians' organizations, hospital administrators, quality consultants, and legislators. One such flurry of dialogue prompted Congress to create the Agency for Health Care Policy and Research (AHCPR) in the late 1980s. The new government agency was given a mission to invest in research on the effectiveness of health care services (what exists) and to develop guidelines to assist providers in improving quality (what should be).

As numerous studies directed by the AHCPR were completed through the 1990s, it became evident that our nation's health care system continued to experience serious and widespread quality problems. When problems related to quality surface, patients are usually harmed. If one simplifies the extensive body of research, three types of quality problems can be categorized. The first group refers to patients who do not receive beneficial health services. This is an example of underuse. Underuse of appropriate and effective interventions is pervasive across the delivery of preventive care, acute care, and chronic care. Underuse problems are often aligned with financial barriers (lack of insurance or coverage for specific preventive care). Some segments of our population also distrust physicians in general and may avoid the health care system altogether.

The second category of quality problems is apparent when patients are subjected to treatments or special procedures from which they most likely will not benefit. This is called overuse. One probable cause of overuse stems from the fact that doctors are overly enthusiastic about the potential value in new medical and surgical interventions. These newly developed technologies are sometimes used in absence of reliable evidence supporting their effectiveness. As noted many times in our textbook, Americans are infatuated with technology and may implore their physicians to use whatever it takes to cure their maladies.

Finally, a third type of quality problem emerges when patients receive appropriate medical services, but those services are delivered poorly, directly exposing recipients to additional risk of what are preventable complications. This is misuse. Most of you have, at some point in your lives, heard about one

or more medical malpractice suits. In fact, Academy Award nominated films have documented in a dramatic fashion the health problems incurred as a result of misuse in the medical profession. Litigation in the actual (not dramatized) court battles commonly is focused on errors made in hospitals, a percentage of which includes patients who are injured as a result of negligence. A significant proportion of quality problems due to misuse occurs when competent professionals make mistakes and the systems fail to prevent their mistakes from causing serious harm. Doctors, after all, are as human as the rest of us, but it is easy for us to forget that when we are counting on them to save our lives.

Solving this vast array of quality problems has become an imperative and timely challenge in our current health care system. Over the past few decades, American automobile manufacturers and producers of consumer electronics goods have made great progress in improving quality of their products and services. The extent to which similar improvements have been made in health care is discouraging. One obstacle to improvement is the lack of competition due to the overwhelmingly local nature of health care and wellness in our society. In many ways, our automobile and consumer electronics industries were transformed by competitive pressures from higher quality products built in Japan and other nations. It seems unlikely that hospitals and clinics in St. Louis will lose patients to competitors in Los Angeles (let alone Tokyo) just because those other institutions have earned higher quality ratings. Competition in primary health care is local, much like competition in the grocery store business.

Other obstacles to improvement include an absence of quality guidelines backed by empirical evidence; a lack of return on investment; a lack of demand for improvement, since consumers often just want to feel better right away; and a void in leadership, with regard to solving quality problems on a united front. While the government might be looked to as a source of leadership, as when it created the AHCPR, we have witnessed a political trend against government intervention in society—especially toward efforts at health care reform. It is interesting to note here that ten years after the creation of the AHCPR, its authorizing legislation was rewritten specifically to *exclude* guidelines development and policy from its mission.

Medical Care Options

The steady rise in health care costs, coupled with the availability of sophisticated medical technologies and advanced drug therapies, has produced a variety of medical plans. The major payers of health insurance premiums are the government and private employers. Each of these has initiated steps to eliminate the incentives for excessive treatment, utilization, and abuse of technology.

Diagnosis-Related Groups (DRGs)

> **DIAGNOSIS-RELATED GROUP (DRG)**
> A payment scheme, in which the hospital receives a fixed fee reflecting an average cost of curing the patient's condition. If the patient is worse than average, the hospital must pay any cost beyond the DRG allowance. If the patient is better than average, the hospital keeps any money left.

The diagnosis-related groups (DRGs) are a patient classification scheme originally developed as a method of reviewing the relationship between the types of patients a hospital treats and their incurred costs over a year. In the early 1980s, Congress enacted a DRG-based prospective payment system for all Medicare patients. Instead of being reimbursed for Medicare patients on the basis of charges or costs, hospitals are paid fixed sums for each type of patient by diagnosis. The Health Care Financing Administration (HCFA) holds the responsibility for the maintenance and modification of DRG definitions. An operating division within the Department of Health and Human Services, the HCFA updates the Medicare DRGs annually. A number of states and large insurance programs have adopted the DRG structure for non-Medicare patients. In addition, DRGs are used as the basis for global budget allocation and payment in several European countries and Australia.

DRGs classify patients into clinically cohesive groups demonstrating a similar consumption pattern of hospital resources and duration of stay in the medical facility. Besides reimbursement, DRGs also evaluate the quality of care, and they assist in assessing the utilization of services. Each DRG represents the average resources needed to treat patients assigned to that DRG, relative to the national average of resources to treat all Medicare patients. In the late 1990s, there were 492 Medicare DRGs in use. For example, the payment for simple pneumonia (DRG 89) is based on nationwide average costs for treatment and an average length of stay of 5.7 days. Hospitals that treat patients for less than the DRG allocation can keep the difference. If the treatment costs more, the hospitals must absorb the loss.

Fee-for-Service Payment

Fee-for-service medical care was standard practice when health insurance was introduced in the middle of the twentieth century. Insurance programs were designed to pay the participant's health care bills. In this system, an annual premium is paid. In return, the insurance company pays for at least a portion of an individual's medical bills. Through the years, these plans have become more complete, and health insurance has often been buried in the benefits packages offered by employers. Subsequently, employees have seemingly lost sight of the fact that the medical treatments they receive are costing money. Some observers contend that this situation has given patients and doctors alike an inclination to buy or recommend excessive amounts of medical care regardless of expense.

The bulk of fee-for-service coverage is provided by the following:

❑ Medicare. This is a federal program for retired people.

❑ Medicaid. This is a state program for low-income people.

❑ Blue Cross and Blue Shield. These are nonprofit organizations offering coverage for hospitalization and doctors' fees, respectively.

The advantage of fee-for-service coverage is that the patient can choose any desired provider. Any qualified physician who offers a recognized health care service can submit a bill to the insurer. If any emergency occurs, care can be sought at any site with the expectation of at least partial reimbursement. Virtually anyone can buy individual insurance to cover fee-for-service care, but the cost for an individual policy may not be affordable for the average, middle-income wage earner.

Pre-Paid Managed Care Plans

The most notable alternatives to traditional fee-for-service providers are known as *health maintenance organizations (HMOs).* HMOs, like the DRG system, utilize a fixed payment plan. In this type of medical plan, the provider offers a binding estimate of what a given amount of health care will cost the average patient for a one-year period. Individuals who enroll in the plan pay that amount annually. If total costs go higher than

the sum of these prepayments, the HMO loses money. If the costs stay below the estimate, the HMO operates in the black.

In actuality, the HMO is a merger of an insurer and a provider of health care. As an insurance company, it collects annual premiums from subscribers and then absorbs the risk that this cumulative income will cover its expenses. As a provider of medical services, it has its own staff of physicians and may also contract services out to other facilities. Enrollees must choose a primary care physician, who is authorized to refer patients to medical specialists. For example, if a person develops a skin rash, he or she must first see the primary care physician (sometimes called the gatekeeper) in order to receive an approved referral to a dermatologist. The theory is that HMOs have an incentive to deliver quality health care and also stay within their budget.

Critics of pre-paid plans hold that HMOs can often place too much pressure on doctors to limit care for the sake of protecting their profit margin. They usually make reference to a desirable feature of traditional fee-for-service plans in which the price of medical treatment is not as likely to affect its selection or recommendation. Physicians employed within this type of structure may be more convinced that it is financially worthwhile to care for difficult cases.

Another form of managed care seeking to preserve this aspect of traditional health plans is known as the *preferred provider organization (PPO).* PPOs are hospitals or groups of physicians offering their services at a fixed price in exchange for a guaranteed supply of patients. The insurer (business firm, production facility, state agency) represents its clients by initiating a cost comparison search within the medical marketplace. It next contracts with a group of providers who agree on a predetermined list of charges for all services rendered. After the rate is paid, the patients are covered for the medical care services offered by the list of preferred providers. Unlike an HMO, if you want to see a doctor of your choice not connected with the PPO, the plan may pay a percentage of the medical bills much like an indemnity plan. These types of arrangements represent an alternative to traditional health care packages.

Upon first glance, it may seem that the options for health care coverage have become too confusing. We may ultimately look back on this time as a period of experimentation. In any event, it is evident that our society is determined to develop a system in which the highest quality of medical care is provided at the most reasonable price. The medical profession cannot allow the pressures to cut costs to have an adverse effect on the quality of treatment administered. It is also essential to recognize that theoretical arguments can be made in favor of each of the programs described in the previous section.

Health Priorities

Economists tell us that increase in demand is one of the main factors contributing to rising costs of health care in this country. Several agents have fueled this increase, including the widespread availability of health insurance, changes in our lifestyle, an awareness of the effects of the environment on our physical well-being, a larger and more specialized population of health care providers, and technological progress. Advances in medical technology have increased demand by offering diagnostic and therapeutic procedures previously not available. As noted above, the quality problems related to overuse are often a result of the "if the technology fits, then use it" syndrome.

The benefits associated with technological advances in medicine are usually accompanied by a list of social costs. Some of the

recent developments have actually lowered the cost of health care. The polio vaccine eliminated the need for the expensive iron lung, and the rubella vaccine reduced the incidence of disease among pregnant women and cut down on the birth of mentally retarded babies. The use of mind-stimulating drugs in the treatment of mentally ill patients has also yielded expenditure-cutting results. On the flip side of the coin, the conquest of infectious diseases, such as pneumonia, has increased our life expectancy. The most dramatic improvements can be seen for those who have chronic conditions. These people seemingly remain alive to incur high health care costs long after they have been "cured." Advances in surgical techniques have also influenced health care expenses.

Continued research efforts will add to the complexity of the technology, cost, and human need problem. The concept of technology overkill will not disappear in the near future. In a general sense, technological progress in medicine often makes what can be done much clearer than what should be done. The numerous technical manipulations and procedures can be defined with much conviction. It is far more difficult to decide when and where such technological interventions *should* be utilized.

If advanced technology continues to inflate the cost of health care, it may no longer be accessible to all people who may be able to benefit from its use. Some medical experts believe we may be approaching a day when we will be forced to confront the rationing of costly high-tech procedures. Making life-and-death decisions about the use of costly medical technology is an exercise in which few legislators and even fewer doctors want to engage.

Over the past several decades, technological success in medicine has intensified political pressure for public access to the advances. The subsequent expansion of health insurance coverage increased access to new technology. Insurance coverage also provided incentives to develop newer, more effective, and inevitably more costly technological devices. Coming to grips with the cyclical nature of this dilemma may be the central challenge for the present generation of medical professionals.

In a recent study, it became apparent to several Harvard researchers that Americans really do have a concise set of health priorities. Cost of care is clearly a top issue, as are the affordability of prescription drugs and the percentage of people in our nation who have no medical insurance. Americans are also very concerned about emergent deadly diseases spreading internationally, which they believe threaten their long-term health and well-being. Respondents to questions posed in this Harvard study expressed great anxiety and concern about infectious diseases like the Ebola virus, mad cow disease, and the West Nile virus. They ranked these problems as high priority items for government legislative action. Now, in the aftermath of the 2001 terrorist attacks in our nation, concerns with bioterrorism will most likely be a stimulus for strong public support for major initiatives to improve our public health system. In summation, the Harvard study made the following conclusions:

> In the absence of a perception of crisis or large scale efforts to raise the salience of other health and health care problems, the public agenda will be dominated on the one hand by concerns with cancer, HIV/AIDS, and heart disease, and on the other hand by cost, prescription drugs, and the uninsured. However, one set of issues that may be growing in public concern is the threat of emerging international infectious diseases that

may affect health in the United States. After September 11, concerns about this in regard to biological terrorism are likely to grow.[1]

Contemporary Technology in Medicine

A number of recent medical technology breakthroughs were mentioned in previous sections of this chapter. The next few sections endeavor to summarize the technical explanations and the social ramifications of just a few of these "miraculous feats." Extensive research and development (R & D) is ongoing across numerous fronts in the medical world (for example, hospitals, universities, industrial centers, and government research labs). The examples cited in this section are accompanied by a vast array of equally compelling diagnostic and therapeutic techniques, all of which are totally reshaping the way we practice medicine today and the way we might be practicing it in the near future.

At the onset of the twentieth century, the average life expectancy for American men and women was less than fifty years. Today, these figures exceed seventy-five years for men and women alike in this country. Changes in lifestyle certainly account for a portion of this gain, but many recent discoveries in medicine have also played a major role. As with other fields of specialization, the capacity for scientific research and technological applications to change the way our minds and bodies feel and function continues to accelerate at a breakneck pace. Medical knowledge doubles at a rate that defies physicians to remain current. (You should by now realize, however, this rate of change in critical facts and figures is not unique to the medical industry—it is characteristic of

---

[1]Blendon, R. J., K. Scoles, C. DesRoches, J. T. Young, M. J. Herrmann, J. L. Schmidt, and M. Kim. 2001. Americans' health priorities: Curing cancer and controlling costs. *Health Affairs* 20 (6): 231.

most contemporary technology fields.)

Today's medical researchers and doctors keep working to make inroads toward developing treatments for various forms of cancer. Current developments in biotechnology were discussed in Chapter 2, many of which are driving emerging and future breakthroughs in medicine for humans. The Human Genome Project is largely complete, and it entails decoding the sequence of one complete set of human genes (previously discussed in Chapter 2). Emphasis in research is now shifting to the study of how deoxyribonucleic acid (DNA) varies from one person to another. These variations, however minute they might be, may help determine which patients may be harmed by particular drugs and which groups will benefit most from them. Slight genetic variations between individuals are called single nucleotide polymorphisms (SNPs), or more simply, they are the substitution of one letter of the DNA alphabet for another in a person's genes. Amazingly, these small differences can make a huge impact on how a person responds to a given medication.

This field of study carries the label *pharmacogenomics,* a term that arrived in the medical literature just a few years ago. We are truly living through an era of transition in medical therapies. We are moving out of the "one size fits all" drugs venue, where all patients receive the same prescription, and toward a time where drugs are personalized and geared to the specific genetic makeup of groups or individuals. Eventually, it may be possible for you to review a readout of your own genome with your doctor in her or his private office. This information will provide indications of whether a particular medication may be dangerous for you to take and which health care options will be optimal for you.

DNA microarrays, more commonly referred to as *DNA chips,* are DNA-covered wafers manufactured out of silicon, glass, or plastic, capable of analyzing thousands of genes at a time. These chips have great potential to join our ongoing quest to eliminate all forms of cancer in our lifetime. DNA chips potentially provide a whole new way of detecting cancerous cells in a person's body—one also promising to be earlier, easier, and more precise. In the very near future, pathologists expect to be using DNA chip–based tools to spot genetic differences between cells. It is there that illuminating differences can be used to detect cancerous cells long before symptoms develop and it is too late for treatment (for example, symptoms like a chronic cough, intense pain, or an internal lump). Further, DNA chips will also be employed to provide a genetic profile of cancerous cells, allowing medical professionals to distinguish one form of cancer from another.

The dream of a bionic body is seemingly becoming a reality. Doctors have been able to replace the human heart, pancreas, kidneys, limbs, blood vessels, and hip joints with synthetic devices. See Figure 5-3. In some cases, the technological substitute may actually function better than its organic counterpart.

In an even more incredible fashion, the field of tissue engineering promises that we may soon be able to design and grow living replacement body parts. A number of biotech firms are aggressively engaged in basic and applied research, which may lead to ready availability of replacement heart ventricles and bladders. See Figure 5-4. Already, there are a small number of tissue-engineered products on the market, including skin, bone, and cartilage implants. These breakthroughs are aligned with recent advances in biomaterials compatible with living cells and a much better understanding of cellular behavior. Groundbreaking work completed in the 1990s by a medical team from Boston's Children's Hospital shows great possibilities for human trials to begin soon, in efforts to tissue engineer the bladder. These researchers successfully engineered new bladders for beagle dogs; the altered animals had almost as much bladder capacity as dogs with their original equipment. What remains unknown is how long the tissue engineered organ will last inside the dogs' bodies.

Balloon angioplasty, also called balloon dilation therapy, is one of three standard medical interventions for coronary artery disease; the other two are medication and surgery. Balloon angioplasty can sometimes replace coronary bypass surgery to open clogged arteries. The goal of the procedure is to push the fatty plaque back against the walls of the artery to allow more room for blood to flow. This procedure can be performed by snaking a thin tube with an uninflated balloon at the tip through a patient's femoral artery (which is found in the upper thigh or groin area) into the heart and the pulmonary artery. Once the balloon-tipped catheter is at the site of the blockage, it is inflated for several seconds to clear the clogged area of fatty deposits. Once clear, the catheter is removed and replaced with a stent, which is a wire mesh tube to keep the artery open. This medical intervention can be done for a fraction of the cost of bypass surgery, and the period required for recovery is significantly shorter (most patients are free to return home after just twenty-four hours). Another form of balloon angioplasty

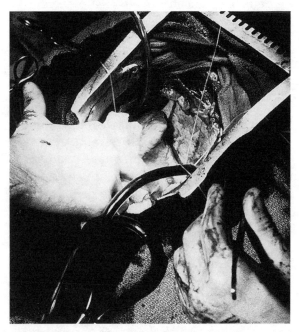

Figure 5-3.
Using synthetic devices, doctors can now perform surgical procedures, such as this one, involving a hip joint replacement. (El Camino Hospital, Mountain View, California)

Tissue Engineering in Industry		
Company	**Location**	**Products in the Pipeline**
Advanced Tissue Sciences	La Jolla, CA	Skin, (TransCyte, Dermagraft); cartilage, ligaments, and tendons; blood vessels and heart valves
Genzyme Biosurgery	Cambridge, MA	Cartilage cells (Carticel); cartilage graft (Carticel II)
CryoLife	Kennesaw, GA	Heart valves and blood vessels; ligaments
Curis	Cambridge, MA	Cartilage gel to prevent urinary reflux (Chondrogel); bladder
LifeCell	Branchburg, NJ	Skin (Alloderm); blood vessels; ligaments and tendons
Organogenesis	Canton, MA	Skin (Apligraf, Vitrix); blood vessels

Figure 5-4.
Here is a partial list of biotechnology firms working to build new tissues and organs. They are leaders in a field of medical research called tissue engineering.

has the capacity to treat clogged arteries in other areas of the body, including the brain, neck, kidney, and abdomen.

Surgeons replace organs with plastic parts and vaporize tumors with laser rays. They know how to perform delicate lifesaving operations on human fetuses in utero. Arthroscopic surgical techniques facilitate the treatment of joint problems and sports injuries and have nearly perfected the diagnostic accuracy of knee pain conditions. Haptic simulators are being used in medical schools to train physician candidates to perform traditional procedures (such as catheter insertion), and to learn new procedures (such as deep brain stimulation). The term *haptic* is from the Greek *haptikos* meaning "to grasp or perceive." Haptic simulation exercises provide a realistic way for doctors and nurses to practice their techniques without causing harm to patients.

Try to picture a collaborative medical scene where you are the patient lying awake on an operating table, answering questions posed to you, while doctors perform deep brain stimulation procedures. The objective of the prodding is to enable the patient to help doctors properly implant a neurological pacemaker inside the brain. If this sounds futuristic, think again. Similar to heart pacemakers, which have been in use since the late 1950s, brain pacemakers consist of an electrode permanently implanted in the brain to help maintain neural equilibrium. Heart pacemakers use electrical stimulation to maintain optimal cardiac rhythm. The U.S. Food and Drug Administration (FDA) approved the first use of brain pacemakers in the late 1990s to treat a few forms of tremor, including the one associated with Parkinson's disease. Other candidates for treatment using this form of electronic brain stimulation include people who suffer from severe depression and others afflicted with obsessive-compulsive disorder.

Cutting edge laboratory research involving stem cells derived from human embryos, as well as adult bone marrow, is taking place in institutions across the country. As you may recall from your biology courses, most cells in the human body are specialized to perform specific functions within specific tissues. In the past decade, we have discovered that stem cells can form a number of different tissues. In theory, this area of study may lead to our ability to rebuild human hearts after heart attacks, reverse the effects of Parkinson's disease, and improve the lives of people who suffer from Alzheimer's disease. Efforts are underway to combat these and other diseases using stem cells found in adult tissues (sources of cells include skin, bone marrow, the brain, the liver, and the pancreas). Research using embryonic stem cells shows great medical promise as well, but ethical and political debates continue, which will undoubtedly influence the future of this technique in the United States and other nations around the world.

Given just this short list of breakthroughs, it appears we live in an era when some new spectacular medical discovery is reported on a monthly basis. Perhaps some of us look forward to a world where safe and effectively proven medical tools and techniques are evident to combat any organism that causes disease among humans. Still, medical progress is continually slowed by its share of bottlenecks. Doctors are making great headway, but are stymied by the complexities of acquired immune deficiency syndrome (AIDS). AIDS patients are using various prescriptive "cocktails," but new cases keep emerging around the world—an effective vaccine to prevent the disease is not yet available. Many other diseases, from Alzheimer's disease to lung cancer, still elude the healing touch of medical specialists.

To complicate matters further, medical breakthroughs generally do not emerge overnight. They are often forced to go through a series of political obstacle courses. The FDA is the regulatory agency that decides which drugs and devices will be allowed to be used. Some researchers view the FDA as a bureaucratic barrier to innovation. Others hold that it does not require sufficient laboratory tests before human patients are allowed to serve as guinea pigs.

Then, too, there is the issue of the medical entrepreneur. As with genetic engineering and the entire field of biotechnology, research on artificial body parts has spawned a number of private industrial start-up firms. The flagship technologies developed in these centers may prove to be controversial amidst cost-conscious groups and those who are skeptical about the quality of the research conducted. Nevertheless, as illustrated in the following paragraphs dealing with surgery, procreation, and home testing kits, medical science is persistent in the face of these obstructions.

Minimally Invasive Surgery

During the last quarter of the twentieth century, the medical profession underwent a paradigm shift with regard to methods used to perform surgery. For many procedures, the invasive nature of operating room surgeries was greatly curtailed. In some settings, the term *noninvasive surgery* was even introduced. Advancements in video imaging, endoscope technology, and instrumentation have combined with computers and robotics to reduce the need for open surgeries performed with a scalpel. A few medical interventions falling under the heading of *minimally invasive surgery* include the use of lasers, arthroscopes, shock waves, and robotics to achieve a specific end result inside the patient's body.

MINIMALLY INVASIVE SURGERY

A group of new medical technologies employing endoscopic devices, laser beams, or shock waves instead of a scalpel to penetrate the damaged or affected area of the patient's body. Robotics sometimes assist in these procedures.

Laser is an acronym for Light Amplification by Stimulated Emission of Radiation. A laser consists of a glass rod or tube filled with a gas whose molecules can be stimulated to produce an intensely powerful, concentrated beam of light. Today's sophisticated lasers use many different gases (for example, argon, carbon dioxide, and krypton) that shine at varying wavelengths. During several kinds of surgical procedures, the laser beam simultaneously cuts and coagulates the tissue at which it is directed. In laser stitching inside the body, the beam of light fuses the protein of the tissues. As the wound heals, the coagulated tissue is absorbed by the body and replaced by new proteins.

For women, gynecological laser surgery has been shown to be a safe and effective alternative to having a hysterectomy. In performing this minimally invasive technique, the physician uses a ten-inch long telescopic instrument called a hysteroscope. This tool is inserted into the endometrial cavity through the cervical canal. At the tip of the scope are optical fibers and lenses, which enable the physician to view the walls of the cavity and locate the problem. Another optical fiber that transmits the laser beam is next threaded through a small channel in the hysteroscope.

The cost of hysteroscopic surgery is significantly less than the standard charges for conventional open surgery. It also eliminates most of the postoperative problems, such as pain from incisions, intestinal cramping, and extensive recuperation periods. Since the laser cauterizes as it vaporizes tissue, it stops bleeding and reduces the chances of infection and hemorrhaging.

Hysteroscopic laser surgery represents one of many applications for the lasers. Laser beams can be transmitted through optical fibers in thin tubes (technically called endoscopes) that are inserted into hollow organs of the body. This characteristic makes the technique equally useful in pulmonary medicine, urology, and gastroenterology.

Another form of minimally invasive surgery employs the use of an arthroscope to repair injured joints. The arthroscope is a sophisticated needle-like metal device housing a fiber optic light source. It is only two to six millimeters wide. See Figure 5-5. It can easily be inserted, under local anesthesia, into a knee, ankle, hip, shoulder, or wrist joint through a small puncture. Conventional joint surgeries require general anesthesia, longer stays in the hospital, and extended recovery periods. For example, a surgeon can use the arthroscope to repair damaged knee cartilage. The physician inserts this small viewing instrument into the knee joint through a one-quarter-inch incision. Working through the scope and two additional one-half-inch incisions, the surgeon can insert slim surgical implements to repair the damaged cartilage. Although some people may require general anesthesia, the entire procedure can be viewed on a television with the patient wide-awake.

An even more futuristic-sounding minimally invasive surgical procedure is called electromagnetic shock wave emission, which involves a machine called a *lithotripter.* This expensive piece of equipment uses shock waves to fragment kidney stones inside the body. Manufactured by Dornier Systems, it was approved by the FDA in the mid-1980s. To make effective use of this

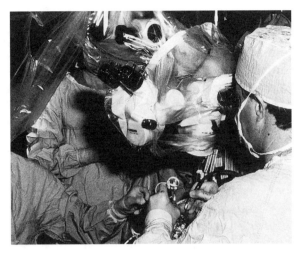

Figure 5-5.
Arthroscopic knee surgery is one type of minimally invasive surgery. A small viewing instrument, called an arthroscope, is inserted into the patient's body through a one-quarter-inch incision. An arthroscope is a needle-like metal device with a fiber-optic light source inside. It is two to six millimeters wide. It can be inserted into a knee, ankle, hip, shoulder, or wrist joint through a small puncture under local anesthesia. The surgeon works through the scope and two additional one-half-inch incisions, through which slim surgical tools are inserted to repair the damaged knee cartilage in the procedure shown here. (El Camino Hospital, Mountain View, California)

technology, it is first necessary to perform a series of exploratory tests to determine the position of the stones and how well the patient's kidneys are functioning.

Following this pretreatment phase, if the technique is deemed appropriate, the patient is prepped for the procedure and given general anesthesia. The person's lower body is immersed in a water bath and an underwater electrode is used to emit bursts of shock waves to the site of the kidney stones. The waves are administered at a rate and rhythm corresponding to the patient's heartbeat. Throughout the procedure, two fluoroscopic X-ray units are attached to television monitors to help the doctor pinpoint the exact location of the stones. The procedure may take an hour or more, but the remnants of the stones can usually be passed through the patient's ureter (flushed out of the body) without difficulty the following day. Although lithotripsy costs more than the standard operation, the patient normally saves time and money in the long run because the hospital stay is much shorter.

Technology associated with robotics has great potential to make minimally invasive surgical procedures available to a much larger group of patients. Already, the process used to remove a person's gallbladder has been influenced by robotics technology. Doctors who know how to perform robotic-assisted surgery equate their control of long-stemmed, narrow instruments to using chopsticks to complete the operation. In the case of cardiac operations, advancements in computer-assisted procedures mean the patient can be spared the pain and trauma involved in cracking open the chest and using a heart-lung machine while the heart is stopped. At the turn of the twenty-first century, this form of surgery was being performed by less than one-third of U.S. surgeons, due to the highly specialized training required.

Medical regulators in our government, however, are enthusiastic about robotically assisted surgery. The FDA recently approved the million dollar *da Vinci*™ Robotic Surgical System manufactured by Intuitive Surgical® of Mountain View, California. See Figure 5-6. Other robotic surgical devices are likely to be cleared for market release, and medical students will increasingly receive training in the use of these new tools of their trade. The potential advantages for robotics in the operating room are many. The physicians being assisted by robots are more successful, as the robotic arm holds and positions the video camera with great accuracy and steadiness and patients endure less trauma since the procedure is less invasive to their bodies.

In light of the fact that minimally invasive surgical techniques are becoming more widely accepted and available, prospective patients should be aware of the drawbacks. In lithotripsy, if the shock waves are improperly focused, they can damage the kidneys and other internal organs. The fluoroscopes used in this technique emit radiation equivalent to the dosage of standard diagnostic X rays. Other side effects, such as internal bleeding and blockages, may also occur. To date, arthroscopy and laser surgery appear to have a low incidence of side effects. They are not foolproof, however. Laser surgery is heat-intensive, and repeated treatments may cause permanent tissue damage. The person who opts for arthroscopy may ultimately need to undergo further treatment or even open surgery if the ligaments or tendons need more work.

Individuals are also advised to review their medical insurance coverage to determine policies regarding minimally invasive and robotics-assisted surgery. Insurance is presently available for most laser and arthroscopic techniques. The average insurance policy may not cover the newer procedures.

Procreation

"Where do babies come from?" is a timeless childhood inquiry having some new and somewhat provocative answers these days. Advances in medical technology have essentially revolutionized the process of human reproduction. Around the world, childless couples have started to investigate a variety of reproductive-assisting options, including IVF, artificial insemination, surrogate motherhood, embryonic transfer, and cryopreservation.

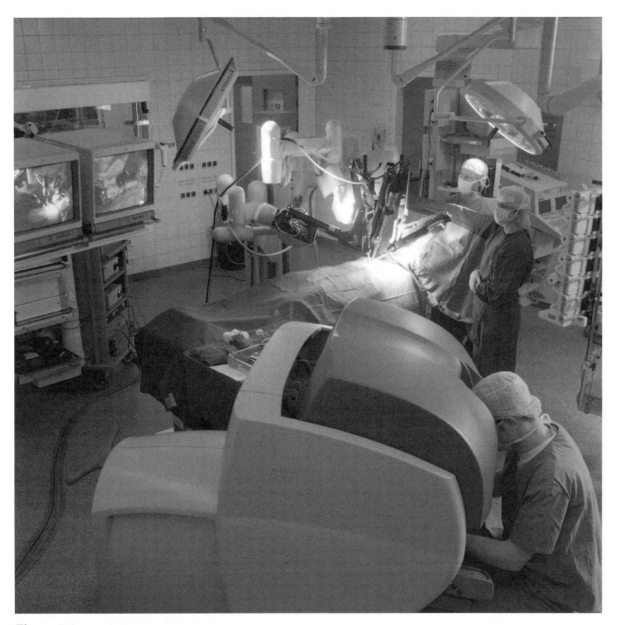

Figure 5-6.
The Food and Drug Administration approved the *da Vinci*™ Robotic Surgical System, manufactured by Intuitive Surgical of Mountain View, California, for use in operating rooms. (©1999 Intuitive Surgical, Inc.)

In vitro fertilization (IVF) is a procedure in which eggs are removed from a woman's ovaries and fertilized outside her body by the father's sperm in a laboratory dish. Once conception occurs, the fertilized egg is reimplanted in the woman's uterus.

In order to have any chance for success, IVF requires a number of well-timed and well-executed procedural steps. The woman who cannot conceive is first given hormones to induce her ovaries to produce more than one egg during her next cycle. Hormonal levels are closely monitored to detect the precise moment at which the eggs are ripening. To extract an egg from the ovary, a small incision is made in the patient's abdomen, through which a fine hollow needle is inserted. The mature egg is deftly retrieved and placed in a medium that further enhances maturation prior to fertilization.

Comparatively, the father's sperm is obtained and placed in a carrier solution that prolongs its ability to fertilize the ovum. The two mediums containing the eggs and sperm are combined and diluted to simulate the normal conditions found in the fallopian tubes. The resulting zygote is transferred to another solution that supports cell division and allows the embryo to mature. Some forty-eight to seventy-two hours later, the embryo is transferred into the mother's body. A large percentage of the embryos transferred by IVF medical teams fail to implant in the uterus, and repeated attempts are usually necessary. The first "test-tube" baby was born in England in the late 1970s.

One social innovation that has culminated from technological experiments in human procreation is surrogate motherhood. A typical situation occurs when an infertile woman and man enter into a contractual agreement with another woman. The surrogate agrees to be artificially inseminated with the father's sperm. After carrying the resulting fetus to term, she is expected to return the child to the original couple. In these instances, the baby has the genetic characteristics of the father and the surrogate mother. Another variation of surrogate motherhood arises when the female partner is not infertile, but incapable of carrying a fetus to term. In this type of case, an egg from the woman is fertilized with a man's sperm in vitro, and the resulting embryo is implanted in the surrogate's uterus.

As you can well imagine, the physical and emotional commitment of the surrogate mother is substantial. Much of the anguish associated with pregnancy cannot be offset by financial remuneration. The couple is obliged to the surrogate and must generally rely on her good faith to relinquish the child.

Embryo transfer is another reproduction-assisting technology that has become available to couples who are unable to conceive a child on their own. This technology may be employed for women who cannot produce eggs, but can carry a fetus to term. Once again, the couple pays a fee to a fertile woman who agrees to be inseminated with the man's sperm. It is critical for this egg donor to be matched with the woman, in terms of their ovulation cycles.

Also, the hormone levels must be the same in both women for a successful transfer to be possible. Four to five days after fertilization, the embryo is flushed out of the female egg donor and inserted into the woman's uterus. A variation of this procedure is known as embryo adoption, in which a male donor other than the woman's husband provides the sperm. As with IVF procedures, the success ratio for these procreation options is not that high.

Through a technology known as cryopreservation, a woman's eggs can be fertilized and cooled to −321 degrees Fahrenheit in liquid nitrogen. It is plausible that embryos could remain frozen in that state for centuries, awaiting implantation in a human uterus. This technology could enable a young couple in their twenties to preserve their own genetic offspring to be born at a later point in their marriage. In other words, neither partner would need to be fearful of infertility, and the proverbial "biological time clock" might not seem as threatening. Frozen embryos present a host of major ethical and legal concerns; some of these are reviewed in the final section of this chapter.

Complementing these scientific procedures designed to assist in the reproductive process are several prenatal diagnostic technologies. Medical scientists have made monumental efforts to identify and eventually treat a wide array of chromosomal abnormalities, metabolic disorders, and other hereditary diseases.

Among others, their research has developed and popularized the following procedures:

❏ *Amniocentesis.* A long, thin needle attached to a syringe is inserted through the lower wall of the pregnant woman's abdomen. The amniotic fluid, which contains some live body cells shed by the fetus, is withdrawn. The chromosomes are karyotyped (tested for specific characteristics of a cell nucleus, including chromosome number, form, size, and points of spindle attachment) in order to identify any abnormalities and determine the sex of the fetus. If the fetus is found to have a severe chromosomal or metabolic disorder, therapeutic abortion may be offered to the mother. This procedure is normally performed early in the second trimester of pregnancy.

❏ Ultrasound. This procedure uses sound waves directed into the abdomen of the pregnant woman. "Pulse echo" sonography presents a visual image of the fetus, uterus, placenta, and other internal features. This noninvasive technology is painless for the patient and reduces the need for X-ray scanning procedures. Ultrasound is one of the medical breakthroughs that significantly impacts the possibility of performing in utero fetal treatments.

❏ Fetoscopy. This application of fiber-optic technology enables the doctor to study the fetus inside the uterus. The fetoscope is inserted through an incision in the woman's abdomen under the direction of ultrasound. Fetoscopy has been used to give blood transfusions to diseased fetuses. Fetoscopy has considerable potential for introducing medicines, cell transplants, or genetic materials into developing fetuses to treat a number of diseases in utero.

Home Health Testing

Less than a decade after the first home pregnancy tests were introduced on the market in the late 1970s, a whole new generation of sophisticated self-diagnostic products began to appear on drugstore shelves. A walk down the aisles of the local convenience store might reveal do-it-yourself devices for pinpointing ovulation and monitoring blood sugar levels. Detection tests are also available for numerous conditions, such as high cholesterol levels, the presence of illegal drugs, venereal disease, urinary tract infections, colorectal cancer, menopause, and hypertension.

A fair amount of self-diagnosis was being practiced even before there were hospitals. Ancient Egyptian societies used papyrus and eucalyptus leaves for diagnosis. To determine whether or not they were pregnant, Egyptian women smeared eucalyptus oil on their arms and breasts before they retired in the evening. If their skin had a greenish tinge in the morning, they were confident that they were pregnant. Another pregnancy test entailed urinating on the leaves of a papyrus plant. If the plant died, the test was negative; if it lived, they began to make plans for motherhood. There does not seem to be any record of how accurate these tests were, but they were interesting nonetheless.

The contemporary movement toward increased self-diagnosis has been stimulated by several distinct social and technological changes. The first reason relates to lifestyle and a trend toward wellness. Generally speaking, our society is much more health and fitness conscious than it has been in the past. Along with the emergence of several new journal and magazine titles, jogging, fitness classes, nutrition awareness, and regimens of megavitamins, it seems people are beginning to take much more interest in their own medical care. Some exhibit an almost insatiable desire to know more about how their bodies function. Subsequently, many of us know all about resting heart rates, lean body weight, metabolic rates, various muscle groups, and the importance of personalized exercise programs. It is probably safe to assume that the average person who you might have approached in a 1960s neighborhood grocery store would not be able to tell you much about his or her optimal target heart rate zone. The availability of diagnostic kits for at-home use fit perfectly into society's significant interests in its physical well-being.

Another reason for the increased use of these products involves the soaring costs of medical treatment. Getting sick or suspecting a medical problem used to mean a trip to the doctor's office. These visits were often accompanied by a lengthy wait for laboratory test results. The evolution of the test-kit industry is providing people who may be ill with a different set of options, often at considerable cost savings, greater convenience, and more privacy. For instance, for a nominal sum of money, a person can purchase a detection test kit to determine the presence of blood in the feces. According to the American Cancer Society, hidden blood in the stool may be an early indication of colorectal cancer. Some people believe it is no longer cost-effective to pay for a rectal exam with a physician unless this type of self-test is positive. Also available, and recently approved by the FDA, is a very accurate and affordable home test for people to check

two types of blood cholesterol levels. Coronary heart disease remains a leading cause of death in America, and elevated levels of cholesterol play a significant role in as high as 50 percent of those deaths.

A third development that has strengthened the growth of the self-diagnosis movement involves a series of successful technology R & D projects. As computer chips have become less expensive and more miniaturized, self-diagnostic kits have become more accurate and easier to use. The development of monoclonal antibodies (MAbs) has also contributed to the substantial increase in the accuracy of the tests performed at home.

MAbs are hybrid cells created through genetic engineering techniques. MAbs have the potential to recognize a particular chemical structure. They can therefore act as sensitive probes for illnesses or physical conditions from pregnancy to cancer. For example, the over-the-counter pregnancy tests first marketed in the late 1970s were found to be about 75 percent reliable. They did not make use of biotechnology. Pregnancy test kits that became available during the mid-1980s were shown, on the average, to be 95 percent reliable—a substantial increase in accuracy due to this technological breakthrough. The cholesterol detection kits mentioned above are 97 percent accurate.

Taking a bird's-eye view of the drugstore shelves, there appear to be three categories of self-testing products. The first group is used to help diagnose a condition or disease in people who possess given symptoms. The most common example here is the pregnancy test kit, which is normally used after a woman notices a change in menstruation. The second type includes screening tests that identify indications of diseases in people without symptoms. Tests designed to detect hidden fecal blood fall into this category. A third classification includes monitoring devices to provide continual checkups on an existing condition. These are often purchased under the advice of a physician. Good examples of this type are kits used by diabetics to measure the level of glucose in their blood and those used by women to determine when they are ovulating. All tests that examine body specimens such as blood, urine, saliva, and so forth are called in vitro tests. Figure 5-7 illustrates several typical self-diagnostic products available without a prescription.

An unfortunate situation may arise if users translate *self-test* to mean *self-diagnosis.* Essentially, some of these test kits could shift the diagnosis of disease from the doctor's office to the home medicine cabinet. This may be construed as a quiet revolution in health care, which has received mixed reviews from practicing physicians.

Some health professionals believe properly administered self-tests show great potential for improving public health, as long as they are performed in conjunction with medical guidance. Home diagnostic testing devices are most valuable when they plug major gaps in current medical care. In other words, people who are concerned enough to take the first step and spend a reasonable sum of money on a test kit may be equally concerned to follow up and make an appointment to see their doctor. There is a

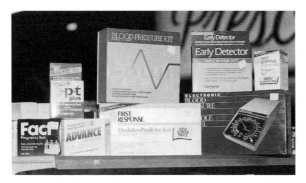

Figure 5-7.
These self-diagnostic products and kits are available over-the-counter at your local drugstore. (Michael Dafferner)

significant benefit procured when the results of an off-the-shelf test prompts a person to get necessary medical care earlier than would have occurred without the test. Furthermore, there is certainly the possibility for more effective treatment when a doctor can say to the patient, "Here is an explanation of your condition. I would like you to go home, test yourself periodically with this test kit, and report the results back to me in three weeks." In some cases, the test kit being recommended might actually be covered as a prescription by the person's insurance.

On the downside, other physicians feel quite uneasy about this self-test health phenomenon. Some fear patients may substitute the kits for a doctor's opinion. Diagnosis by a physician normally involves an evaluation of the patient's medical history, a physical examination, a series of tests, and possible consultation with other medical experts. Another concern expressed involves the fact that people may only test for one disease without knowing that the same symptom could be signaling an even more serious malady.

There is also the chance kit users may perform the test incorrectly or misinterpret the results attained. When this happens, the patient receives what are called *false positives* or *false negatives.* For instance, a woman may perform a test for pregnancy and conclude that she is not pregnant, only to find out three or four weeks later via a doctor's examination that she is pregnant (false negative). You can well imagine the severity of this type of event if the woman had, for example, ingested some form of prescription drug not recommended for use during pregnancy. Similarly, a false positive result in this case may cause premature elation or even unnecessary anxiety. Other causes for concern involve the shelf-life of the test kit and its estimated reliability.

In any event, self-test kits are probably here to stay. We can most likely expect further R & D pursuits to improve the existing products and to also expand the market considerably. The FDA's Center for Devices and Radiological Health regulates medical devices. It requires that devices developed for use by a layperson comply with the same regulations as the devices used in professional laboratories to ensure safety and effectiveness.

While this sounds reassuring in print, self-tests present several variables that are not as evident in the professional sector. For one thing, professional labs are designed, managed, and maintained specifically for scientific testing. Although medical labs are not infallible, their test results are generally more reliable and uniform, since they are acquired in a controlled environment (proper lighting, chemical storage requirements, sanitary conditions, and instrument maintenance). Another factor stems from the fact most laboratory technicians have had formal training and experience that have enabled them to develop expertise in conducting the tests. When we purchase a test kit, most of us must rely entirely on the written instructions or diagrams provided inside the package.

In sum, there appears to be a gradual shift taking place in the home diagnostic market from monitoring chronic illnesses to diagnosing serious or potentially fatal diseases. The FDA continues to warn consumers that they should be wary about buying and using kits on their own. More importantly, the FDA routinely finds unapproved home tests being sold on the Internet. In a recent investigation, none of the bogus human immunodeficiency virus (HIV) kits they confiscated (via the Internet) produced accurate results. The FDA affirms consumers can feel more confident that home test kits purchased from a reputable drugstore or pharmacy

have been cleared by their regulatory agency. For peace of mind, before you decide to buy a self-diagnostic kit or monitoring device on-line, the FDA encourages you to consult their website.

Quality of Life Ethics

Several times throughout this chapter we have made note of the fact that Americans are living longer, which is indeed a key mark of success in public health and medical care. On the other hand, more people are living the last few years of their lives with progressive illness and disability. Better medical interventions, coupled with healthier lifestyles, are giving more individuals the opportunity to grow old, accumulate a myriad of chronic conditions, and die slowly. Until a little over half a century ago, most Americans died of infections, accidents, and organ system failures usually resulting in a quick end to life. In contemporary society, by the time most Americans reach their mid- to late forties, they have already experienced the loss of a loved one living out the end of life with unrelieved pain or an incurable medical condition. People say they feel lucky when a person close to them lives well despite fatal illness, and then dies peacefully. These end-of-life scenarios are replete with emotional reactions related to quality of life ethics. They illustrate the demands made on family members, who are often subjected to difficult care burdens for which they are not adequately trained, and the expectations we have from the nation's health care industry.

It is difficult, if not impossible, to begin to understand the health care system in our society today without recognizing the pervasive influence of technology. Medical professionals, patients, and business leaders have attributed positive values to numerous technological procedures and devices used to treat illness and prolong life. Technology has become the core of modern medical diagnosis and treatment. In achieving that stature, technology often has the power to divert the attention of health care providers from issues of human value to the expedient selection of technological solutions. In other words, many of the scientific miracles in our midst are raising difficult ethical questions about how these lifesaving tools should be used. The decisions regarding who should pay for this treatment also loom over the heads of our medical policymakers.

Some researchers are suggesting that one of the most basic precepts of the medical profession may no longer be useful or even valid. For centuries, physicians abided by the philosophy that they should do everything in their power to sustain the lives of their patients. This basic belief has become confused by new medical technologies and the economic realities of contemporary medical practice. We have reached a point where living longer may not necessarily mean living better. Quality of existence may not be an issue receiving much attention when technological decisions are being made.

Legal Aspects

A simple definition for the elusive term *quality of life* is given by the authors in Chapter 11 of this textbook. One of many quality of life indicators is linked to a person's physiological, psychological, and mental well-being. Quality of life is an interdisciplinary concept with application and relevance in several areas, including medicine, law, and philosophy. In legal settings, quality of life issues surface when it is necessary to make decisions about a person's life, especially in the context of withholding or withdrawing life-sustaining medical technology or other interventions.

An array of powerful new drugs and prescriptive "cocktails" being used to combat serious diseases may cause psychological damage to the patient who places too much hope in their healing potential or curative capabilities. A traditional interpretation of healing has been the process of helping people lead a full existence by allowing them to retain autonomy and maintain a greater measure of control over their bodies. Modern life-support technologies may be truly irresistible, but they can serve to evade these goals. New pharmaceutical technologies have the ability to keep various tissues disease free, but often at great financial expense to the recipient. The machines of technology are often used to sustain life that many people would not deem worth living. Death is no longer viewed simply as part of the natural order of human existence, but as failure of medical technology. In such situations, the legal system may be called upon for ruling.

Courts utilize the quality of life standard most commonly in the cases of terminally ill patients. Years ago, courts used the patient's autonomy to preserve quality of life. In other words, a person's self-determination constituted protection against artificial prolongation of life, and his or her own wishes represented informed consent. People who reject life-sustaining treatment base their decision on personal preferences and knowledge about whether a prospective future medical state is very painful or degrading. In their minds, this is an unacceptable quality of living.

The ability of courts to rule in this manner breaks down considerably when the patients in question do not possess decision-making capacity. Quality of life judgements for people who do not have autonomy to decide for themselves whether life support is or is not desirable are extremely complicated. As you can well imagine, it is difficult, if not impossible, to determine how much discretion should be given to the family of an incompetent person for deciding to terminate her or his life. These situations are only made slightly less agonizing for jurists, judges, and legislators when the incompetent individual has exercised personal autonomy when he or she was competent. Two standards are used: living wills and advance directives.

A living will is a formal written declaration by a competent adult stating a wish or expectation that if she or he becomes so mentally or physically ill that there is no prospect of recovery, any procedures designed to prolong life should be withheld. Individuals often assemble living wills with the help of an attorney. This type of document takes the decision out of the hands of a medical team who may be inclined to administer treatment under the doctrine of necessity. In the United States, the first legislation to recognize the living will was the Natural Death Act in California in the mid-1970s. Since then, many states have enacted natural death legislation serving the following purposes:

❏ Prescribe conditions for the execution of living wills.

❏ Endorse the validity of living wills.

❏ Release medical professionals and institutions from civil and criminal liability.

An advance directive enables a competent person to give instructions about what their wishes are or who should be able to make decisions for them. These advance instructions are given in the context of medical treatment and relate to a person's rights to refuse or change treatment in a disabling chronic or terminal illness. These types of directives must be written and clearly articulated by the person while in a competent state of mind. In other words, if you simply mention to your significant other that you would not want to live in a comatose state,

your statement would not be sufficient to hold in court.

Modern medical technology can and will continue to save more lives each year as we move further into the twenty-first century. For many people, quality of life issues and personal values may lead to their decisions to craft living wills or advance directives. These documents are just another side effect of highly advanced technology in a discipline existing to preserve human life.

Issues of Cost Associated with Medical Technologies

High-tech medicine may provide society with a reduced incidence of disease, but the costs are enormous. Technology in medicine often alienates the patient from the health care professionals themselves. The beeps and squeals of the cardiac monitor, the printed output of a CAT scan, or the sound of a bedside respirator can obscure issues of human value. High-tech medical procedures can often interfere with family support mechanisms that have historically served as a source of encouragement for a person's well-being and sense of self-esteem.

On the economic front, the scenario looks even more complicated and disturbing. It has become quite apparent that there is a definite limit to the economic resources we can devote to health care. Governmental officials have repeated that we may have exceeded those limits for the past several years. In this setting, how can we decide who should receive these new forms of treatment, and in some cases, who should not get them? Major organ replacement is both technology-intensive and cost-intensive; it provides a visible example of this ethical quagmire.

Organ transplantation is unique among modern medical technologies because it is dependent upon human tissue obtained through consent. This characteristic places constraints on the use of this potentially life-saving procedure when the supply of suitable donors falls short of the demand. Some medical observers and professionals advocate a system of presumed consent. This means the organs of an individual who dies may be used, unless he or she has explicitly forbidden that use. This type of system has already been instituted in several European countries. A policy with these parameters places the burden of dissension on the individual. Some individuals and groups resist this approach and view it as a legislative intrusion. Think about this for a moment. Have you ever contemplated being an organ donor? Do you feel that this is a decision you alone should be allowed to make?

In 1999, according to the United Network for Organ Sharing, there were more than seventy-two thousand people in the United States who had their names on transplant waiting lists. By the end of that year, more than sixty-one hundred of them had died while waiting for an appropriate organ. Avowedly, the demand for human organs exceeds the supply, and the cost for the necessary surgical procedures is off the chart. The process of establishing the criteria for selecting the beneficiaries of new transplant technologies seemingly poses another ethical issue. It often seems the patients with families who have the ability to attract public attention and can also afford the steep price of transplantation are the first ones chosen for available organ donations. While it is still debatable as to whether or not total expenditures for organ replacement should be a matter of public policy, an individual's ability to pay should not affect the decision to replace a failing organ. It seems both moral and logical to base the choice on the person's medical need and the extent to which the results are projected to be beneficial for the long term.

These criteria still would require an accounting of the cause of organ failure, the age of the patient, the availability of a suitable donor, and the ability of the patient to recover. For instance, a good candidate for a heart transplant might be a previously healthy young adult who has a terminal heart disease unresponsive to therapy. With this in mind, is the middle-aged adult who has a history of alcoholism a less desirable candidate for a costly liver transplant? How would you respond to this question if you were on a hospital patient selection committee? Future decisions concerning organ replacement or any new technology will not be any easier to make from the ethical or moral perspective. Economic constraints may inevitably lead to mandatory rationing of certain medical technologies. The medical profession will have to agree on a credible list of criteria for the selection of patients who are "good risks" for organ transplant surgery.

The subject of technology and procreation also prompts an array of difficult legal and ethical questions. Who ultimately has legal and moral responsibility to care for the offspring of artificial procreation? Who decides what is best for a child when the traditional mother-father roles are no longer precise? Will surrogate motherhood become commercialized in the next several years? What sorts of emotional reactions might occur when a person learns that he or she is the product of the fertilization of an anonymous donor sperm and donor egg in a petri dish? What becomes of unused embryos?

This last question is crucial when, for example, an IVF team creates more embryos in the laboratory than can be safely reimplanted in the mother's womb. Disposing of these surplus embryos or using them for further research purposes opposes the view held by some that an embryo is a human being who has a right to life. Others say that until the embryo develops to the point at which it is capable of experiencing something, it is not plausible to view it as a human being. Current stem cell research activities, coupled with the scientific feats in cryo-preservation, may lead to an increased practice of freezing embryos for later use. This area of study is fraught with much debate and controversy. To date, a very small number of frozen embryos have been successfully thawed, implanted in the mother's womb, and carried to term. The necessary data is insufficient to evaluate the degree of risk to children who are created in this manner. An even more disturbing question arises if the parents die without leaving specific instructions about the disposal of the embryos.

Medical technology will probably not cease to advance due to an absence of answers for these and other ethical inquiries. The responsibility for directing the course of its development and utilization remains with all of society. While a small percentage of the population will join the ranks of tomorrow's medical specialists, most will continue to play the role of consumer or beneficiary. Value-laden decisions about new technology in medicine must be shared by both sectors.

Summary

Technology has had a positive influence on the practice of medicine in our society. Amidst an impressive array of computer-assisted diagnostics and life-sustaining pharmaceuticals and devices, the cost of health care remains an urgent matter of international concern. The advent of DRGs and HMOs is evidence that some progress is being made toward solving this problem.

Current demographic figures indicate the average life expectancy for American men and women has surpassed the seventy-year mark and is rapidly approaching eighty. Recent technological breakthroughs in the medical field have played a major role in this development. A review of several feats in minimally invasive surgery and the use of technology to enable infertile couples to bear children have been described in this chapter. These areas of specialization were selected to illustrate that science and technology in medicine are here to stay—there may be no turning back. A cost to benefit analysis of these revolutionary techniques must emerge to continually assess social ethics and the moral responsibilities of our physicians. Several factors have stimulated a definite trend toward self-diagnosis in the United States. A diversified collection of self-diagnostic home health test kits and medical devices have become available on the shelves of local drugstores. They have been met with mixed reactions among practicing physicians.

The greatest challenge for contemporary and future health care professionals will revolve around the following areas:

❏ Curtailing the rampant increase in the cost of medical care.

❏ Appropriate and efficient use of medical technology.

❏ Continued monitoring of medical procedures integral to the trend toward self-diagnosis.

❏ Preservation of social ethics related to quality of life.

Discussion Questions

1. With reference to the cost of health care, explain what the statement "Conventional financing plans and governmental policies have stimulated the use of technology at any cost" means.

2. What are the three categories into which the huge number of quality

problems in our nation's health care system can be classified? Give a detailed example of a patient condition or treatment to illustrate each of these limitations in contemporary health care.

3. Identify the primary features characterizing a modern HMO. Provide a specific example of an HMO in your area and retrieve a copy of an information brochure from them, if possible. Be ready to present your findings during an upcoming class discussion.

4. How does the relationship among technology, human needs, and medical costs present a cyclical problem for the health care industry?

5. Explain what the term *minimally invasive surgery* means. How does this type of medical treatment differ from a conventional surgical procedure? Provide at least three contemporary examples in your explanation.

6. What are the three categories into which most self-diagnostic products can be classified? Make a trip to your local drugstore to take a survey of the self-diagnostic kits available without a prescription. Provide a detailed description of the function and usage of at least one of the products located. Identify reasons you believe these types of home health testing kits have become so popular in our society.

7. Do you feel that today's medical professionals are overly dependent on technology? Give reasons to support your position. What effect does the use of technology have on legal and ethical standards in society?

Suggested Readings for Further Study

AMA officer looks at issues in future of health care industry. 1985. *Industrial Engineering* 17 (3): 46–48, 50, 54–56.

Amato, I. 2001. Helping doctors feel better. *Technology Review* 104 (3): 64–71.

Banta, H. D., ed. 1982. *Resources for health: Technology assessment policy making.* New York: Praeger.

Banta, H. D., C. Behney, and J. S. Williams. 1981. *Toward rational technology in medicine: Considerations for health policy.* New York: Springer.

Barger-Lux, M. J. and R. P. Heaney. 1986. For better and worse: The technological imperative in health care. *Social Science and Medicine* 22 (12): 1313–1320.

Becher, E. C. and M. R. Chassin. 2001. Improving the quality of health care: Who will lead? *Health Affairs* 20 (5): 164–179.

Blank, R. H. 1985. Making babies: The state of the art. *The Futurist* 19 (l): 11–17.

Blendon, R. J., K. Scoles, C. DesRoches, J. T. Young, M. J. Herrmann, J. L. Schmidt, and M. Kim. (2001a). Americans' health priorities revisited after September 11. *Health Affairs.* [Internet]. (13 November): 3 pages. <http://www.healthaffairs.org/Blendon_Web_Excl_111301.htm> [7 Dec 2001].

———. 2001b. America's health care priorities: Curing cancer and controlling costs. *Health Affairs* 20 (6): 222–232.

Boyle, J. 1985. The challenges of health care in America: On the side of the patient. *Vital Speeches* 51 (1 May): 426–428.

Brown, H. G. 1990. The twenty-first century and the control of cancer? *Oncology Nursing Forum* 17 (4): 497–501.

Burton, S. L., L. Randel, K. Titlow, and E. J. Emanuel. 2001. The ethics of pharmaceutical benefit management. *Health Affairs* 20 (5): 150–163.

Bush, G. W. 2001 Stem cell research: Cautious, ethical, and moral research and development. *Vital Speeches of the Day* 62 (22): 674–676.

Carey, J. 1985. Medicine chest labs. *Newsweek* 85 (27 May): 87.

Centers for Disease Control and Prevention. *Assisted reproductive technology.* [Internet]. <http://www.cdc.gov/nccdphp/drh/art.htm>.

Clark, M. 1986. Lasers for medical wars. *Newsweek* (2 June): 68.

Consumer's guide to home medical tests, A. 1986. *FDA Consumer* 20 (l): 25–27.

Corporate Rx for medical costs, The: A push for revolutionary changes in the health care industry. 1984. *Business Week* (15 October): 138–141, 144–146, 148.

Cutler, D. M. and M. McClellan. 2001. Is technological change in medicine worth it? *Health Affairs* 20 (5): 11–29.

DelCano, A. M. M. 2001. The concept of quality of life: Legal aspects. *Medicine, Health Care & Philosophy* 4 (1): 91–95.

Dickey, B. 1997. Assessing cost and utilization in managed mental health care in the United States. *Health Policy* 41: S163–S174.

Ditlea, S. 2000. Robosurgeons. *Technology Review* 103 (6): 74–81.

Eckenfels, E. J. 2001. Learning about ethics: the cardinal rule of the clinical experience. *Medical Education* 35 (8): 716–717.

Elkowitz, A. 1986. Physicians at the bedside: Thoughts and actions with regard to health care allocation. *Journal of the National Medical Association* 78 (5): 423–427.

Farley, D. 1986. Do-it-yourself medical testing. *FDA Consumer* 20 (l): 22–24.

Filley, R. D. 1985. Health care costs take spotlight as technological advances collide with issues and practicalities. *Industrial Engineering* (March): 34–37, 40, 42–45.

Fineberg, H. V. 1985. The costs of health: Irresistible medical technologies. *Current* 271 (March/April): 5–7.

Freifeld, K. 1984. Obsoleting the scalpel. *Forbes* 134 (27 August): 130.

Galton, L. 1986. A new age of medical miracles. *Consumers Digest* 25 (2): 57–60.

Garr, D. 2001. The human body shop. *Technology Review* 104 (3): 72–79.

Gold, M. 1985. The baby makers. *Science 85* 6 (3): 26–38.

Goldman, J., P. Schwartz, and P. Tang. 2000. Roundtable: Medical privacy. *Issues in Science and Technology* (summer): 77–78.

Hall, S. S. 2001a. Adult stem cells. *Technology Review* 104 (9): 42–49.

———. 2001b. Brain pacemakers. *Technology Review* 104 (7): 35–43.

Harwood, C. T. 1985. Pulverizing kidney stones: What you should know about lithotripsy, *RN* 48 (l): 32–37.

Health care and cost. 1986. *Harvard Medical School Health Letter* 11 (4): 1–4.

Hellerstein, D. 1983. Overdosing on medical technology. *Technology Review* 86 (6): 13–17.

Herzlinger, R. 1985. Corporate America's "Mission Impossible": Containing health care costs. *Technology Review* 88 (8): 40–49.

[Internet]. (1995). *The International Health Care Cost Crisis.* <http://www.mum.edu/msvs/6195Herron1.html> [6 Dec 2001].

Jacob, M. 1985. The light fantastic: Lasers brighten the future. *The Futurist* 19 (6): 36–38.

Kanner, B. 1985. Test thyself. *New York* (20 May): 18, 20, 23–24.

Knudson, M. 1998. The hunt is on for new ways to overcome bacterial resistance. *Technology Review* 100 (9): 23–29.

Kwan, J. W. 1989. High-technology infusion devices. *American Journal of Hospital Pharmacology* 46 (2): 320–335.

Lane, H. C. and A. S. Fauci. 2001. Bioterrorism on the home front: A new challenge for American medicine. *JAMA* 286 (20): 2595–2597.

Langer, L. F., H. Brem, and R. Langer. 1991. New technologies for fighting brain disease. *Technology Review* 94 (2): 62–71.

Langer, R. 2001. Drugs on target. *Science* 293 (6 July): 58–59.

Larkin, T. 1984. Computers may be good for your health. *FDA Consumer* 18 (9): 8–11.

Laufman, H. 1990. Environmental concerns in surgery in the 1990s. *Today's OR Nurse* 12 (10): 41–48.

Lewis, C. (2001). Home diagnostic tests: The ultimate house call? *FDA Consumer Magazine*. [Internet]. (November/December): 6 pages. <http://www.fda.gov/fdac/features/2001/602_home.html> [13 December 2001].

Loannou, L. 1985. Bloodless operations. *Savvy* 6 (10): 84–85.

Lunzer, F. 1985. A lot to learn. *Forbes* 136 (1 July): 102–103.

Lynn, J. and J. H. Forlini. 2001. Serious and complex illness in quality improvement and policy reform for end-of-life care. *Journal of General Internal Medicine* 16 (5): 315–319.

Mack, M. J. 2001. Minimally invasive and robotic surgery. *JAMA* 285 (5): 568–572.

Mann, R. 1991. Scalpel! Clamp! Floppy disk! *Technology Review* 94 (7): 66–71.

McKillop, T. 2001. Healthcare and society: The role of the global pharmaceutical industry. *Vital Speeches of the Day* 67 (21): 650–653.

Medicine's new triumphs. 1985. *U.S. News & World Report* (11 November): 46–48, 53–54, 56, 58.

Merchant, J. 1996. Biogenetics, artificial procreation, and public policy in the United States and France. *Technology in Society* 18 (1): 1–15.

Mereson, A. and A. Dorfman. 1985. New at-home diagnostics: Over-the-counter detection kits offset soaring medical costs. *Science Digest* 93 (3): 16.

Misener, J. H. 1990. The impact of technology on quality health care. *Quality Review Bulletin* 16 (6): 209–213.

Mulder, J. T. 2001. Ailing hospitals want cash transfusion. *Syracuse Post-Standard Newspaper* (6 December): A-1, A-6.

Potera, C. 1998. Making needles needless. *Technology Review* 101 (5): 67–70.

Randel, L., S. D. Pearson, J. E. Sabin, T. Hyams, and E. J. Emanuel. 2001. How managed care can be ethical. *Health Affairs* 20 (4): 43–57.

Rawls, R. 1984. Laser heart surgery teams chemistry and medicine. *Chemical and Engineering News* 62 (12 March): 14–15.

Reiser, S. J. 1978. *Medicine and the reign of technology.* Cambridge: Cambridge Univ. Press.

Reiser, S. J. and M. Nabar, eds. 1984. *The machine at the bedside: Strategies for using technology in patient care.* Cambridge: Cambridge Univ. Press.

Relman, A. S. 1985. Technology and medical care: Cost control, doctors' ethics and patient care. *Current* 271 (March/April): 8–13.

Ricardo-Campbell, R. 1986. Economics and health: The medical mystique. *Vital Speeches* 52 (15 March): 345–349.

Robinson, R. 2000. Managed care in the United States: A dilemma for evidence-based policy? *Health Economics* 9 (1): 1–9.

Runde, R. 1985. Take two tests and call me in the morning. *Money* 14 (3): 173–174.

Schatz, R. 1985. No place like home: Self-diagnostic kits are a hot item. *Barrons* 16 (8 July): 18.

Sciaky, P. (1997). *What's Wrong with our Health Care System?* [Internet]. <http://bcn.boulder.co.us/healthwatch/wrong.html> [7 Dec 2001].

Seppa, N. 2000. Edibile vaccine spawns antibodies to virus. *Science News* 58 (22 July): 54.

Singer, P. 1985. Technology and procreation: How far should we go? *Technology Review* 88 (2): 23–30.

Sisk, J. E. 1984. Effects of competition in health care on the use and innovation of medical technology. *Health Care Management Review* 9 (3): 21–34.

Spiegel, A. D., D. Rubin, and S. Frost, eds. 1981. *Medical technology, health care and the consumer.* New York: Human Sciences.

Stewart, C. F. and Fleming, R. A. 1989. Biotechnology products: New opportunities and responsibilities for the pharmacist. *American Journal of Hospital Pharmacology* 46 (l): 4–8.

Stikeman, A. 2001. The programmable pill. *Technology Review* 104 (4): 78–83.

Tracy, E. J. 1985. High-tech diagnosis at home. *Fortune* 112 (9 December): 121–122.

Voelker, R. 2001. Will focus on terrorism overshadow the fight against AIDS? *JAMA* 286 (17): 2081–2083.

Weil, T. 1985. Procompetition or more regulation? *Health Care Management Review* 10 (3): 27–35.

Weinhold, B. 2001. Making health care healthier: A prescription for change. *Environmental Health Perspectives* 109 (8): A370–A377.

Weisbrod, B. A. 1985. America's health care dilemma. *Challenge* 28 (4): 30–34.

Wilson, J. 2000. Beating-heart surgery. *Popular Mechanics* 177 (4): 50, 52.

Wooten. J. O. 2000. Health care in 2025: A patient's encounter. *The Futurist* 34 (4): 18–22.

Wortman, M. 2001a. DNA chips target cancer even before tumors form. *Technology Review* 104 (6): 50–55.

———. 2001b. Medicine gets personal. *Technology Review* 104 (1): 72–78.

Chapter 6

Manufacturing and Production Enterprises

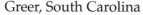
Introduction

During the last twenty years, the United States experienced the erosion of its steadfast leadership position in the manufacturing arena. The nation, faced with numerous challenges to rebuild antiquated factories, retool production lines, and rethink industrial management paradigms, lost a significant portion of its competitive edge in manufacturing. The casual reader need only pick up yesterday's *Wall Street Journal,* this morning's *New York Times,* or an issue of *Fortune* from last year to review yet another bleak forecast regarding the decline of America's manufacturing base. Workers around the country relate their stories about being laid off unexpectedly, as their employers decide to relocate production facilities to overseas sites with lower hourly wages. Overall, the manufacturing share of total employment in the United States has been cut in half since 1970.

Without question, all American manufacturing firms have been confronted with a dramatic increase in foreign competition. In one attempt after another to match foreign competitors' low labor costs and more relaxed regulations, garment makers, appliance manufacturers, and electronics components companies moved their operations overseas.

These decisions were made to enable the home facilities to focus on research, design and development, and the production of fewer products. In most instances, these companies took this action to remain competitive, but they ultimately became marketing and distribution centers for products *not* made in the United States.

There appears to be yet another series of revolutionary trends emerging across the playing field of American manufacturing. Companies nearly paralyzed by unfair trade practices, reverse engineering feats, and a lack of long-range strategic planning in the 1980s, have come charging back. Across the nation, manufacturing employers have embraced an attitude of "can-do optimism," as they breathe new life into manufacturing and production enterprises that had all but been pronounced dead just a few years ago.

Some examples of America's manufacturing resurgence are displayed where manufacturing firms recognized that an overdependence on large-scale flexible manufacturing systems (FMSs) and across-the-floor automation is potentially detrimental to their success. A contemporary automation paradigm or pattern is being spotlighted and applauded as corporate executive officers pursue an ingenious balance between software packages and computer networks—human operators and automation. En route, they have discovered it is often much more cost-efficient to use hand labor alongside software networks versus sophisticated robots. Too much automation results in lower profitability, especially when dealing with odd-sized components, a large variety of complicated machine tools, and tight tolerances. By taking some of the automation out of a facility, a U.S. manufacturer can ironically outdistance foreign competition in such critical measures as time-to-market and manufacturing flexibility.

General Electric (GE) is gaining an impressive market share abroad because of the exceptional time-to-market and customizing capability of its Erie, Pennsylvania locomotive plant. The company retained traditional automation at the start of the production line to complete harsh repetitive work. At the end of the line, where customized cabs, motors, and paint schemes become a factor, GE has installed flexible cells that can be programmed. The plant has reduced the time needed to make a locomotive from ninety-two days in the early 1990s to twenty-five days today.

Another example of flexible automation can be viewed at Alcan Aluminum's Oswego Works (Oswego, New York), which recycles aluminum cans. In the process of recycling cans into aluminum, unmanned vehicles dart back and forth between the machines to deliver and pick up material. These office desk–sized vehicles are not like the carts in Japanese FMS installations that travel in rigid patterns along wire routes imbedded in the factory floor. Innovative technologies allow the automated guided vehicles (AGVs) at Alcan to run around anywhere. Each vehicle has a range finder that bounces laser beams off targets placed around the factory floor to let it know where it is headed. See Figure 6-1. A series of computer-directed radio signals from a control center dictate its pickups and deliveries.

Figure 6-1.
This is an illustration of laser guidance for an automated guided vehicle (AGV). Laser guidance works by use of a laser scanner mounted onboard the AGV. By referencing an onboard map of the layout of the factory, the vehicle determines its position by sweeping its beam across reflectors placed on walls or columns. (Stefan Karlsson, AGV Electronics)

As one culls the literature associated with manufacturing and production enterprises, a huge assortment of technical acronyms becomes evident. Perhaps more than any other discipline, the array of alphanumeric abbreviations in manufacturing is overwhelming. Most of you are at least somewhat familiar with terms like *CIM, CAD, CAM, FMS, JIT, MRP,* and *SPC.* Recent years have brought us new constructs like lean manufacturing, agile manufacturing, rapid prototyping (RP), and concurrent engineering (CE). We are also being challenged to decipher a host of new abbreviations, including *MRP II, GT, CE,* and *DFMA.* Each of these technical terms is identified in the Key Concepts section, but recognize that this is only a small sampling of what is everyday jargon in industry. Manufacturing managers are continually being charged to learn about, evaluate, and potentially implement one or more of what might be labeled *integrated automated process technologies* to launch their firms into future successful ventures.

The focus of this chapter will be to discuss the new and emerging technologies affecting the manufacturing environment in the United States, as well as overseas. Some of the questions we will address include the following:

❑ Will integrated automated process technologies enable manufacturing enterprises to grasp and hold on to leadership positions in the global arena?

❑ Why do some corporations remain reluctant to recognize and enthusiastically support the procedures central to CE?

❑ When should the decision to automate be made? Is there such a thing as too much automation on a production floor?

❑ Assuming that lean manufacturing is more efficient, what strategies can an organization use to become leaner and more agile?

❑ What is the difference between material requirements planning and just-in-time? What are the advantages and disadvantages of each in a manufacturing environment?

❑ How do present-day workers respond to the presence of robots in their daily work environments?

❑ What new manufacturing and production enterprises are on the horizon, and in which disciplines do they show the most promising applications?

Key Concepts

As you review the material contained in this chapter, you should learn the meaning of each of the following technical terms and phrases. Be especially attentive to the many manufacturing acronyms as they are mentioned throughout this discussion.

agile manufacturing
bar coding
computer-aided design (CAD)
computer-aided manufacturing (CAM)
computer-integrated manufacturing (CIM)
concurrent engineering (CE)
design for manufacture and assembly (DFMA)
flexible manufacturing system (FMS)
group technology (GT)
just-in-time (JIT)
lean manufacturing
manufacturing resource planning (MRP II)
material requirements planning (MRP)
quality function deployment (QFD)
rapid prototyping (RP)
statistical process control (SPC)

Organization of the Work Environment

In one way or another, every new change on the production floor challenges traditional assumptions and contemporary procedures. Old ways in manufacturing are slow to die, but the horizon seems much brighter than it did just a few years ago. Is manufacturing on the rise in this country? Although the spread of new technology and innovative strategies across the nation is uneven, the following discussion will reveal numerous examples of an exciting U.S. manufacturing renaissance.

As the average consumer attempts to make sense of manufacturing buzzwords, it becomes apparent that manufacturing technology systems generally share three common features: integration, automation, and computerization. A simple explanation of any manufacturing system is an entity providing concept implementation, starting with design, continuing through realization of the product, and culminating with customer satisfaction. A computer-integrated manufacturing (CIM) setting implies a situation in which all components essential to the production of an item are integrated.

COMPUTER-INTEGRATED MANUFACTURING (CIM)

The umbrella acronym (pronounced "sim") for a host of automation technologies in the manufacturing environment. It does not refer to one specific technology, but to the integrated use of computers in all sections of the enterprise, from the planning of production, through the design and manufacture of a product, to the assurance of good quality.

In the broadest sense, CIM encompasses a diverse array of manufacturing strategies in use today, including the initial stages of planning and design, all the way through the final stages of manufacturing, packaging, and shipping. In actuality, CIM is not a specific technology that can be purchased one week and installed the next. Even though much has been written about the CIM factory of the future, it is not as much a technology as it is a management philosophy of operation. It involves strategic and aggressive efforts to combine all available technologies to manage and control the entire business of bringing new products to market. Several of the manufacturing strategies that can make for a successful CIM enterprise include computer-aided design (CAD), computer-aided manufacturing (CAM), artificial intelligence (AI), expert systems, group technology (GT), rapid prototyping (RP), automated materials handling (AMH), robotics, manufacturing planning and control systems, automated inspection methods, and continuous quality improvement.

Although CIM is a unifying philosophy of manufacturing production, almost every company using CIM implements it differently. Also, the tools each company uses in its CIM implementation are dependent upon many diverse factors. From this ever-expanding list of technology-based activities, it should be obvious that company-specific planning is imperative. While CIM might be regarded as a unifying force among the basic functional areas of design, production, and management, it is essential to note that management is the central integrating force, and the computer is a tool for execution. Basic to an effective CIM strategy is a strong commitment to intercompany communication. Success in the long-term requires a spirit of cooperation from the highest level of management, through each level of production, on down to the shop floor, sustained by employees who are not afraid to accept new technology.

The term *lean manufacturing* was initially popularized by researchers involved with Massachusetts Institute of Technology's International Motor Vehicle Program.[1] To become leaner, manufacturers must remove obstacles preventing them from manufacturing with high velocity. Obstacles, such as complicated setups, excessive material handling, poor physical flow, and production interruptions, interfere with an organization's ability to design and build the best quality products in the shortest time possible. Lean manufacturing often begins with lean design. Although lean manufacturing began in Japan, American and European manufacturers

---

[1]Womack, J., D. Jones, and D. Roos. 1991. *The machine that changed the world: The story of lean production.* New York: Harper Perrenial.

have adapted this philosophy into their manufacturing plants. One such plant is New United Motor Manufacturing Inc. (NUMMI), a joint venture between General Motors (GM) and Toyota, located in Fremont, California. See Figure 6-2.

An essential element of a successful lean manufacturer is the use of *concurrent engineering (CE).* This innovative approach to product development is a method of simultaneously integrating the many aspects of product design, development, and manufacture.

CE contrasts with the traditional linear approach, in which marketing experts conceptualize the product, design engineers design the prototype, and the manufacturing and purchasing departments make all decisions regarding the production process and parts suppliers. Fundamentally, the results of decisions made at each stage, in the traditional linear approach to product design, are thrown over the wall metaphorically existing between the various departments in an organization, with little, if any, interaction

Figure 6-2.
New United Motor Manufacturing, Inc.'s production method is based on Toyota's lean production system. The system is an integrated approach to production using machinery, material, and labor as efficiently as possible.

among departments. In contrast, CE increases the interaction between designers and manufacturers before the product goes to the manufacturing floor. Under CE, once a new product is designed, there should be no problems in its manufacturing. As companies discover the advantages of getting things right the first time, the benefits of CE are hard to dispute. ***Design for manufacture and assembly (DFMA)*** is a tool used for product simplification. It is an analysis technique requiring a strict evaluation of product complexity and systematic development of design alternatives that simplify the product and improve the incidence of defect free manufacturing.

JUST-IN-TIME (JIT)

A manufacturing philosophy attempting to eliminate waste throughout the system, including inventory at both ends of production and all machinery and manpower not adding value directly to the product. JIT has its roots in the Japanese automobile industry, which sought to get rid of excess, waste, and unevenness. Even companies in the United States and Europe often use the Japanese terms for these three problems—*muri* (excess), *muda* (waste), and *mura* (unevenness).

The core objective of JIT is to achieve low cost, high quality, on-time production by minimizing instances of idle equipment, facilities, or workers and reducing excess work in the process. Instead of making parts to stock, JIT emphasizes having the right parts, at the right time, in the right quantities, on the manufacturing floor. JIT keeps inventory costs down, but also demands well structured supply lines and very cooperative employees.

In addition to the manufacturing changes in the last twenty years, organizations worldwide have undergone restructuring to reduce the layers of management and increase profits. These leaner manufacturers are fundamentally different from their historical counterparts. Traditionally, U.S. manufacturers would have many layers of management separating the assembly line, or blue-collar workers, from the company's president. Each layer of management would report to and create reports for a higher layer. Most of the middle management existed to report and summarize the work of the employees below them and to give this analyzed information to the employees above them. Beginning in the 1970s, companies began restructuring the management hierarchy, and the organizational structures of companies began to flatten. Instead of six or seven layers of middle management, companies reduced middle management to one to three layers. This flattening of the organizational structure has a direct effect on manufacturing enterprises because it puts more responsibility and decision-making in the hands of the assembly line workers and their first line managers.

These are not the only challenges for the modern manufacturer. There has been an increasing amount of competition in manufacturing, not only from traditional economic competitors like Japan and European countries, but also from developing countries. There is a growing focus on the quality of products and quality assurance strategies that remain in the forefront of operations planning. New and improved engineering materials are being introduced on a routine basis. Selecting the most effective manufacturing system takes time, and start-up costs are often exorbitant. Companies making systematic and informed choices will remain competitive and emerge as international manufacturing leaders well into the next century. These agile organizations will most probably be the ones contributing to the global market with a host of functional products that *are* made in the United States. Manufacturing managerial decisions are constantly reshaping the profile of tomorrow's factories. The implementation of integrated, automated systems will most likely have a visible impact on the U.S. lead in production, as well as on the credentials of the present and future work forces of the United States.

Computer-Integrated Manufacturing (CIM)

Forming the backbone of a computer-integrated manufacturing (CIM) system is an integration of the mechanical, electrical, and informational subsystems. A successful CIM operation makes it possible for the computer workstations used by management, design, and manufacturing personnel to communicate with one another. Data can be entered from the operations of the entire plant, and the resultant information is continuously processed, updated, and sent on demand to any employee for efficient decision-making. CIM allows in-plant design systems to obtain data from management, on the current cost of raw materials, and also from manufacturing, regarding ways to adapt design for more efficient production. Likewise, manufacturing cells on the factory floor have a direct link to design data in order to more effectively plan the steps for making products. Finally, managers and operators at all levels can work from their computer terminals to acquire up-to-date information from both design and manufacturing databases. Ostensibly, this feature of CIM allows for maximum coordination and centralized control of many industrial engineering operations. See Figure 6-3.

If one takes a bird's-eye view of a CIM enterprise, computers are found being used throughout to accomplish the following:

❑ Provide design assistance.
❑ Translate specifications into drawings, parts lists, and routings.
❑ Schedule work based on machine availability and promised delivery dates.
❑ Reassign personnel to various tasks to keep in-process inventories at a minimum.
❑ Control machine tools.
❑ Provide vital, up-to-the-minute information for management.
❑ Execute routine clerical tasks.

None of the departments in a modern manufacturing facility can be allowed to consider itself a closed cell in its daily operations. In order to realize all the potential benefits, all employees—in design, engineering, manufacturing, and administration—must be able to work in unison from the same manufacturing database. Among other items, such a centralized database might consist of the following elements:

❑ Computerized drawings of parts (CAD).
❑ Files for parts programs to be used in numerically controlled machines *(computer-aided manufacturing [CAM]).*
❑ Files on routing that detail the workstations through which various parts must be processed.
❑ Files for all workstations listing production schedules, tools required, and operators assigned for each shift.
❑ Files on all tools and equipment detailing maintenance and depreciation time lines.

A large variety of tangible benefits have been reported among CIM users. Several of these benefits follow:

❑ A reduction in the duplication of data entry concerning product specifications, tolerances, order quantities, inventory levels, and raw materials.
❑ Fewer discrepancies in the data that may occur when a number of people gather information from different sources at different times in the production cycle.
❑ Prototypes can be prepared and tested more quickly, resulting in more precise

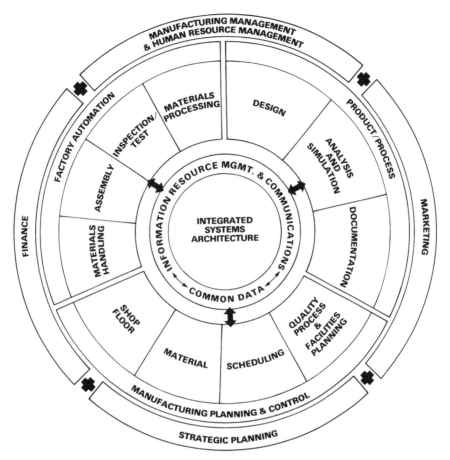

Figure 6-3.
This schematic from the Society of Manufacturing Engineers (SME) illustrates one perspective via a logical diagram identifying many of the foundational components of computer integrated manufacturing (CIM). This framework implies the need for critical integration among all subsystems of a manufacturing enterprise. CIM can include many other manufacturing processes in its umbrella of terms, including *computer-aided drafting (CAD), computer-aided manufacturing (CAM), flexible machining systems (FMS),* and *automated storage and retrieval systems (ASRS).* (Technical Council of the Computer and Automated Systems Association of SME)

estimates for customers as to when they can expect delivery of their orders.

❏ Lower energy bills.

❏ Less scrap and rework problems.

Manufacturing managers are more inclined to implement CIM strategies when they realize a sizable list of intangibles, such as better quality control, higher competitive standing, faster product introduction, and increased flexibility in design, product mix, production volumes, and process routings.

The increasing use and availability of communication technologies, particularly the Internet, has expanded the possibilities of CIM. One new CIM product, SpinDFS, is an Internet portal that allows electronics designers and engineers to work directly with component suppliers during the design process, thereby reducing the time-to-market for a new electronics product.[2] The ability to work with suppliers during the design stage dramatically reduces the cost of the finished product.

Most successful manufacturers have recognized a need to automate at some level and improve their productivity to stay in business. To remain both solvent and competitive, manufacturers are learning to formulate ways to get new products into the marketplace at an ever-increasing rate of speed.

Concurrent Engineering (CE) Design Environment

The concurrent engineering (CE) approach to developing a new product is a way of integrating the many aspects of product design, development, and manufacture. The progression of steps central to product design must proceed concurrently within the boundaries of the manufacturing systems infrastructure. Contemporary design engineers must regard their design function as a pervasive activity influencing such labor functions as research, development, process planning, manufacturing, assembly, quality assurance, packaging, distribution, and marketing. At the bottom line, design must truly represent the dedicated delivery of a solution to a problem within the parameters of profitability, customer satisfaction, and the prosperity of the social community.

There is no one set way for a new product to come to market. Because of the competition existing today, the time allowed to develop new products is shrinking. Four years used to be common from the problem definition stage to release of the new product. Now, we are seeing time-to-market periods of six to eighteen months.

Perhaps you have heard the phrase, "If you can design it, we can build it!" The unspoken caveat or concern here has to do with cost and time-to-market cycles. When the sky is *not* the limit, DFMA becomes an important aspect of CE. Also referred to as design for manufacturability (DFM), its goal is to ensure a design inherently efficient and economical to make. The overall economy of manufacture is directly proportionate to the firm's success in controlling the manufacturability of the design. Design engineers must be articulate and knowledgeable in the fundamentals of the manufacturing processes applicable to their areas of expertise, including properties of materials; materials forming and processing; machining; fastening techniques; tool design and selection; electronics packaging, fabrication, and distribution requirements; and geometric dimensioning and tolerancing.

How does CE differ from traditional design engineering? The key difference is increased communication between the different stages of the design process. In traditional design, engineers would design a new product with little or no input from manufacturers. Today, with CE, the focus is on increased input and feedback in the design stage. The goal is to reduce the product costs by designing a product that can be manufactured from the beginning without any problems, but it is not only manufacturers that get involved in CE. Quality assurance supports all functions in the cycle by verifying design reliability, maintaining supplier assurance programs, creating and maintaining inspection procedures, and monitoring processes. Other departments that might be involved include material engineers, marketing, and personnel. Usually, the departments work together in cross-functional teams with the goal of producing a product with a short time-to-market and a low development cost.

---

[2]SpinDFS is a product developed by SpinCircuit (2000) in a collaboration among Cadence Design Systems, Flextronics International, and Hewlett Packard (HP).

Individual companies differ as to the composition of their product development teams, but no matter what the team composition, the use of CE has significantly reduced the time and cost in bringing a new product to market.

Lean Manufacturing and Agile Manufacturing

Sports teams may be referred to as "lean and mean." The connotation is one of streamlined fitness and competitive aggression. When the same word *lean* is used as a modifier for terms like *production* and *manufacturing,* we are inclined to think of these same descriptors. Lean production, in contrast with craft production and mass production, relies on teams of multiskilled workers at all levels of the firm who operate flexible, automated workstations to produce items in varying lot sizes. The craft producer employs highly skilled workers or artisans to build very high quality products customized for the consumer. The mass production methodology employs narrowly skilled people to design standardized products that semiskilled laborers will crank out in large volumes.

Lean manufacturing environments tend to use fewer resources—less space, less inventory, and less workers—to manufacture a given product than do traditional assembly lines. Also, workers in a lean system are given more authority. Instead of quality control occurring after the product is completely built, the emphasis is on prevention. Workers are encouraged to stop production if they find a defect and work together to solve any problems.

Lean manufacturing was developed by Toyota in Japan and is part of the Toyota Production System this company developed. Using lean manufacturing, a company can implement JIT and move away from a made-to-stock organization to a made-to-order organization. Since its perfection by Toyota, thousands of manufacturers have implemented lean manufacturing techniques. For these manufacturers, the result has been an increase in product quality and increased productivity. *Industry Week*, a weekly business magazine, has been giving "Best Plant" awards since the early 1990s; an analysis of the winners show that the majority use lean manufacturing techniques in their plants.

Two of the *Industry Week* "Best Plant" winners are Senco and Donnelly. Senco, which manufactures fastener systems, including nails and screws, began its work with lean manufacturing in the early 1990s. Senco reorganized its manufacturing floor and converted it into production cells. This company used to have months of inventory in storage at its plant because it built parts to a pre-designed schedule not necessarily agreeing with the demand. Since implementing lean manufacturing, the 1997 "Best Plant" winner has built what is needed on a day-to-day basis, has less than a week's worth of inventory in stock, has reduced air emissions by 99 percent, and at the same time, has increased its volume from two hundred thousand to three hundred thousand units annually.

Donnelly Corporation, the largest producer of automobile mirrors in the world, also increased its productivity through lean manufacturing. Donnelly implemented lean manufacturing into all of its plants in the mid-1990s. Its results since that time have been extraordinary: a 35 percent reduction in scrap, a 30 percent improvement in productivity, and a 90 percent reduction in customer defects.

Lean manufacturing is not, however, the only new manufacturing technique. In recent years, the word *agile* has been added to the manufacturing lexicon. It puts a slightly different spin on the decade-long focus on leanness. If a person is described as being agile or having agility, you most likely envision one who moves quickly and easily, with great balance, in a seemingly effortless manner. How does this relate to manufacturing enterprises? With more companies competing worldwide, the pressure is on for organizations to design and build the best quality products in the shortest time possible. Delivering a top performance today requires agility, as well as ability.

In industry, the agile manufacturer is fastest to market with the lowest total cost and the greatest ability to meet varied customer requirements. The ultimate measure of its agility is its ability to delight and satisfy its customers. With this explanation in mind, it is logical to conclude that agility is industry specific. The criteria for being agile in the textile industry are quite a bit different than those specified by the automotive industry. Even still, both types of manufacturing enterprises must adopt strategies that will enable them to streamline the physical flow of materials, integrate processes, and close the distances between supply, production, assembly, distribution, and customer fulfillment.

Although there are similarities, *agile manufacturing* and lean manufacturing are not the same. Lean manufacturers see themselves in partnerships with their suppliers, and they generally cultivate long-term relationships with suppliers to ensure quality. Agile manufacturers focus on the customer and meeting the customer's needs. They will continually find and switch suppliers, depending on their product needs.

Whether a company is lean or agile, several of the current technologies underlying a firm's transition are also beneficial for reducing product defects. On a short list of management techniques used to reduce product defects you might find quality function deployment (QFD), risk management,

DFMA, design for defect reduction, and statistical process control (SPC). While several of these concepts will be explained a bit further in a later section of this chapter, it is essential to note their particular influences on a company's success in achieving agile manufacturing. First and foremost, if the products being shipped are not acceptable to the customer, those customers will quickly take their orders elsewhere.

Supply Chain Management

In the 1920s, the Ford Motor Company River Rouge Complex was the ultimate manufacturing plant. Ford put all the steps in automobile production in one location, including blast furnaces, a glass manufacturing plant, a tire plant, and the assembly line. At River Rouge, Henry Ford reached a level of total self-sufficiency and a complete vertical integration in automobile production. Each part of the Model T® vehicle was built at the River Rouge complex of plants. Today, companies no longer build all the assemblies and parts going into their products. Today, manufacturers are streamlining their operations and outsourcing their manufacturing to reduce the overall cost of manufacturing.

Companies are now using supply chain management to make significant changes in the way they manufacture a product. Most companies, in contrast to the River Rouge plant, have always used suppliers to provide parts. The goal of supply chain management is to manage these suppliers so production is maximized. Inventory is a key aspect of all supply chains. Too much inventory has a negative effect on the bottom line, while too little inventory causes production delays. It is not only the amount of inventory that matters, however. It is the relationship between the manufacturer and its suppliers that changes.

How does supply chain management work? Companies today, in their quests to be lean and agile, are specializing as never before. As specialization increases, there is more dependence on external suppliers or the outsourcing of an increased percent of manufactured components. Since other companies are now supplying critical components, a manufacturer becomes dependent on these external suppliers for the overall quality of the product. Part of supply chain management includes the certification of suppliers who then are considered part of the product team.

Chrysler, for example, used supply chain management to change the way it designed and manufactured cars. Chrysler shrunk its number of suppliers from 2,500 to 1,140. It also has changed the way it works with this smaller group of suppliers. Previously, Chrysler followed the traditional American model of choosing a supplier based only on cost, and Chrysler, as most other manufacturers, would switch suppliers on a year-to-year basis. Chrysler began selecting its suppliers using the additional criteria of quality and on-time delivery. Instead of making suppliers compete for business every year or so, Chrysler chose them for the life of the car model. In addition, suppliers had become far more involved in the design process. Instead of designing the entire car itself, Chrysler had its suppliers get involved during the design phase by taking responsibility for designing the parts they were supplying. This created a cooperative, rather than an adversarial relationship, between Chrysler and its part suppliers. The results? For one of Chrysler's models—the LH car—the company was able to reduce the development time from 234 to 183 weeks, a savings of almost a year. Supply chain management affected Chrysler's bottom line as well. Its profit per vehicle skyrocketed from $250 in the 1980s to $2,110 by the mid-1990s.

Boeing is using supply chain management to achieve lean production. The company wanted to reduce the time a plane is at each workstation. Also, they desired to reduce the immense amount of inventory they had in stock. One technique was to have their parts suppliers do more work on the subassemblies before delivery. This way the subassemblies could be delivered right to the assembly line where they could be installed faster. Boeing implemented lean manufacturing in its development and manufacture of the two X-32A Joint Strike Fighter® concept demonstrator aircraft for the Department of Defense. See Figure 6-4. The results of this lean manufacturing are staggering. Boeing was able to build these aircraft in half the time of previous comparable aircraft with almost no rework, which is practically unheard of when building a plane for the first time. Also, the project came in under

Figure 6-4.
The structurally complete X-32A Joint Strike Fighter concept demonstrator is moved from final assembly to structural proof testing. (NASA)

budget—overall assembly costs were 30 to 40 percent below the company's estimates.

Chrysler and Boeing are not the only American companies utilizing supply chain management. Companies from Campbell Soup to Hewlett-Packard to Wal-Mart have used these techniques to reduce costs. Although their implementations of supply chain management differed, each of these companies built new and improved supplier relationships to increase their efficiency and, in most cases, profitability.

Just-in-Time (JIT)

Just-in-time (JIT) is a management philosophy that has been in practice in Japan's production facilities for nearly five decades. JIT was a crucial part of the systematic changes developed at Toyota beginning in the 1950s. The overall system is called the Toyota Production System, and many other companies across the globe have imitated it since its popularization in the 1980s. Interestingly, the model for the Toyota Production System was the American supermarket. Taiichi Ohno, a Toyota executive, traveled to the United States in the mid-1950s to visit American automobile plants. During his trip, he was impressed with American supermarkets and the way shoppers got the *products* they wanted in the *amount* they wanted *when* they wanted them. Also, he noticed that the shelves were restocked quickly, either from a stockroom or through deliveries. This observation led him to think of the Toyota car factory as a type of a supermarket where each workstation would get its supplies from an upstream workstation—an ultimate expression of the "supply and demand" principle at work.

JIT is an attempt to achieve agility by reducing work-in-process (WIP) to an absolute minimum. It has gained notoriety and praise in a number of American firms

in the past decade. A core objective of JIT is the elimination of waste that interferes with agile performance. Waste appears in the manufacturing arena in many forms, including lead times for start-up, setup, and changeover periods; defective parts; excess inventory; unsatisfactory raw materials; and unnecessary material handling.

Within a JIT operation, production processes are viewed as the only means of adding value to the product being manufactured. The indirect, but necessary tasks of transportation, inspection, and storage are looked upon as wastes that should be minimized or eliminated wherever possible. JIT managers recognize the dynamic nature of manufacturing and subscribe to a philosophy of continuous improvement, synchronization, and simplicity. In practice, JIT focuses on achieving integrated, highly consistent, short cycle operations requiring minimal WIP inventory. The JIT approach is a departure from conventional systems in which the emphasis was on automating indirect operations that do not add value to the product, rather than eliminating the indirect operations. These conventional efforts may lead to "islands of automation," where there is a lack of continuity in the flow of materials between production stages.

JIT assumes the production rate at the final assembly stage is even. External to the JIT plant, the company's suppliers are expected to provide parts in smaller batches on a continuous basis. A manufacturer's close rapport with its suppliers may ultimately ensure quality materials, as well as patience from the suppliers when they are asked to deliver smaller amounts more often. The hallmark of JIT purchasing seems to require the steady purchase of parts in small lot sizes (just in time for use on the line), as opposed to conventional purchasing practices in which raw materials are ordered from suppliers in anticipation of future production.

A major drawback of this approach occurs if a key supplier's plant shuts down due to a labor strike, natural disaster, or financial failure. Also, since there are reduced inventories, there must be increased quality, both for the parts suppliers and on the manufacturing floor. Since parts are ordered at the exact quantity needed, defects become more disruptive to the manufacturing line.

Another feature of JIT is the reduction or elimination of setup times at the individual machines. Ideally, all setup should be done off-line. If a plant is making two different products on a JIT line, then the setup for product 2 will be occurring off-line, while product 1 is being assembled. Then, it is a quick switch to product 2 on the main line. Although this is a fairly simple example, it illustrates a fundamental difference of JIT from traditional manufacturing. Since the setup time does not add value to the product, it should not be done on the line. The line can be more efficiently used to produce the product. The reduced setup times lead to another feature of JIT—reduced lot sizes. Since setup times are reduced or moved off-line, a smaller lot size of each product can be produced at one time. This allows for

more flexibility on the production floor and also allows the company to produce the products needed. Since the production floor has more flexibility, a company can switch from one product to another if the sales demand changes.

One critical component of JIT production control is the use of the *kanban*. The word *kanban* translates to mean "signboard." An item called a kanban card or ticket usually accompanies WIP parts. A withdrawal kanban reveals the type and amount of product the next process should withdraw from the preceding one. A production kanban specifies the type and quantity of product that the upcoming process must produce. See Figure 6-5. The kanban tickets, illustrated in Figure 6-6, are used in place of job orders and routing sheets, and they emphasize small lot sizes. Subsequently, less paperwork is required to coordinate planning and control operations. JIT is a pull system, in which the user department pulls the parts or subassemblies from supplier departments. The items necessary to keep production on schedule are pulled from the preceding workstation only as required. This contrasts with the push system

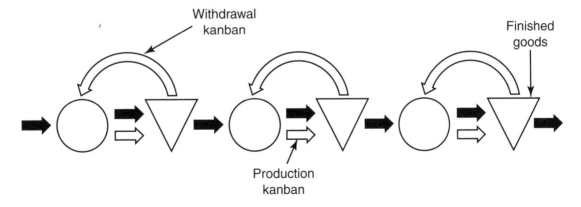

Figure 6-5.
This is a simple illustration of kanban control. The dark arrows represent movement of parts, while the light arrows show the circulation of kanban. Machines are shown as circles, and buffers are shown as triangles. The last buffer is the finished goods inventory.

Figure 6-6.
Here is an example of kanban move tickets, which are an essential ingredient in just-in-time production. A typical kanban card will contain information such as part number or name, quantity, previous operation, and production routing and schedule. (New United Motor Manufacturing, Inc.)

characteristic of traditional manufacturing, in which items are pushed indiscriminately onto the succeeding workstations, even if that center is not ready to receive and process them. The basic difference between a pull system and a push system is their relationships to demand. A pull system like JIT starts production as a reaction to current demand, while a push system starts production in anticipation of future demand.

Most companies that have adopted the JIT philosophy have also established quality circles in their work forces. These employee groups are encouraged to make suggestions that may lead to improvements in the company's operations. They strive toward the goal of simplified synchronous production and often help to improve design methods, cut down lot sizes, reduce lead and setup times, minimize scrap losses, rearrange process flow, and solve vendor problems.

Through most of the last two decades, JIT strategies have been used for mass-produced items in the United States, Japan, and parts of Europe. Success stories usually make claims about increased labor productivity, reduced inventories, reduced quality rejection rates, and reduced need for plant space. GM implemented the JIT strategy in the early 1980s and reduced its annual inventory-related expenses from $8 billion to $2 billion in a few short years. Black and Decker reported in the mid-1980s that, with the application of JIT concepts, it was able to increase its volume of throughput by 300 percent and reduce manufacturing lead times by 50 percent. Harley-Davidson took its top management team to Japan in the early 1980s to study JIT factory operations. The company subsequently regained lost market share, and the motorcycles it built in the early 1990s were 99 percent defect free, as compared to 50 percent prior to the trip to the Far East. Dupont's May Company, a textile producer in South Carolina, utilized a variety of JIT principles to improve its manufacturing performance in the early 1990s. The company reports that its WIP declined 96 percent, while quality improved and significant savings in the operational budget were realized. Appliance makers, including General Electric (GE), Westinghouse, and Motorola, have also made great strides in the implementation of JIT manufacturing and purchasing methods in their plants. Despite the numerous advantages cited for using JIT, firms still report a variety of limitations and shortcomings, including faraway suppliers, unreliable freight systems, loss of individual autonomy, poor quality of delivered parts, and worker resistance to change. This simply verifies that no philosophy, JIT or otherwise, is a panacea or cure-all for the many challenges confronting contemporary manufacturing and production enterprises on a daily basis.

Material Handling Procedures

Material handling refers to the manual, mechanical, or automatic treatment of material when the intended purpose is to move, hold, or control objects either individually or collectively. Material handling may entail controlling material quantities, location, timing, sequence, orientation, alignment, or condition. *Bar coding* is the most widely used and accepted automatic identification technology, particularly in material handling. Bar coding is an excellent example of a technological tool that has revolutionized the movement of material through all areas of a CIM facility. The familiar black bars and white spaces all appear to look pretty much the same. There are, however, a variety of code types, called symbologies, created to handle different data sets for specific applications. The most common symbology is the Universal Product Code (UPC) widely used in retail functions. Data gathered by bar code scanners is being used to control and drive complex conveyor systems. Critical data is available instantly and automatically for sorting, order verification, distribution, and shipping. Virtually every facet of the material handling sequence is affected by bar coding.

It has been estimated that, in traditional manufacturing environments, more than 90 percent of a product's time spent in the manufacturing cycle is consumed by material handling and associated activities. For this reason alone, the effects of material handling on product quality demand serious attention. The major negative effect of material handling on product quality may cause damage to the point where the product is no longer fit for use. Inappropriate material handling methods may lead to product quality degradation as a result of positioning errors in fixtures or jigs, weather, dust, static electricity, contaminating vapors or obsolescence.

The selection of suitable material handling procedures requires the thorough analysis of materials and process requirements. Ideally, material handling systems should be viewed as an integral part of process and facilities design. Specifically, quality assurance specialists must take into account the number of times each product is handled, the distances over which moves occur, and the length of time the product spends being handled. Analysis and control of these three main variables should result in reduced incidence of product quality degradation.

Some of the most common material quality related characteristics a manufacturing firm must deal with are fragility, surface finish, sensitivity to electrostatic discharge, sensitivity to magnetic fields, environmental sensitivities, and shelf life. From a material handling perspective, fragility may translate into impact, vibration, temperature changes, friction, and weight support tolerances. Furthermore, a product's sensitivities to weather exposure, moisture, temperature conditions, ambient pressures, dust, and contaminating vapors demand specific tolerances. These tolerances subsequently lead to decisions regarding appropriate material handling methods, movement paths, and storage locations.

Two direct strategies used to enhance product quality throughout the manufacturing process are to *minimize* the number of times materials must be handled and to *shorten* the distances of travel between workstations. Several of the CIM and lean manufacturing strategies described earlier seem to respond directly to these recommendations.

Material Requirements Planning

Material requirements planning (MRP) is a technique for planning future purchase orders and manufacturing lots according to what is required to complete a master production schedule. *Manufacturing resource planning (MRP II)* is a dominant application software structure being used by manufacturing managers. MRP II commonly includes planning applications, customer service applications, production control, purchasing, inventory product data management, and various financial functions. Essentially, MRP II creates a dynamic closed loop management system integrating all of the major subsystems of the organization.

While Japanese manufacturers were adopting JIT in the 1970s, American manufacturers were pursuing a different approach to production control. This approach was MRP. MRP is a computer-based system designed to handle the ordering of supplies and the scheduling of work in the manufacturing plant. An MRP system builds on a product's list of materials, called a bill of materials (BOM). The MRP system integrates all the individual BOMs and the forecasted demand for the individual products into an overall master production schedule. See Figure 6-7.

Demand forecasting provides an estimate of the demand for each type of product sold. Long-term forecasting is done by an organization to develop an overall plan for one to five years, to allow the organization to acquire new facilities, hire people, and buy new equipment. Short-term forecasting is usually done from one month to one year and focuses instead on the manufacturing of the company's products.

Once demand is determined, an aggregate production plan can be developed. Typically, the sales and marketing departments make demand forecasts that are a combination of orders on hand and anticipated forecasts. These estimates must be reconciled with the manufacturing constraints of the plant (such as plant capacity and work force). Since a firm can make thousands of products, this data is usually combined at some manageable aggregate level—therefore, the term

aggregate planning. One of the major uses of an aggregate plan is to level the production schedule so production costs can be minimized. Since the aggregate plan does not provide a plan for individual manufactured products, this plan must be subdivided into another plan specifying quantities of manufactured items by time period. This resulting plan is the master production schedule. The master schedule states which items are going to be manufactured, the quantities of these items, and when these items are needed. The master schedule divides the planning into several time periods. The duration of these time periods vary greatly from company to company.

After the master production schedule has been developed, rough cut capacity planning is performed to determine if the master production schedule is feasible. In rough cut planning, some broad guidelines are used, with the guidelines usually relating to a key resource (such as equipment or direct labor). If the master production schedule meets the rough cut capacity, the detailed production requirements can be planned.

After MRP is completed, a detailed capacity analysis is done to determine whether the production schedule exceeds the capacity of the plant. If the capacity limitations cannot be resolved, the master production schedule will be altered, and the process will repeat itself. The process of evaluating the capacity requirements begins with the due dates of each order. Using lead times, BOMs, and routings, each order is back scheduled from the due date through the required operations to determine when individual workstations will be used.

As time passed, various operational functions were added to extend the range of tasks in MRP systems. These extensions include master production scheduling (MPS), rough cut capacity planning (RCCP), capacity

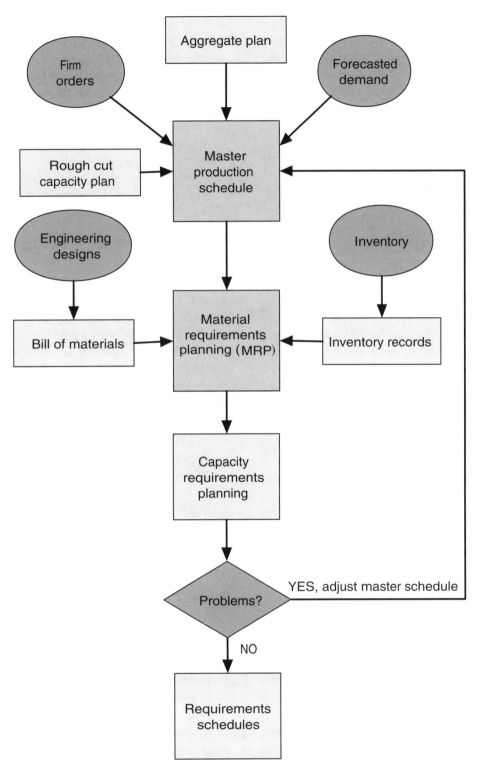

Figure 6-7.
This is an overview of material requirements planning.

requirements planning (CRP), production activity control (PAC), and purchasing. The combination of these functions was termed *closed loop MRP*. With the addition of certain financial modules, as well as the extension of MPS to deal with master planning and the support of business planning in financial terms, this extended MRP was labeled manufacturing resource planning (MRP II). MRP can be used with JIT, although the structured environment of MRP makes JIT more difficult to accomplish. In these cases, JIT is used for parts ordering, while the schedule is maintained through MRP.

Which is better: JIT or MRP? This is an impossible question to answer. In an assembly line where production is constant, MRP provides a stable way to manage and plan for production. In a factory where different products are produced on a daily or weekly basis, JIT is generally more successful. Either approach requires extensive training and expertise to operate efficiently.

Emergent Manufacturing Technologies

The future state of the manufacturing environment, especially for those firms that have already implemented lean production technologies and JIT strategies in all phases of their operation, will most assuredly feature aspects of soft, custom production. Soft manufacturing brings yet another approach to the automation mix, and it delivers with it much greater agility to the plant. Sophisticated software-driven computer networks used and operated by a skillful and dexterous human work force represent the backbone of what some experts refer to as the digital factory. Companies are discovering they can customize products literally in quantities of one, while churning them out at mass production speeds.

Robots, if they are found at all, play a much less omnipresent role in the soft manufacturing firm. Over the last twenty years, a number of U.S. manufacturers realized that an overdependence on automation through the installation of huge *flexible manufacturing systems (FMSs)* actually cost them profits. Large, expensive, complicated systems may be more vulnerable to failure than smaller, more manageable cells of machines. In places where dexterity and judgment are critical, human operators are back in assembly. Now this is a paradigm shift— "Humans replace robots!" Several well recognized companies are currently reaping the rewards of soft or custom manufacturing based upon their abilities to outdistance foreign competition in such critical measures as time-to-market and manufacturing flexibility.

As CAD equipment and RP machines continue to evolve, the world will enter yet another new age of custom manufacturing. Where can an avid cyclist find a replacement part for an upscale Italian Ducati motorcycle when it breaks down on a country road in northern Maine? In a few short years, the rider may simply catch a ride to the nearest town and hand over a floppy disk containing all pertinent design files for the motorcycle to the operator at the town's corner parts factory. And you thought twenty-four hour copy shops were a convenience!

Contemporary manufacturing is moving toward an era of mass customization. While this sounds like an oxymoron, the earlier examples of agile manufacturing hint at a trend toward customized fabrication. The entire suite of sciences and technologies we have referred to as *rapid prototyping (RP)* provides the foundation to this developing approach to manufacturing. This is indeed new territory, but it is ground where U.S. research and development teams have already established solid footing. Researchers are presently experimenting with and creating intelligent materials that can anticipate failure, repair themselves, and adapt immediately to changes in the environment.

Rapid Prototyping (RP)

Before any firm commits to the mass production of a new item, it builds different prototypes for design, ergonomics, safety, ease of assembly, and fitness for use (quality). In recent years, rapid prototyping (RP) has emerged as a well-regarded manufacturing technology in the CE design environment. The aim of RP systems is to make full use of prototypes early in the development stage to identify errors in design and make necessary modifications. This expanding technology has the potential to allow designers to produce a prototype within minutes of completing a *computer-aided design (CAD)* drawing of the part, thus obtaining a physical model of a proposed design, while avoiding the lengthy and costly use of conventional tooling and casting processes. See Figure 6-8.

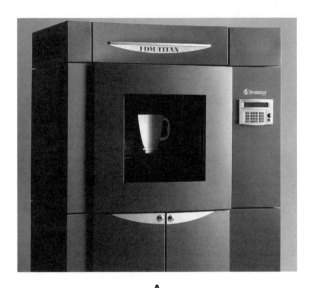

A   B

Figure 6-8.
Rapid prototyping (RP) machines use high performance engineering materials to create parts with superior strength, heat resistance, and chemical resistance. A— The FDM Titan from Stratasys is the only RP system that can build parts from high-temperature, high performance, and durable engineering plastics, such as polycarbonate, polyphenylsulfone (PPSF), and ABS plastic. (Joe Hiemenz, Stratasys) B—This partially completed prototype of a model airplane was built using a RP machine. (Phoenix Analysis & Design Technologies)

RP systems use data from a three-dimensional (3-D) CAD file to construct a model. Charles Hull patented one of the first RP systems in the mid-1980s, with the founding of 3-D Systems, Inc., to develop commercial applications for the process he called stereolithography. The number of commercially available RP systems has increased considerably to include laser modeling systems, solid ground curing, fused deposition modeling, fast casting, and laminated object manufacturing. The purchase of these highly sophisticated systems exclusively for internal use is often prohibitively expensive for many companies. For this reason, a large number of companies outsource their rapid prototype manufacturing requirements.

Remember our broken motorcycle? In an example of the use of prototyping to replace the motorcycle part using fused deposition modeling, a machine tool would receive a geometric description of the broken part from the CAD file on the disk. The program then divides the model into evenly spaced layers, each as small as a few thousandths of an inch. The model is built layer by layer from the bottom up. The program instructs the tool to deposit thin layers of liquid, one layer at a time, which subsequently fuse together to build a complete part.

Another RP technique is solid ground curing, developed by Cubital America. In this technique, the models are made from a material that hardens after being exposed to ultraviolet light. Ford Motor Company used solid ground curing to build an air intake model for its new 4.6 liter V-8 engine. Ford reduced the cost of prototyping this part by 33 percent because the company was able to create the prototype directly from the CAD file.

Almost every issue of *Machine Design*, a trade weekly, contains an advertisement for new and improved types of product prototyping, and there is no denying the benefits of this technology. Prototype parts can save costs because they allow the engineer and manufacturer to see the final product early in the design stage, thereby saving expensive revisions and rework. RP itself has expanded into two additional areas: rapid tooling and rapid manufacturing. Rapid tooling refers to the use of RP to develop molds for use in production. Rapid manufacturing is the use of RP for low volume production. There is no doubt that its use will continue to expand in the future as more companies exploit this technology to decrease costs and reduce development time.

Cellular Manufacturing

Cellular manufacturing is a type of equipment layout in which the machines are grouped into cells rather than being placed on an assembly line or divided into different functions (for example, all drills together or all lathes together). The parts that will be produced in a particular cell determine the layout of a cell. In order to have an effective cellular arrangement, a company has to group their products that use similar manufacturing processes. All parts in one group (called a family) will follow the same route in the cell, although individual products may spend more time at a particular machine than other products in the same family.

Historically, the layout of manufacturing facilities was classified as a job shop, flow shop, or fixed layout. Cellular manufacturing is a new type of production layout. *Group technology (GT)* is used in order to achieve cellular manufacturing. GT is an approach to manufacturing that seeks to maximize production efficiency by grouping together similar and recurring tasks, procedures, problems, and bottlenecks. A key feature of GT is the segregation of parts according to their designs, manufacturing features, or a combination of these. When similar parts are grouped together, each collection can ultimately share setups and machine tools, thus reducing production costs. GT is applicable to both automated and nonautomated manufacturing and can be used in new or existing facilities.

GT has attracted a great deal of interest from manufacturing firms because of its proven capacity to simplify material flow on the production floor. The GT approach is a marked improvement over traditional batch processing methods. Experts estimate that most manufacturing is still done in small batches ranging from a single workpiece to several thousand pieces. In many cases, these parts cannot flow smoothly through the manufacturing process since different parts require different setups or must be transferred to another machine. The application of computerization to manufacturing more than twenty-five years ago enabled managers to improve the production of both small and large batches through the use of scheduling software, sequencing software, and MRP systems.

It also became feasible for companies to identify and track the thousands of different parts being produced through the use of GT methods. Design engineers have found they can use GT systems to determine whether an existing part could be used in a new application, thus eliminating the need to design a new one. This potential to eliminate design duplication and the parallel need to build a new jig or fixture can yield significant economic benefits.

Basically, a GT database is a computerized filing system that speeds up the retrieval of parts information, facilitates the design process, and enhances the communication between individual functional areas inside the plant. GT systems also improve the accuracy of process planning and aid in the creation, layout, and operation manufacturing cells. GT is a critical building block for

CIM. A GT database can contain detailed information about the parts a company produces, the machines and equipment available to produce those parts, and the best processing methods to complete the job on-time as promised. Other advantages of adopting GT strategies include reduced materials handling costs, improved flow of materials, small in-process inventory, reduced changes in production planning and control, less floor space needed, reduced design modifications and easier design retrieval, more accurate cost and delivery estimates, and improved quality and productivity through the workplace.

One of the salient benefits of GT is a direct result of using a formal coding system in which each part receives a numeric or alphanumeric code describing specific characteristics or attributes. To be most useful, the code should be able to describe the part from both a design *and* a manufacturing perspective. Such physical features as the external and internal shapes, dimensions, threads, grooves, and splines describe the geometric form (morphology). The structure and chemistry of the raw material, the surface finish, various tolerance requirements, and the need for heat treatment are examples of manufacturing specifications. It is easy to visualize a code with a large number of characters if one tries to capture all significant attributes. A number of GT coding schemes have been developed, including MICLASS (twelve digits), DCLASS (eight digits), Vuoso Praha (four digits), and Opitz (five plus four digits).

Figure 6-9 provides an example of a principal GT coding system called Opitz serving to illustrate this key aspect of GT. Once the parts are coded, they are further grouped into families. As you can see, the Opitz code classifies a part according to its external shape, internal shape, the machining of its surface, and its additional holes

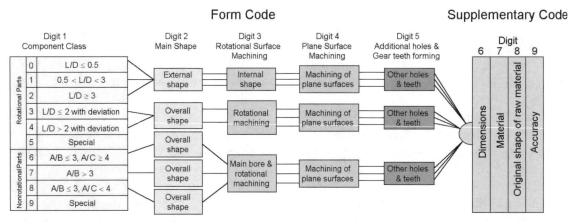

Figure 6-9.
This chart shows the basic structure of the Opitz code.

and teeth. The part receives a score from zero to nine for each digit in the Opitz code. Under Digit 2, Main Shape, a code of zero would represent a smooth external shape, while a code of eight would indicate the part is threaded like a screw. The combination of codes for each digit then gives a complete description of the part. The special codes following identify the specific design attributes of the part, including its dimension and material. Using GT codes, it is far easier to locate one part fitting the current needs at hand. Then the engineering department can avoid designing a brand new part. As a rule, GT codes are assigned to both purchased items and fabricated parts.

A simplified version of GT in a job shop setting involves the sequencing of similar parts on a machine or series of machines. This job shop facility can process a wide variety of parts since it is still a job shop, but the parts are processed in families (using the GT code) to realize some of the benefits of GT.

A more sophisticated approach to GT entails the creation of manufacturing cells. A manufacturing cell is a collection of machine tools and materials handling equipment grouped together to process one or several part families. Transfer of the piece from one process step to another within the cell and possibly on to a different cell can be automated. Essentially, a manufacturing cell is a hybrid production system meeting the needs of a specific firm. The development and application of manufacturing cells are dependent on the type of manufacturing operations performed, the life cycle of products fabricated, the product mix, and projected customer demand. Cells are a blend of job shops producing a large variety of parts and flow shops dedicated to the mass production of one product. Figure 6-10 portrays one view of the differences between manufacturing parts in a job shop using a conventional layout and a cell shop subscribing to GT principles. A fully operational flexible manufacturing cell speeds up the manufacturing process faster, since parts are moved quickly and systematically from one workstation to the next. This allows the manufacturer to reduce inventories of partially finished parts representing significant cost savings.

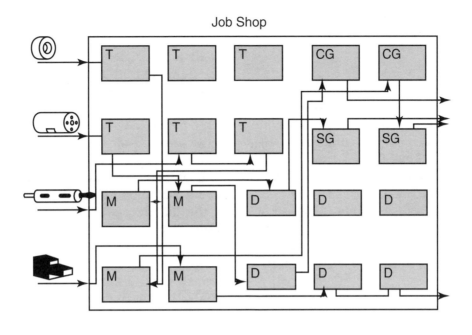

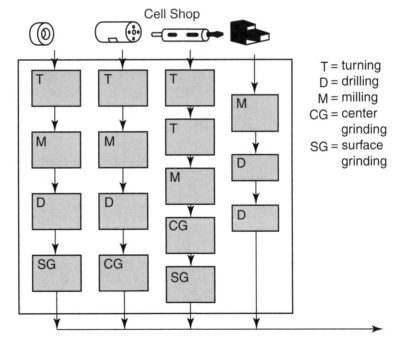

Figure 6-10.
This simplified representation illustrates the movement of parts through a traditional job shop compared to the movement through a group technology–oriented cell shop.

Total Quality Management (TQM)

Along with the strategic implementation of manufacturing technologies, the management of productivity and quality has emerged as a major business strategy in numerous organizations. There appear to be several forces driving this ongoing trend. The first force is related to an intensive amount of global competition in all industrial sectors. There has been an increasing demand for better quality goods and services at either the existing price or at a significantly lower one.

A second force stems from the fact that production costs have gone up considerably over the past several years. To offset this situation, manufacturers are investigating new ways to improve the productivity of labor, as well as to apply a variety of innovative manufacturing technologies. Third, the level of consumer education has intensified, leading to a more acute awareness of quality. Public acceptance of manufactured goods is based on strict conformance to specifications. People commonly expect the products they purchase to be 100 percent defect free, and they quickly lose confidence in a manufacturer when these expectations are not satisfied.

A widely accepted generic definition of quality is fitness for use. The manufacturer views fitness for use from the perspective of its own ability to process and produce finished goods with less rework, less scrap, minimal downtime, and high output. Customers view fitness for use whenever they consider things like product durability, availability of spare parts, identity, and comfort. Two main aspects of quality are the quality of design and the quality of conformance.

Quality of design involves the features obtained through changes in or manipulation of design parameters. Differences (not always improvements) in quality may be achieved by changing elements, such as the size of the item, the materials used in the product, the equipment used in production, and the tolerances during manufacturing.

The quality of conformance is a measure of the extent to which the product conforms to the specifications and tolerances required by the design. Many factors influence this measurement, encompassing training, employee motivation levels, complexity of the production process, and the quality assurance system being used.

Consumers evaluate a product's fitness for use through a review of certain quality characteristics. Depending on the item, any number of features may be critically considered. If the buyer expects the product to last for a certain length of time, quality elements, such as its warranty, serviceability, reliability, and maintainability, are important. Ergonomic features, including comfort, size, and ease of use, fall into another category. Finally, one should not overlook the sensory-oriented qualities, such as the product's color, taste, fragrance, beauty, and appearance. If you have ever been involved with the selection and purchase of a new automobile, you should remember these things well; there is nothing quite like the new car smell!

There are a large number of quality management functions that must be executed at different levels of the manufacturing process. The corresponding control activities and the types of data collected at each level tend to vary considerably. Quality planning is a function parallel to process planning. For each workpiece handled, the measurement parameters, tolerances, and test sequences must be determined. It is also necessary to establish sampling plans, process capability, studies, and the amount and type of quality data to be accumulated and stored. This planning operation is normally completed by the company's quality expert with or without the use of a computerized data storage and retrieval system.

As previously described, quality is the sum of all attributes and characteristics of a product or service contributing to its usefulness or its ability to perform certain functions. Quality management is therefore a regulatory process through which performance is first measured and then compared with preset standards. If necessary, corrective action is taken. The backbone of an organization's quality management system is the internal, national, and international standards it aspires to guarantee. These standards are essentially akin to a contractual agreement between the manufacturer and the customer. Adherence to publicly accepted standards is the aim of a quality assurance system. In recent years, many organizations have adopted the International Organization for Standardization ISO 9000 standards as the basic foundation for their TQM system.

An efficient quality management system may be described as a complex adaptive control loop. All critical properties and product performance standards are identified and checked throughout the factory:

1. At receiving.
2. At the location where the parts originate in the manufacturing process.
3. At subassembly stations.
4. During the final acceptance test of the finished product.

Within the system, when the cause of any defect and its place of occurrence are known, measures for its correction or elimination are initiated. Requisite corrective actions may take place at the factory floor level or may extend back into product design. An important aspect of quality management is reaction time. This is the length of time (seconds, minutes, hours, or days) between the instant the defect is recognized and the instant it is corrected. Coordinated efforts between plant managers and shop personnel, combined with advanced computerization, will eventually lead to shorter reaction times and more efficient quality assurance programs. JIT production goes hand in hand with good quality. Poor quality requires buffers of inventory ready for use when bad or unacceptable parts are found during assembly. As stated earlier, one goal of lean production and JIT is to operate with greatly reduced inventory.

A new technique pioneered by Motorola has been added to the TQM movement—Six Sigma. Six Sigma refers to the amount of variation existing in a product. The average product at a company usually has a variation of four sigma—this equates to more than six thousand defects per million. In the late 1980s, Motorola started a campaign to reduce defects in its products. The company used a quality system review to assess the effectiveness of all the major business units in Motorola. It developed the Six Sigma concept to improve its processes and increase product and process quality. Six Sigma is equivalent to products that are 99.9997 percent acceptable or, in other words, have only 3.4 defects per million opportunities. In 1988, Motorola was the winner of the first Malcolm Baldrige National Quality Award.

The Six Sigma concept has spread to dozens of leading companies since its establishment by Motorola. At Seagate, a disk drive manufacturer in Scotts Valley, California, engineers used Six Sigma to reduce the turnaround time on disk drive repairs. GE estimated that its profits in the late 1990s increased by $6 million because of its use of Six Sigma, while AlliedSignal recorded more than $800 million in savings. Six Sigma represents a higher level of commitment by an organization because it requires a paradigm shift in thinking about quality. Six Sigma focuses on selecting projects carefully targeted to improve selected business operations. The goal is to eliminate defects before they occur in an organization. Six Sigma is often implemented in conjunction with lean manufacturing.

Statistical Process Control (SPC)

While quality assurance has traditionally been accomplished through product control (such as inspection), the focus of process control is on individual operations and the roles they perform in manufacturing. Process control strategies are prevention-oriented as opposed to inspection-driven. The ultimate goal of process control is to have each operation functioning within its normal capability limits. This type of control is often referred to as *statistical process control (SPC)*.

STATISTICAL PROCESS CONTROL (SPC)

An approach to quality assurance differing from the traditional policy of product control via inspection. The focus of process control is on individual operations and the roles they perform in manufacturing. SPC is prevention-oriented versus inspection-driven. It involves statistical analysis and increased personnel responsibility to ensure all equipment and processes are operating within acceptable limits. SPC implementation focuses initially on two or three critical parameters in each process that must be controlled.

As a theory, SPC is over sixty years old. It is based on work done by Walter Shewhart during the 1930s in Bell Telephone Laboratories. SPC principles were widely adopted only about twenty years ago. However, since that time, they have become a pervasive feature of manufacturing plants, as well as more commonly used in other industries.

Process control involves both statistical analysis and increased personnel responsibility for making sure equipment and processes are operating within acceptable limits. It provides a unifying and validating factor for the overall quality program. Process control provides management with critical feedback regarding the condition of the plant's manufacturing equipment and machinery and may pave the path toward excellence in timely machine tool replacement. Managers who are committed to their quality assurance programs are beginning to realize the "run-it-til-it-drops" strategy for machine replacement is not the way to stay ahead in the international manufacturing race.

The decision to implement process control should focus initially on the identification of two or three critical parameters or characteristics in each process that must be controlled in order for the process to adequately fulfill its function. Next, the operating ranges of these critical parameters must be specified. After this step, it is possible to go back and evaluate how each process can be improved to tighten the acceptable operating ranges and enhance overall product consistency. Managers can systematically utilize this input to plan and rationalize future capital improvements.

Process control is fundamentally different from traditional product inspection systems. The basic philosophy that seems to permeate an inspection-based system is first you make the product, then you check it to see how good it is. In many cases, if you make enough of it, you ship it out the door unless it is totally disastrous. This type of attitude and approach to manufacturing lies at the heart of the problems many American firms have had trying to maintain quality. As an alternative, the basic premise of process control is that the manufacturer knows enough about the operation to keep the system up and running and how to keep it within certain limits. Also, under a process control system, the quality monitoring is delegated to the worker on the production floor. In this system, the worker knows how to continually monitor his process to ensure quality. The worker measures the critical parameters at his station in order to assure that all products are built with an acceptable quality standard.

The process control system demands quality by design. CE, discussed previously in this chapter, has been effectively used to increase the quality of the product through the design process. Design for defect reduction can be an effective way to quantify and minimize the probability of product defects. This requires an analysis of the defects that will most seriously affect customer satisfaction and an examination of the manufacturing processes to define which ones might cause defects. Those manufacturing operations unable to stay within tolerances should be corrected or not be allowed to keep running.

Quality Function Deployment (QFD)

As stated earlier, today's lean and agile manufacturers are already thinking about quality outcomes during the product design and development stage. A technique known as *quality function deployment (QFD)* has emerged in recent years. QFD is a systematic way for the manufacturing firm to identify customer requirements and convert them into design and manufacturing needs. Essentially, it is a management-planning tool that can be used in any phase of production. QFD helps all concerned personnel identify the critical design parameters that are then optimized through quality engineering to minimize variation during production. To ensure customer expectations are met in an economical fashion, adequate resources are focused on any areas that could cause failure.

The main feature of QFD is its focus on meeting customer needs, called the "Voice of the Customer." The House of Quality is the most recognized tool used in QFD. A multifunctional team takes the customer requirements obtained from market research and benchmarking and converts these requirements into a new product design using the House of Quality. After the customers' needs are translated into technical and design requirements, the team looks at the relationship between these two categories. Say, for example, we were designing a new pencil. Two customer requirements might be that the pencil was easy to sharpen and that it was hard and lasted a long time. After translating these two customer needs into technical requirements, the team might determine that these two needs conflict. If a pencil is harder, then the material used has to be stronger, and this would make it harder to sharpen. This type of information would go in the top of the House of Quality, in the technical correlation matrix. Figure 6-11 shows a sample House of Quality, but most companies adapt this form to serve their specific needs. A key to all implementations is the focus on the customer.

International Organization for Standardization (ISO) 9000 and Standards-Based Manufacturing

Another significant change in the past twenty years has been the rise of the quality standards movement. International Organization for Standardization (ISO) 9000 is the most popular quality standard in the world, and it has been adopted by thousands of organizations. ISO 9000 is actually a set of standards. The most recent standard, ISO 9001: 2000, applies to manufacturing, service firms, and public agencies. ISO 9001: 2000 is a series of three interrelated standards.

A company would first develop a quality system meeting the standards. There are

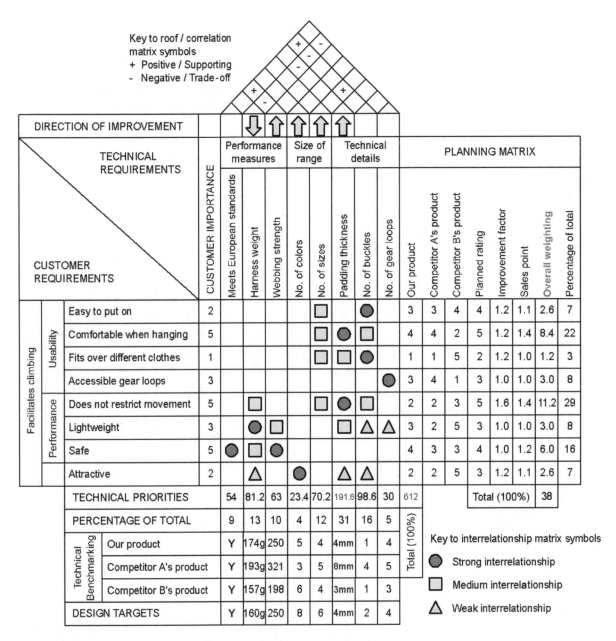

Figure 6-11.
The House of Quality is the most recognized tool in quality function deployment. (Dr. A. J. Lowe MEng PhD)

eight sections to ISO 9001: 2000. The first three sections provide background. Section 4—Quality Management System—requires a company to establish and document its quality system. The focus is on continuous improvement, both of the product and of internal processes in the company. Section 5 focuses on the responsibilities of management and requires the company to satisfy their customers, establish a quality policy, and perform management reviews. Resource Management is Section 6—the standards state that without adequate resources, a quality plan will not be successful. Section 7 relates to product design and development, while Section 8 focuses on measurement, analysis, and improvement.

Once the quality system has been developed, the company carries out an internal audit to make sure the quality program is working. The last step is to have a visit by an external evaluator. The evaluators audit the company's quality system. If the auditor approves the new system, then the company receives an official certificate that the company is ISO 9000 registered.

Since ISO 9000 certification indicates that a company meets international quality standards, companies are using their achievement of ISO certification as a marketing strategy. Even small companies are jumping on the ISO 9000 bandwagon. See Figure 6-12. Also, for some companies, ISO 9000 certification is now required to get contracts, particularly in Europe, where the European Community (EC) pushes ISO certification. In certain industries, the EC requires ISO 9000 certification because the products have stringent requirements for safety and reliability. Even if a small company does not sell directly in Europe, ISO certification can still be useful because the small company might sell to other companies that sell in Europe. Also, as a large company achieves ISO 9000 certification, often it pushes down this requirement to its suppliers.

With the globalization of business, ISO certification creates a standard companies can rely on as they buy goods and services from across the world. Not surprisingly, the number of ISO 9000 certified companies is growing each year. In the late 1990s, the number of ISO 9000 certificates increased by over seventy thousand—bringing the total number of ISO companies to almost four hundred thousand. There is no doubt that

Figure 6-12.
ALOM Technologies, a small manufacturer and distribution center in Milpitas, California proudly shows it has received ISO 9002 certification. ALOM helps high-tech and e-commerce companies get their products manufactured and distributed to customers. ISO 9002 is the ISO standard for companies only participating in production, installation, and servicing.

ISO certification will remain a large part of the manufacturing world in the future.

Advances in Automation

It's hard to believe, but the robot is well into its middle age. In 2002, it turned forty! According to Joseph Engleberger, who is often referred to as the father of robotics,

Unimation installed the first industrial robot in the early 1960s on a production line at GM. Since that time, more than five hundred thousand robotic units have been set to work in factories around the world. Robots are being used for a variety of operations. They are familiar entities in chemical processing plants, automobile assembly lines, and electronics manufacturing facilities. The most popular applications worldwide include arc welding, spot welding, spray painting, parts and tool handling, and assembly. Robots are commonly used to replace human labor in repetitive and potentially dangerous operations. Robots can work in hazardous conditions in which high temperatures, toxic chemicals, or radioactive substances are present. Robotics has also moved into warehousing, shipping, and receiving functions. Robots are installed in some offices and used in several service industries, including hotels, hospitals, restaurants, and landscaping.

Industrial robots are a critical component of factory automation. Alongside peripheral technologies like CAD, numerical control, and automatic identification, these ever improving factory floor helpers come in all shapes and sizes. Very few of them bear any resemblance to the humanlike androids pictured in science fiction movies. Most of them can be described as computer mechanisms to which limbs, tools, and other appendages have been attached. Although somewhat less exotic, a more technical term for a present-day robot is an *automatically controlled programmable manipulator*. The most commonly accepted definition still in use today for an industrial robot is "a reprogrammable, multifunctional manipulator designed to move materials, parts, tools, or specialized devices through variable motions for the performance of tasks."[3] Robots operate with several degrees of freedom and may be either fixed in place or mobile. The key word in this definition is

reprogrammable. The reprogrammable robotic feature differentiates industrial robots from other forms of so-called hard automation that do not allow for easy changes. Research efforts in the fields of AI and expert systems will continually expand the capabilities of these steel collar employees.

Regardless of their size or functions, industrial robots have three main parts, as shown in Figure 6-13. The main body of the robot is called the manipulator. Its base may be fastened to the floor, modified to move on tracks, or hung from an overhead support. The power mechanism moving the arm of the robot is contingent on its size. Hydraulic power is used for heavier arms; electric or pneumatic power is used on lighter units. Electrical robots can return to a given position repetitively and do it quicker and more accurately than can hydraulic or pneumatic robots. On the other hand, these smaller units have a limited load capacity. The number of joints, or degrees of freedom of the arm, will determine the robot's dexterity, as well as its cost. For example, it takes six degrees of freedom to emulate human arm motion (two degrees of freedom at the shoulder, one at the elbow, and three at the wrist).

The second essential part of the robot is an end effector. This is the gripper, welding gun, or other tool allowing the arms to execute assigned tasks. Grippers are normally custom-made, and they vary a great deal from one application to another. The gripper designed to pick up a book will look quite different than one used to grasp an egg. The grasping of an item by a robot emulates human finger motions. Some grippers can lift several heavy objects at one time. Other mechanisms can grasp a fragile component without damaging it. Examples of grippers include vacuum cups, hooks, electromagnets, clamps, scoops, or finger-like devices.

The third essential part of a robot, the controller, is the program and computer

[3]Robot Institute of America. 1981. *Worldwide survey and directory on industrial robots.* Dearborn, Mich., 1.

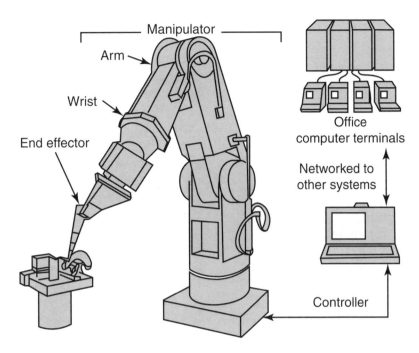

Figure 6-13.
The three main systems of an industrial robot are the controller, the manipulator, and the end effector.

used to activate the robot, guide it, and direct its movements. One very common way to program the unit is to lead it through the desired sequence of operations. A human operator moves the robotic arm physically or with switches on a control panel, while the controller records the path. This is an effective programming method for tasks such as recording the curved motions needed to produce an even coat of spray paint or to pick up parts from one location and move them to another. On the other hand, the path is difficult to edit or revise without completely rerecording or reprogramming the task.

Off-line programming is when an operator writes a new program at a terminal. Off-line programs can be written without interrupting production. They can be used to issue a different set of instructions to the robot dictated by the situation at hand (for example, if no part is available for treatment, stay on hold and wait for the next cycle). In the mid-1990s it became possible to program industrial robots with Windows® operating system–type technology on a personal computer (PC) in an office with interactive video. In recent years, computers have not only made robots smarter, but they have made it possible to control, program, and monitor these machines from remote sites anywhere.

Robot manufacturers classify robots with respect to their abilities. A point-to-point robot uses mechanical stops or limit switches to go from one predefined point to another. This type of robot is less expensive and generally used for simple maneuvers. A controlled path robot continuously follows a preselected geometric path. A servo-controlled robot uses software to control its movements. A servo-controlled robot can sense points on a path and feed back the information so the controller can take alternative actions. Due to their greater flexibility and the controllability of the sequence of motions, most contemporary industrial robots are servo-controlled.

Applications in Industry and Elsewhere

Whether robots are simple or sophisticated, the decision to implement robotics into manufacturing and production operations should be based on a comprehensive analysis of the entire operation. Until the mid-1980s, the robotics industry grew fast, as companies invested heavily in them to automate their factories and replace blue-collar employees. Industrial managers discovered industrial robots were not a panacea or solution for all manufacturing jobs. Not everything can be automated. In some cases, robots can be a major disappointment. As assemblers, they may dumbly attempt to jam a nut into an opening, even if it does not fit; as surface finishers, they may actually paint themselves on the paint line. It is essential to recognize robots for their realistic potential:

❏ Robots are not mechanical people.

❏ Robots do not replace human labor, but enhance it.

❏ Robots are tools within an integrated manufacturing system.

The U.S. robotics industry has only recently recovered from the downswings of the 1980s. There is only one U.S. company, Adept, still producing industrial robots. Most of the other companies went out of business, were consolidated, or were sold to European and Japanese companies.

Robots work as an integral component of a production system. Environmental conditions such as heat, humidity, static electricity, and dust determine the surroundings and should therefore influence the selection of the robot. The range of motions and speed at which they are required to operate determine the capabilities demanded of the robot. If robots are used to perform material handling tasks, sensors are necessary to make them aware of the work environment at any given moment. Improvements are being made to robotic devices on a regular basis, and these changes usually involve speed, precision, and remote control.

Where is the functionality of today's industrial robots? Many early robots were used (and still are) for monotonous or dangerous jobs and for loading and unloading heavy parts. More sophisticated robots are capable of a variety of value-added jobs, including assembling electronics, sanding missile wings, and moving parts from one machine tool to another. In some cases, robots almost appear to be replacing machine tools. To summarize the most common industrial operations assigned to robotized devices in factories worldwide, the list would minimally contain these tasks: materials handling, assembly operations, spraying, cutting and deburring, and inspection. The need for sensing devices is most acute when robots are used in inspection processes. This area may involve some form of monitoring, detection, analysis, or calibration. Robots are used to gauge and measure tolerances on manufactured parts, as well as used with ultrasound to detect leaks in assemblies.

As "senses" are added to industrial robots, they become ever more valuable to manufacturing enterprises. The cost of technologies like sight, touch, and voice is decreasing. Equipped with camera eyes or optical scanners, robots are able to select, inspect, assemble, and place parts or products on machines, trolleys, and warehouse shelves. The BMW automation machine from the *Dateline* section of this chapter is an example of this type of automated system. After a car body is painted, it moves by conveyor to an inspection station. Here, a SmartEye® reader—an optical scanner—checks the number and sends the car body to its correct location.

Look away from the factory floor for just a moment. In the near future, household robots may provide companionship and assistance for the elderly or for people with physical disabilities. Using electronics, servomechanisms, controllers, sensors, communication equipment, active and passive beacons, and a receiver for the Global Positioning System will enable household robots to navigate living spaces in an effortless manner. These robots may soon be purchased by consumers for their capacities to give ambulatory aid, fetch and carry, cook, clean, monitor vital signs, entertain, and even get in touch with relatives in case of emergency.

As discussed in our chapter on space exploration (Chapter 4), robotic explorers have already gone to Mars to collect critical data. In the medical field, robots hold the promise of increased accessibility to medical care for people living some distance away from specialists. The promise of medicine at a distance—telemedicine—allows a person to receive specialized care from a surgeon using a robot. For example, people requiring medical assistance may find robotic devices have the capacity to save their lives, even when human surgeons are miles away. With the use of robots, it might be possible to perform complicated surgery at remote sites or even on the battlefield! See Figure 6-14.

Implications for Workers and Their Job Environments

Robots and other automation have been successfully integrated into the American manufacturing and production scene without the catastrophic consequences projected by some in the early 1980s. It was theorized that robots would migrate through factories and displace people. This problem never became a reality. The role of robotics in reducing the percentage of the work force employed in manufacturing has not been as dramatic as initially projected. Industrial robots, like other forms of technology, are tools allowing people to become more productive and skillful. Industrial robots have not replaced shifts of workers, and "lights-out" factories are not expected to become a reality in the near future.

Employees may express concern about job security when confronted with a discussion on automation. The extent to which automation is successfully implemented is a reflection on the amount of cooperation and communication among management, supervisors, employees, and the design engineering staff. In industries where automation is being implemented, employees must be open-minded in their reactions to changes in their work environment.

An industrial organization's future success is linked directly to its ability to manage and support a manufacturing system focusing on producing a high quality product at a competitive price. Labor costs continue to rise, while robot equipment prices drop, and second hand robots are a viable option. In instances where it is strategically and financially determined a corporation's needs can best be served through the use of industrial robots, the decision to invest in the appropriate equipment is a wise one. The primary consideration for robotics implementation must take into account product characteristics plus the manufacturing processes required. As previously stated, not everything can be successfully automated through the use of robots. A number of companies have found they can be more competitive if they use a softer, and perhaps even more traditional, mix of human labor and machine technology.

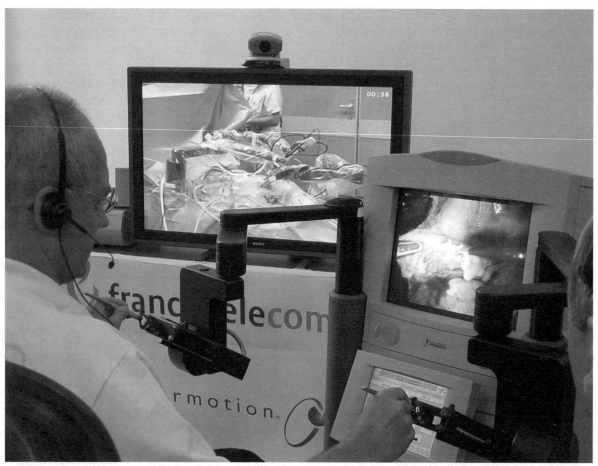

Figure 6-14.
A surgical team operating at a console in New York sent instructions to a set of robotic arms that removed the gallbladder of a patient across the Atlantic in Strasbourg, France. "Operation Lindbergh" was a world first in telesurgery. (IRCAD)

Summary

The executive suites of manufacturing firms in the United States look to be happier places than they were less than a decade ago. Managers in industries as diverse as construction equipment, motorcycles, electronic pagers, automobiles, and personal computers (PCs) have analyzed and adopted a striking array of integrated automated process technologies. By studying each step in the manufacturing process and cyclical fluctuations in workloads, companies have found ways to reduce variation, eliminate bottlenecks, improve quality, and speed up time-to-market. Many of the strategies being applauded fall under the computer integrated manufacturing (CIM) umbrella, where a company-specific blend of computers, machining centers or cells, industrial robots, and human labor is fundamental to success.

The concurrent engineering (CE) design environment has enabled manufacturing and production enterprises to discover the advantages of getting it right the first time.

The benefits of simultaneous product design, development, and manufacture are hard to dispute. The growing ability of competitive firms to gain a larger share of their markets has, to a great extent, been accomplished through their implementations of lean manufacturing techniques. There is a trend toward increased customization of products. For a manufacturing company, this puts increased pressure on being able to respond to market demand in a quick and efficient manner, and on the road to becoming leaner and meaner, just-in-time (JIT) philosophies have flourished among manufacturers across the nation.

A whole new collection of "intelligent materials" that can anticipate breakdowns and repair themselves is on the horizon. Industrial robots are being renovated, and industrial organizations are being more cautious about investments in large-scale flexible manufacturing systems (FMSs). The future of manufacturing will be structured within a process manufacturing paradigm that will undeniably include rapid prototyping (RP), customized production, and delivery to consumers just in time.

Discussion Questions

1. Develop an argument to either defend or refute the following statement: "Computer integrated manufacturing (CIM) is not a specific technology."
2. Explain some of the benefits manufacturing firms have realized as a result of implementing concurrent engineering (CE) strategies. Use a specific industry to illustrate your response.
3. How does the just-in-time (JIT) management philosophy differ from traditional manufacturing philosophies practiced by manufacturing firms in this country? What potential problems might be associated with its adoption in, for example, a shoe manufacturing plant in Syracuse, New York?
4. Briefly describe what the term *lean manufacturing* means. Select one industry and give examples of how it can make itself leaner and subsequently more competitive.
5. Provide a description of a "personal service robot" that sets it apart from an "industrial robot." What uses might you project for personal robots in the near future?
6. In a creative way, identify all the similarities one might find between an agile manufacturing enterprise and a seasoned jazz band.

Suggested Readings for Further Study

Adler, Paul S., Avi Mandelbaum, Vien Nguyen, and Elizabeth Schwerer. 1996. Getting the most out of your product development process. *Harvard Business Review* 74 (March–April): 134–152.

Aronson, Robert B. 1995. LEAD winners find CIM is key to improvement. *Manufacturing Engineering* 115 (November): 63–64+.

———. 2000. It's not just RP anymore. *Manufacturing Engineering* 124 (5): 98–112.

Benders, Jos. 1995. Robots: A boon for the working man? *Information & Management* 28 (6): 343–350.

Biekert, Russell. 2000. *CIM Technology Fundamentals and Applications.* Tinley Park, Ill.: Goodheart-Willcox.

Billesbach, Thomas J. 1994. Applying lean production principles to a process facility. *Production & Inventory Management Journal* 35 (3): 40–44.

Bushnell, R. 1994. Where to begin with bar codes. *Modern Materials Handling,* 49 (November): 38.

Bylinsky, Gene. 1994. The digital factory. *Fortune* 130 (14 November): 92–94+.

Cheng, T. C. E. and Susan Podolsky. 1996. *Just-in-time manufacturing.—An introduction.* New York: Kluwer Academic.

Chu, X. and H. Holm. 1994. Product manufacturability control for concurrent engineering. *Computers in Industry* 24 (May): 29–38.

Clay, G. Thomas, and Preston G. Smith. 2000. Rapid prototyping accelerates the design process. *Machine Design* 72 (9 March): 166–171.

Dietrich, Edgar. 2000. SPC or statistics? *Quality Progress* 39 (August): 40–45.

DuVall, J. B. 1996. *Contemporary Manufacturing Processes.* Tinley Park, Ill.: Goodheart-Willcox.

Engelberger, J. F. 1995. Robotics in the 21st century. *Scientific American* 273 (September): 166.

Fisher, Marshall L. 1997. What is the right supply chain for your product? *Harvard Business Review* (March–April): 105–116.

Gardiner, Keith M. 1996. An integrated design strategy for future manufacturing systems. *Journal of Manufacturing Systems* 15 (l): 52–61.

Hales, Robert F. 1995. Adapting Quality Function Deployment to the U.S. culture. *IIE Solutions* 27 (10): 15.

Handfield, Robert B. and Mark D. Pagell. 1995. An analysis of the diffusion of flexible manufacturing systems. *International Journal of Production Economics* 39 (May): 243–253.

Hassan, Mohsen. 1995. Layout design in group technology manufacturing. *International Journal of Production Economics* 38 (2–3): 173–188.

Hopp, Wallace J., and Mark L. Spearman. 2001. *Factory Physics.* New York: Irwin McGraw-Hill.

[Internet]. *American Society for Quality.* <http://www.asq.org/>.

[Internet]. *Association for Manufacturing Excellence.* <http://www.ame.org>.

[Internet]. (2000). International Organization for Standardization. *Annual Report 2000: Harmony for Prosperity.* <http://www.iso.ch/iso/en/aboutiso/annualreports/2000/index.html> [11 March 2002].

[Internet]. *Laboratory for Manufacturing and Productivity.* <http://web.mit.edu/lmp/www/>.

[Internet]. *National Association of Manufacturers.* <http://www.nam.org/>.

[Internet]. *National Center for Manufacturing Sciences.* <http://www.ncms.org/>.

[Internet]. *Society of Manufacturing Engineers.* <http://www.sme.org>.

[Internet]. *Stanford University Alliance for Innovative Manufacturing (AIM).* <http://www.stanford.edu/group/AIM/>.

Jordan, James and Frederick Michel. 2001. *The lean company: Making the right choices.* Dearborn, Mich.: Society of Manufacturing Engineers.

Karmarker, Uday. 1989. Getting control of just-in-time. *Harvard Business Review* (September–October): 1–10.

Kochan, Anna. 1995. Renovation gives robots a new lease of life. *Industrial Robot* 22 (2): 24–26.

Krizner, Ken. 2001. Manufacturers adopt a lean philosophy: Methodologies and tools, mixed with information technology, can allow companies to streamline costs, increase efficiencies. *Frontline Solutions.*

Lang, James S. and Paul B. Hugge. 1995. Lean manufacturing for lean times. *Aerospace America* 33 (May): 28–33.

Ligus, Richard G. 1994. Enterprise agility: Jazz in the factory. *Industrial Engineering* 26 (November): 18–19.

Masterson, James W., Robert L. Towers, and Stephen W. Fardo. 1996. *Robotics Technology.* Tinley Park, Ill.: Goodheart-Willcox.

Min, Hokey and Dooyoung Shin. 1994. A group technology classification and coding system for value-added purchasing. *Production and Inventory Management Journal* 35 (l): 39–42.

Nicholas, John M. 1994. Concurrent engineering: Overcoming obstacles to teamwork. *Production and Inventory Management Journal* 35 (3): 18–22.

Ono, Taiichi and Taiichi Ohno. 1988. *Toyota Production System: Beyond large-scale production.* Portland, Oreg.: Productivity.

Parsell, D. L. (2001). Surgeons in U.S. perform operation in France via robot. *National Geographic News.* [Internet]. <http://news.nationalgeographic.com/news/2001/09/0919_robotsurgery.html>.

Pearch, Clyde and Jill Kitka. 2000. It's here: ISO 9001:2000. *Manufacturing Engineering* 125 (October): 98–110.

Pesch, Michael J., Larry L. Jarvis, and Loren Troyer. 1993. Turning around the rust-belt factory: The $1.98 solution. *Production and Inventory Management Journal* 34 (2): 57–62.

Phillips, T. A. 1996. American manufacturing: The just-in-time solution. Research paper, Department of Technology, State University of New York, Oswego.

Ramaswamy, Ramana and Robert Rowthorn. 2000. Does manufacturing matter? *Harvard Business Review* (November–December): 2–3.

Rogers, Craig A. 1995. Intelligent materials. *Scientific American* 273 (September): 154–161.

Shina, Sammy G., ed. 1994. *Successful implementation of concurrent engineering products and processes.* New York: Van Nostrand Reinhold.

Simmons, P. 1996. Quality outcomes: Determining business value. *IEEE Software* 13 (January): 25–32.

Singh, Nanua. 1995. *Systems approach to computer-integrated design and manufacturing.* New York: John Wiley & Sons.

Smith, Patricia L. 1997. Tales of rapid prototypes. *American Machinist* 141 (July): 52.

———. 1999. Agile manufacturing: The piece maker. *American Machinist* 143 (11): 58–60.

Spear, Steven and H. Kent Bowen. 1999. Decoding the DNA of the Toyota Production System. *Harvard Business Review* 77 (September–October): 96–106.

Stevenson, W. J. 2002. *Operations management.* New York: McGraw-Hill/Irwin.

Strozniak, Peter. 2001. Toyota alters face of production (Toyota production system fuels company's success). *Industry Week,* (13 August).

Sule, Dileep R. 1994. *Manufacturing facilities: Location, planning, and design.* Boston: PWS.

Thorpe, Paul A. 1995. Concurrent engineering: The key to success is today's competitive environment. *HE Solutions* 27 (October): 10–13.

Tracy, Michael J., James N. Murphy, Robert W. Denner, Bruce W. Pince, F. Robert Joseph, Allen R. Pilz, and Michael B. Thompson. 1994. Achieving agile manufacturing in the automotive industry. *Automotive Engineering* (November): 19–24.

Turbide, David A. 1995. MRP II: Still number one! *HE Solutions* 27 (July): 28–31.

Turino, Jon. 1992. *Managing concurrent engineering: Buying time to market.* New York: Van Nostrand Reinhold.

Ward, G. R. and S. R. G. Went. 1995. Robot safety. *Industrial Robot* 22l: 10–13.

Weimer, George. 1995a. Factory robots in 2000: Smart, fast and sensitive. *Material Handling Engineering* 50 (October): 32–35.

————. 1995b. Manufacturing on a roll! *Material Handling Engineering* 50 (November): 27.

Winek, Gary and Vedaraman Sriraman. 1995. Rapid prototyping: The state of the technology. *Journal of Engineering Technology* 12 (fall): 37–43.

Withers, Barbara E. and Maling Ebrahimpour. 1993. A comparison of manufacturing management in JIT and non-JIT firms. *International Journal of Production Economics* 32 (November): 355–364.

Womack, James and Daniel Jones. 1996. *Lean thinking.* New York: Simon & Schuster.

Womack, James, Daniel Jones, and Daniel Roos. 1991. *The machine that changed the world: The story of lean production.* New York: Harper Perrenial.

Wood, Lamont. 1993. *Rapid automated prototyping: An introduction.* New York: Industrial.

Wurz, David. 1995. Bar coding basics. *Plant Engineering* 49 (8 May): 94–96.

Zaciewski, Robert D. 1995. ISO 9000 preparation: The first crucial steps. *Quality Progress* 28 (November): 81–83.

Zayko, Matthew J., Douglas J. Broughman, and Walton M. Hancock. 1997. Lean manufacturing yields world-class improvements for small manufacturer. *IEE Solutions* (April).

Chapter 7

Technology Transfer

Matamoros, Mexico

Mexico's nearly forty-year-old Border Industrialization Program is seemingly responsible for serious social and environmental problems. Overcrowding forces hundreds of thousands of employees working in foreign-owned factories to live in squatters' camps without heat or electricity. Millions of gallons of raw sewage are dumped into the local waterways each day.

Besides overburdening a weak sewer system, owners and managers of these small factories, or maquiladoras, readily admit to moving their operations to Mexico because their hazardous processes are more severely regulated in the United States (especially in operations such as metalworking, tanning, and dyeing). The Mexican government does not have strict regulations, and Mexican employees are subjected to hazardous working conditions....

Introduction

The international technology transfer network is a topic that has received much critical analysis since the end of World War II. At that time, the United States began to export a significantly greater number of manufactured goods to war-torn European nations. On the home front, the U.S. government continued to increase its domestic expenditures on military programs and aerospace research and development (R & D) projects.

TECHNOLOGY TRANSFER
The development of a technology product or process in one setting, which is then transferred for use in another setting.

Technology transfer is the process by which technology developed for one purpose is employed either in a different application or by a new user. To cite one specific historical example, the transfer of U.S. technology helped to speed up Europe's recovery process following the end of World War II. By the mid-1950s, increased European competition emerged in the U.S. export markets. Similarly, significant technological developments occurred in the U.S. commercial sector as a result of sizable government R & D spending. This served to reduce the technical and market risks that would have otherwise been confronted by individual industries. As a result, the 1950s witnessed the introduction of numerically controlled machine tools, integrated circuits, commercial aircraft, and communications satellites.

A more comprehensive explanation of the technology transfer concept evolves throughout a complex and somewhat controversial network of activities. This transfer continuum is marked at one end by the uncritical transfer of technology to reluctant recipients. The opposing end of the same continuum is replete with covert activities akin to high-tech espionage. This chapter discusses a variety of topics and events relevant to the paradoxical nature of the contemporary U.S. technology transfer engine. As you read through the various sections, keep in mind the following inquiries:

❑ What are the key patterns, motives, and channels associated with technology transfer activities?

❑ Should the United States continue to address the "technology gap" perceived to exist between the creators of technology and the borrowers of technology?

❑ To what extent is the increasingly competitive aspect of entrepreneurial science impacting the traditional roles played within institutions of higher education?

❑ What strategies and guidelines can be employed to determine the extent to which a specific technology can or should be transferred to another nation?

❑ How rapidly do industrial innovations spread from one country to another? Which factors determine the relative ease with which the recipients can begin producing these innovations on their own?

Key Concepts

As you study this chapter, there are several key concepts and terms with which you should become familiar:

appropriate technology (A.T.)
demand lag
entrepreneurial science
foreign direct investment
imitation lag
licensing
maquiladora
most favored nation (MFN)
reverse engineering
technology transfer
technology transfer channel
turnkey plant

The International Technology Transfer Profile

It is evident the technology transfer process encompasses a diverse range of activities. There are, however, a few generic elements that may be used to examine a

majority of the activities in a comparative manner.

The first key element involves the transfer item itself. What is being transferred (for example, a product, a process, or information)? Second, who is the technology donor (who possesses ownership rights to the transfer item)? Third, who is the technology recipient (an individual, a manufacturing firm, or the residents of an entire geographic region)? Of central concern is the recipient's capability to effectively adopt and implement the technology once it has been transferred. In some cases, less-developed nations have particularly poor capacities to assimilate new technology. A fourth element for comparison is the type of *technology transfer channel.* How will the transfer item be delivered from the technology donor to the recipient? Typical transfer channels include foreign direct investment, licensing, and the direct purchase of manufactured goods.

TECHNOLOGY TRANSFER CHANNEL
Within the international network of technology exchange and transfer, the channel is the formal or informal route through which technological artifacts travel from the supplier to the end user.

Finally, the technology transfer process may be studied through an analysis of the rate of diffusion of the technology. The concepts of imitation lag and demand lag provide two perspectives for this examination and are discussed later in this chapter.

Transfer Channels

The flow of technological innovations and information through the contemporary international network is a continuous process occurring through many avenues.

See Figure 7-1. The mode of transfer that can be effectively employed is influenced directly by the nature of the technology being transferred; the commercial considerations, as well as the economic incentives associated with the transaction; and the expectations expressed by the recipients of the technology.

Commercial transactions between American and foreign firms seem to form both the foundation and the overarching protocol for the international flow of technology. The underlying motive in a majority of these exchanges is not solely technology transfer. These interchanges are usually associated with the potential for a significant return on investment or the desire to gain market access for future investment opportunities. A "human interest" dimension, regarding the degree to which the recipients will benefit equally or at all from the new technology, may never even be considered.

Several principal channels through which technology is formally transferred from one country to another include foreign direct investments, turnkey plants, licensing, entrepreneurial science, and manufactured exports. Numerous informal transfers occur via student exchange programs, professional meetings, technical seminars, cooperative scientific ventures, dissemination of technical and scientific publications, and the review of online resources. These lists are certainly not exhaustive, but an examination of these types of transnational channels will enhance your understanding of the complexity of the exchange procedures and legislation related to technological products and knowledge.

Formal Channels

Foreign direct investments represent a primary channel for the transfer of technology. These are generally made with commercial

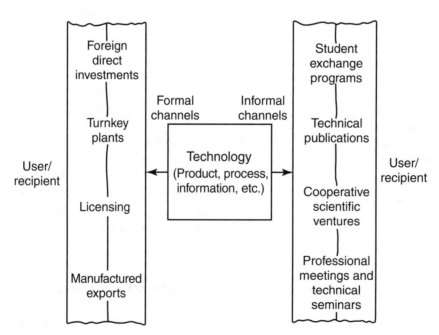

Figure 7-1.
International technology transfer activities typically occur through these channels.

considerations in mind. A manufacturing firm's decision to establish overseas subsidiaries is often related to its desire to reduce current distribution costs, to avoid hefty import duties, or to manufacture with lower labor costs. Nevertheless, the firm's transition from a national to a multinational corporation provides an environment highly conducive to technology transfer.

In this type of endeavor, the technology donor is the parent firm claiming either full or partial ownership of the foreign subsidiary. It is typical for the recipient country to provide material and labor resources, while the parent firm assumes responsibility for the capital, technology, management, and marketing skills.

It seems natural to assume the economic incentive alone should provide motivation for the parent firm to do everything possible to ensure the transfer proceeds efficiently. If they are successful and an effective working relationship develops between donor and recipient, technology transfer occurs in a variety of ways. Most commonly, the technology donors train the resident operatives and managers, communicate with engineers and technicians in order to translate critical information, and may ultimately encourage the local suppliers to upgrade their existing technology.

In a growing number of instances, the recipient nations in these transactions are expressing a desire for greater control over the local enterprises of multinational corporations. This has produced a trend toward joint ventures in which the subsidiary is partially owned by the parent firm and partially owned by public or private interests within the host country. Both parties contribute capital to the project, and the management, control, and profits are shared in proportion to each party's equity in the enterprise.

Turnkey plants provide another example of a channel for technology transfer. In this type of operation, the technology donor constructs a fully operational production plant for the recipient nation. The term *turnkey* refers to the notion that once the facility is completed, the recipient simply needs to "turn a key" to initiate the production process. Since the parent firm retains no equity interest in the plant after its completion, turnkey operations are more accurately described as a contractual sale between the party with the technical and managerial expertise and the party that has requested this technological assistance.

Varying levels of technical knowledge and skill are transferred via this form of exchange. In some cases, the receiver can write a clause into the contract to have the donor thoroughly train its local technicians and managers to be able to operate the facility on their own. In these instances, the technology donor leaves the site upon the completion of the construction or training project.

If, on the other hand, the recipient nation feels it lacks the capacity to execute the skills needed to effectively operate the new technology, the seller may be asked to stay and provide assistance even after the plant is finished. This type of request will generally require a formal management contract or technical service agreement. This type of contractual agreement, combined with the turnkey operation itself, becomes an extremely effective mechanism for bilateral technology transfer. Specifically, a high level of personal interaction between the two countries over an extended period of time is necessary. Visitors to the foreign nation ultimately acquire new language skills and gain greater sensitivity and awareness of cultural differences. Further, they may learn new techniques and methods by teaching others how to safely manage and operate the facility. In sum, a considerable amount of on-site technical information and experience eventually becomes a mutual acquisition.

Licensing is one of the primary mechanisms unaffiliated firms employ for transferring technology among themselves. The license is an agreement specifying the conditions under which the recipient may employ the donor's technology. In other words, the license agreement is technology-specific, and it may cover a patented process, trade secret, manufacturing permit, trademark, or the sales and distribution rights for a particular product. The exact terms of the agreement are arrived at through bargaining between the owner of the technology and the purchaser. In most cases, the licensor receives either a lump sum payment at the time of transfer or later receives a predetermined percentage of resultant sales in the form of royalty payments—sometimes both.

Licensing agreements often specify restrictions on the use of the technology item being transferred in order to protect the donor's competitive position, as well as its current markets. For example, the licensee may be prohibited from reexporting the resulting products or services it develops through the use of the newly acquired technology.

Licensing is most useful as a technology transfer mechanism in situations in which the recipient nation has a high capacity to absorb the technology being transferred. Licensing another entity's technology can be quite beneficial for several reasons: it provides access to current technologies at minimum cost, it reduces the duplication of effort and can save a significant amount on the costs of in-house R & D efforts, and it may help to raise the recipient's level of technical competence if supported with an orientation training program.

Entrepreneurial science is essentially subsumed within the licensing channel we have just described, but its impact deserves individual attention. Entrepreneurial scientists who work in the laboratories of entrepreneurial universities are engaged in

numerous commercial agreements throughout the world in which knowledge is transferred into intellectual property. At leading research academic institutions, faculty members and their graduate students are increasingly learning to assess the commercial, as well as intellectual, potential of their cutting edge research. Essentially, universities have begun in earnest to combine teaching and research with technology transfer and are subsequently starting to play a pivotal role in the economy of our country.

Interestingly, entrepreneurial science represents a convergence between basic and applied research culminating with commercial opportunities linked to the "pure" science. As you learned in Chapter 1, opportunities for commercial utilization of scientific research may have been available to scientists prior to the late 1970s, but disciplinary norms did not permit them to violate the respected boundaries between science and business. In 1980, the Bayh Dole Act permitted U.S. universities to take advantage of opportunities to gain income from licensing their intellectual property rights (most commonly held in the form of patents). This legislation made receipt of federal funds conditional on universities making an effort to put research results to good use in society.

While the entrepreneurial spirit in science sounds promising for young and aspiring scientists, it is perceived by some as a threat to scientific inquiry since it has the tendency to squelch the free flow of ideas at scientific meetings. This reality became quite apparent during the early years of the biotechnology industry when the money madness created a competitive spirit rather than a cooperative one, where all new genetic engineering research was involved. Nevertheless, when scientific knowledge is appropriated to generate income, the discipline of science is transformed. Critics of entrepreneurial universities debate value issues and question whether participation in licensing ventures detracts from the traditional educational and research missions of the universities. This debate is far from over, and this formal channel for the transfer of technology from university research labs to industrial firms to the general public is most definitely here to stay.

Export of manufactured goods represents another principal formal channel for international technology transfer. In terms of dollar volume, the sale of technology-intensive products represents the dominant mode of technology transfer today. The mere existence of a new "gadget" in the recipient nation may result in the transfer of technology in the form of increased knowledge, skill, or convenience.

In many instances, the core technological data and know-how is normally difficult (if not impossible) to extract from the products that have been imported. Subsequently, this channel for transfer may not be perceived as the most efficient choice. On the other hand, numerous recipient nations have demonstrated very astute *reverse engineering* capabilities over the past few decades. This introduces a unique twist into the transfer network, as the imported technological device is carefully disassembled, examined, and eventually copied by the purchasing agent. The resultant reengineered product may not meet or exceed the quality of the original, but these imitators often compete well on the open market.

Informal Channels

Noncommercial avenues also make a significant contribution to the international flow of technology. For example, governmental agreements between the United States and other countries are set up on a regular basis to exchange technical information or to conduct cooperative R & D projects.

These types of endeavors usually involve extensive contact between government representatives, tours of industrial plants, and the establishment of domestic training and educational programs. The information exchange is purported to be mutually beneficial to all participants involved. A few contemporary areas of concern that have initiated or stimulated these types of agreements include environmental protection (especially global warming), the impact of AIDS on Earth's population, global energy resource management, national defense policies (especially regarding terrorism), and space station design and construction.

Scientific and technical publications provide another outlet for the transfer and exchange of information. One of the principal norms of science is to report and disseminate its research findings. Subsequently, journal literature becomes a repository of most scientific knowledge. Technological knowledge may be more difficult to acquire through this channel since it is often proprietary in nature. The review of printed scientific and technical journals has a serious drawback as a means of transferring technology. The information may well be obsolete by the time it is finally published! In recent years, more current and more abundant information, facts, and figures can normally be found on-line.

Technical seminars, conferences, and trade shows help to compensate for the delays inherent to the printed publishing circuit. See Figure 7-2. International meetings provide a convenient forum for the development of new contacts and often stimulate a dynamic interchange of opinions regarding contemporary technological projects still "on the drawing board." Manufacturers who invest huge sums in acquiring a prime location on the trade show floor will usually exhibit their most sophisticated and up-to-date technological wares. Valuable technical information can usually be attained through

Figure 7-2.
Manufacturers on the trade show floor at MacWorld Conference and Expo, held at the Moscone Center in San Francisco in 2002, exhibit their newest technological products. It is the world's largest trade show for Mac enthusiasts and professionals seeking world-class education, peer-to-peer networking, and one-stop comparison shopping. (IDG World Expo)

an informal conversation with one of the booth's representatives.

Technology transfer continues to occur through many different types of channels. Those discussed here represent a few of the principal exchange systems utilized in the current international network. Each of these transfer modes has distinct advantages and disadvantages for both the technology donor and the technology recipient. The next section will examine several of these concerns as they relate to the role of the United States in technology transfer as an operative device to close technology gaps where they may exist in the world.

Closing the Technology Gap

The heading for this section might strike you as being rather presumptuous—it suggests there is indeed a social phenomenon labeled a *technology gap*, and furthermore, the United States is somehow capable of closing it. It forces one to visualize a global scenario in which the United States is leading the world in technological pursuits, while the rest of the world's countries are sluggishly lagging behind. This "gap" concept portrays a world order that can easily be assembled into suitable categories, such as the "techno-leaders" and the "techno-followers."

Based on your intuitive understanding of technology and the technology transfer concept, it is appropriate to develop an interpretation for this elusive "technology gap" idea. Very few people would dispute there is a general, across-the-board technological difference between the United States and several developing nations across the African continent. However, the gap between the United States and the United Kingdom (UK), or other so-called "advanced" countries, or even emerging *Third World* powers is not so clear-cut.[1]

In fact, it has become quite apparent that international technological leadership constantly fluctuates among any number of key players. Present technology leads may be quickly lost or ceded, and they may or may not ever be regained. During the early 1980s, the United States reigned as a leading producer of high-tech products, but its margin of leadership narrowed by the decade's end, as Japan aggressively enhanced its stature in high-tech fields. The U.S. share of global shipments of high-tech manufactured goods declined in the 1980s, while Japan's market share increased nearly 10 percent in the same period. During the 1990s, U.S. high-tech industries regained some of the market share they had lost during the previous decade. By the end of the 1990s, the United States had again become a leading producer of high-tech products, responsible for one-third of the world's production.

Before we proceed any further into the dialogue, it might be useful to set some parameters to the term *high technology*. The Organization for Economic Cooperation and Development (OECD) presently identifies four industries as high technology, based on their high R & D intensities: aerospace, computers and office machinery, electronics-communications, and pharmaceuticals. High technology industries are viewed as being important to nations because they are associated with innovation, are linked with high value added production and success in foreign markets, and perform industrial and science-based R & D that has positive spillover effects (for example, new products and processes might be generated leading to gains in productivity).

As we stated previously, the U.S. high-tech industries faltered, losing market share during the 1980s, and the Asian global market share in high-tech industrial ventures gained steadily. Interestingly, it was the developing Asian nations that made the most dramatic gains during the decade of the 1980s. In the decade following, production by China's high-tech industries was larger than any European nation's. The profile of economic growth in South Korea is remarkably quite similar. If one compares South Korea's high-tech production with that evident in the four largest European countries in the late 1990s, South Korea's share of world production was smaller than Germany's or the UK's, but larger than that produced in both France and Italy. To the west, the U.S. world market share remained steady through the 1990s and moved up slightly at the close of that decade. See Figure 7-3.

In each of the aforementioned industries making up the high-tech cluster, the United States maintained strong, if not leading, market positions for most of the last two decades of the twentieth century. Competitive pressures emanating from countries around the globe contributed to the decline in the U.S. aerospace and communications equipment industries during the 1980s. Since then, both have reversed their downward paths, but the aerospace industry continues to struggle. See Figure 7-4.

Another perspective on the technology transfer network has to do with overall balance of trade. Near the end of the 1990s,

[1]*Third World:* Contemporary writers continue to use the term *Third World* to refer to those countries that have yet to achieve a level of technological expertise comparable to their industrialized neighbors. The authors italicize this term in Chapter 7 because they believe it is difficult to identify which nations conversely belong to the "First" or "Second" Worlds. We encourage readers to understand *Third World* to mean less-developed nations, newly industrialized countries, emerging participants in the international technology transfer arena, or even just simply countries in transition.

Country share of global high-technology output

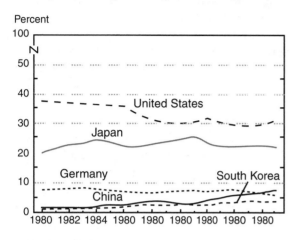

Percent

Figure 7-3.
This graph shows country share of global high-tech output. (Adapted from *Science & Engineering Indicators—2000)*

U.S. global output share, by high-technology industry

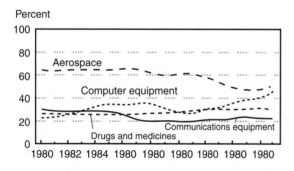

Percent

Figure 7-4.
This graph shows U.S. global output share, by high-tech industry. (Adapted from *Science & Engineering Indicators—2000)*

U.S. trade in technology products accounted for a larger share of U.S. exports than U.S. imports. The United States also remained a net exporter of technological know-how sold as intellectual property. Royalties and fees received from interested foreign firms have averaged three times the amounts paid out to foreign-based companies by U.S. firms—for access to new technology. Japan is the largest consumer of U.S. technology sold as intellectual property, and South Korea is a distant second. You should recognize this technology transfer channel as one labeled *licensing* earlier in this chapter.

The U.S. Bureau of the Census uses a classification system for exports and imports of products that in one way or another embody new or leading edge technologies. This system allows international trade to be examined in ten major technology areas, including biotechnology, life science technologies, optoelectronics (such as optical scanners, optical disc players, and laser printers), computers and telecommunications, electronics, computer integrated manufacturing, material design, aerospace, weapons, and nuclear technology. Using this classification system, with the inclusion of computer software as an eleventh category, the three largest foreign markets for U.S. manufactured technology products can be determined. See Figure 7-5. You will see that Japan and Canada are among the top three technology recipients across a majority of the product categories, with the UK making itself known as well. On the flip side of the transfer scene, the leading foreign suppliers of technology products to the United States can also be determined. See Figure 7-6. Imported technologies from leading economies in Asia and Europe enhance productivity of U.S. firms and workers and offer U.S. consumers more choices.

Until this point in the chapter, the topic of international technology transfer has been discussed from the perspective of a willing technology donor and an anxious technology recipient. By virtue of its leadership position in the realm of technology-intensive products, the United States is either

Three largest export markets for U.S. technology products: 1998

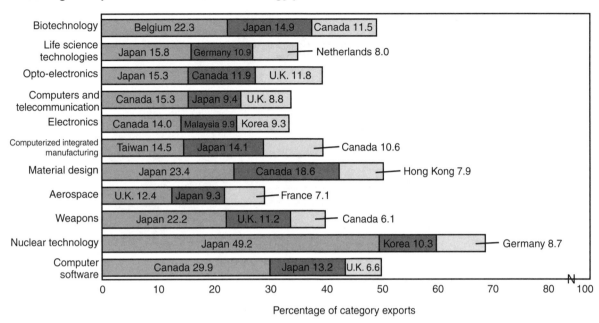

Percentage of category exports

Figure 7-5.
This graph illustrates the three largest export markets for U.S. technology products. (Adapted from *Science & Engineering Indicators—2000*)

revered or coveted for its capacity to reduce technology gaps where they may be evident. Does the United States truly have the technical prowess to share its science and technology (S & T) with those nations who indicate a need or desire for it and subsequently create a new world order?

Creating a New World Order

Sociologists, anthropologists, and government leaders have addressed the issue of international development for several decades. One interpretation of international development suggests a driving force to improve the standard of living and quality of life for people in countries where food supply, nutrition, shelter, and education are inadequate, and where human beings are unable to live up to their full potentials. The gross national product (GNP) per person is the most commonly referenced statistic used to measure development, but it does not always include these other quality of life factors in its calculations and inferences. International development refers then to a combination of improving the quality of life, meeting basic needs, and increasing per capita GNP.

With these parameters or boundaries in mind, what role can the transfer of technological products and knowledge play in the overall international development scheme? Should a primary objective for U.S. technology transfer activities be to equalize the levels of technological development on a global scale? Stated differently, should a humanitarian connotation be ascribed or attributed to a schedule of transfer activities seeming to be

Top three foreign suppliers of technology products to the United States: 1998

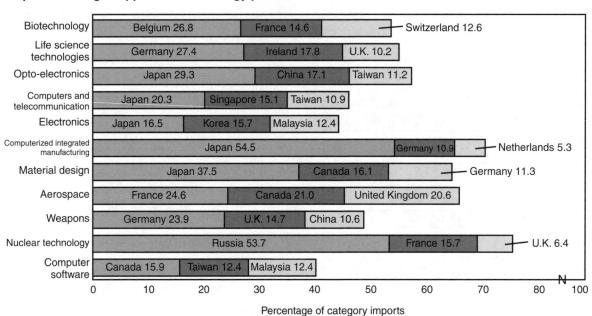

Biotechnology	Belgium 26.8	France 14.6	Switzerland 12.6
Life science technologies	Germany 27.4	Ireland 17.8	U.K. 10.2
Opto-electronics	Japan 29.3	China 17.1	Taiwan 11.2
Computers and telecommunication	Japan 20.3	Singapore 15.1	Taiwan 10.9
Electronics	Japan 16.5	Korea 15.7	Malaysia 12.4
Computerized integrated manufacturing	Japan 54.5	Germany 10.9	Netherlands 5.3
Material design	Japan 37.5	Canada 16.1	Germany 11.3
Aerospace	France 24.6	Canada 21.0	United Kingdom 20.6
Weapons	Germany 23.9	U.K. 14.7	China 10.6
Nuclear technology	Russia 53.7	France 15.7	U.K. 6.4
Computer software	Canada 15.9	Taiwan 12.4	Malaysia 12.4

Percentage of category imports

Figure 7-6.
This graph depicts the top three foreign suppliers of technology products to the United States. (Adapted from *Science & Engineering Indicators—2000*)

heavily commercial, as well as competitive, in nature?

A brief review of the U.S. government policies since the Truman administration (the mid-1940s to the early 1950s) indicates that a general interest in international development has been maintained. The motivations have been mixed, however, and the framework for involvement has changed over the years. A thread of humanitarian concern has endured. The element of competition with the former Soviet Union for international influence in defense lasted several decades and is not expected to completely disappear with the breakup of the Union of Soviet Socialist Republics (USSR). The importance of *Third World* resources to the United States and the awareness of the potential for export markets in those countries emerged as the quest for technological

leadership continued under the guise of development or foreign assistance. Meeting basic needs and human rights became a prominent component of U.S. foreign policy in the *Third World* during the late 1970s. More recently, the principal policy objectives of the United States seem to focus on combating terrorism, providing medical assistance, strengthening the private sector, and supporting U.S. national security interests. Regardless of the administrative tenets espoused over the years, technology transfer has retained a consuming presence throughout.

Economists have consistently highlighted the importance of technology in stimulating a nation's economic growth. Sometimes the creators of the technology benefit. In other cases, the borrowers of technology benefit. In any event, since technology has speeded economic growth and development in industrialized nations, should it not be expected to assist in the development of less industrialized countries?

Transferability and Appropriateness

Two important aspects of technology transfer to be studied are the nature of the transfer item itself and the recipient's capacity to adopt the technology once it has been transferred. An attempt to study how certain transfer items are assimilated and modified to suit local circumstances raises the issues of transferability and appropriateness. Should, for example, aircraft, microwave ovens, fiber optics, computers, or satellite dishes be transferred to India, the former Soviet Union, China, and countless African nations? Are these technology-intensive items useful to or even desired by the citizens of these countries?

Appropriate technology emerged as a discipline in which the objective was to develop a form of technology that meets local needs, uses local resources, involves the local residents in the decision-making process, and does not adversely affect the environment. *Appropriate technology (A.T.)* is discussed in greater detail in Chapter 8, but is defined here as a technology that "fits" the cultural or social situation for which it is intended without causing more problems than it solves. E. F. Schumacher popularized the term in the UK during the early 1970s. The term should not be assumed to describe secondhand or second-rate technology that has become obsolete and is no longer useful.

It seems reasonable to assume that no one is in favor of "inappropriate technology." It is also evident that some capital and technology-intensive products have proven to be unworkable (inappropriate) in certain geographic areas where the level of technological sophistication was not conducive to their implementation. In other words, the introduction of the new technology directly caused more problems than it solved.

Consider the situation in Mexico as described earlier in the *Dateline* segment of this chapter. In the mid-1960s, the Mexican government spearheaded the Border Industrialization Program to lure foreign manufacturing business from the United States into Mexico's border states. By the late 1990s, there were more than three thousand plants in operation, mostly along the border, employing more than nine hundred thousand people. The plants, known as *maquiladoras,* extend along the two thousand mile U.S.-Mexican border from Tijuana in the west to Matamoros on the Gulf of Mexico. Their owners include the likes of IBM, General Motors, Motorola, Eastman Kodak, and Zenith. These manufacturing jobs have been transferred to Mexico (and other *Third World* nations as well) with high social and environmental costs. Mexico's lax monitoring of industrial practices allows haphazard dumping of toxic wastes. Working conditions in these plants are typically inferior to those required by law in the United States. Maquiladora employees continue to work with unsafe machinery, without protective clothing, while being exposed to toxic pollutants, in areas poorly lighted and ventilated. The typical maquiladora worker is a woman in her prime reproductive years between the ages of sixteen and twenty-eight. Sanitation is poor, production quotas are high, and noise levels are often unbearable.

Another unfortunate transfer situation, spurred by the rush to develop an energy reserve in Alaska during the 1970s, involved the introduction of modern technology in the form of snowmobiles and televisions in small Alaskan communities. Technology artifacts uprooted the efficiency of a traditional hunting and fishing lifestyle. Grocery stores eventually supplanted subsistence hunting. Most homes not only became electrified, but had cable television installed as well. Without the emphasis on hunting and fishing that once existed in these communities, *boredom, idleness,* and *low self-esteem* became the operative terms to describe the situation in many Alaskan communities.

Another case of technology transfer's role in international development involved an agricultural project that procured its initial funding from the Rockefeller Foundation. U.S. scientists went to work with local scientists in Mexico and the Philippines to develop new varieties of wheat and rice that could outproduce existing local strains. What became known as the "Green Revolution" was considered to be a great success toward the abatement of famine and starvation.

Conversely, this transfer of agricultural technology hurt the small farmers who could not afford the fertilizer and water required by the new varieties. There is also an increased risk of crop failure through monoculture diseases introduced via the transfer activities. While it is true that recipient nations have been able to increase their crop yields, the problems of starvation still exist. These illustrations are not meant to be representative of all technology transfer activities occurring between the United States (as the technology donor) and less-developed societies (as the technology recipients). There have been, and are currently, many successful projects executed through hundreds of international organizations. Three of these groups include the following:

❑ International Monetary Fund (IMF). Headquartered in Washington, D.C., the IMF was established to promote international monetary cooperation, exchange stability, and orderly exchange arrangements; foster economic growth; and provide temporary financial assistance to countries to help ease balance of payments adjustment. The purposes of the IMF have developed over the years to meet the needs of its 183 member countries in an evolving world economy. It was established in the mid-1940s.

❑ United Nations Conference on Trade and Development (UNCTAD). Headquartered in Geneva, its mission is broadly economic through the promotion of international trade. It plans and implements trade policies favorable to developing nations. It fosters technology transfer and encourages international patent dialogue between industrialized nations and less-developed countries. The UNCTAD seeks to stimulate economic cooperation between member countries, expand world trade, and coordinate aid for less developed countries. It maximizes the trade, investment, and development opportunities of developing countries and assists them in efforts to integrate into the world economy on an equitable basis. Founded in the mid-1960s as a permanent intergovernmental body, it has 191 members.

❑ Organization for Economic Cooperation and Development (OECD). Headquartered in Paris, its thirty members include representatives of West European governments, Australia, Canada, Japan, Korea, Mexico, New Zealand, Scandinavian countries, Turkey, and the United States. These nations share a commitment to democratic government and the market economy. It was founded in the early 1960s.

Ostensibly, the technology transfer recommendations, projections, and assistance these organizations provide for those parties seeking assistance are based on the importance of the transferability and appropriateness ingredients, with reference to the recipient nation's existing "technological mix."

Rates of Diffusion

Once a technological process or product is introduced to the public, the potential for its introduction into the international technology transfer network is inevitably addressed. From the economist's perspective, the diffusion rate of this new technology quickly becomes a prime dimension for statistical analysis. The rate of diffusion of technology is typically examined from two vantage points. First, after the lead nation begins producing a new device, how long does it take for other countries to obtain the production capabilities to produce or reverse engineer the same product? This interval period is called the *imitation lag.*

Second, after the technology is first introduced into the innovating country's domestic market, how long does it take for it to gain acceptance in foreign markets? This is referred to as the *demand lag,* or social readiness, for the technology.

The imitation lag concept is most closely associated with an earlier discussion that focused on the decline during the 1980s in the U.S. share of the export market involving technology-intensive products. Some groups believe this apparent narrowing of the American technological lead is due in large part to the transfer of technology between U.S.-based multinational firms and their overseas subsidiaries. There is the logical assumption these transfers will increase the likelihood that key technological information and know-how will be assimilated into the minds and hands of those who run offshore non-U.S. firms that will eventually use it to their competitive advantages.

Besides the initial introduction of a new technology by the developing agent, there are also other factors seeming to influence the speed with which other countries are able to develop and produce a similar product on their own. The first of these factors involves domestic government expenditures on technology-based R & D activities. Organizations spending relatively large amounts on R & D projects are generally closer to the leading edge of the technological frontier and are more capable of efficient imitation activities.

Nations aspiring to be competent players in the venues of profitable technology transfer agreements recognize the need to promote R & D in order to strengthen their internal technological infrastructures. One of the most widely used indicators of a country's commitment to growth in scientific advancement and technology development is the ratio of R & D spending to gross domestic product (GDP). See Figure 7-7. You will see on these charts that the United States and Japan both peaked at 2.7 and 2.8 percent, respectively, in the early 1990s and dropped for several years before rising again near the close of the decade. Strength in U.S. R & D spending since the mid-1990s is attributed to electrical equipment, telecommunications, and computer services. In Japan, expenditure increases were most significant in the electronics, machinery, and automotive sectors—each associated or aligned with new digital technologies.

A second factor is related to the recipient nation's current level of concentration in the relevant industry. In other words, there may be a greater motivation for a country to venture into reverse engineering strategies to imitate a product that could greatly enhance their own productivity in a related discipline. For instance, a country such as Germany may have more interest in robotics technology than Spain, where the automobile manufacturing industry is not as concentrated. If the recipient country's existing technology mix is ripe for the implementation and full utilization of this new technology, there is no question they will place greater emphasis on imitation strategies.

R & D as a percentage of GDP, G-8 countries

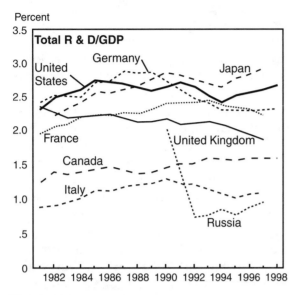

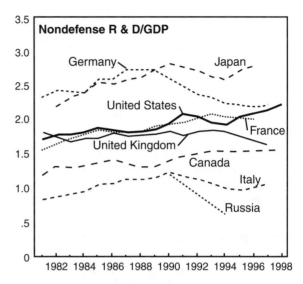

Figure 7-7.
This chart shows research and development funding as a percentage of gross domestic product, by country. (Adapted from *Science and Engineering Indicators—2000*)

The predominant transoceanic export and licensing of American technology, coupled with stronger trends in R & D expenditures in other industrial nations, provide the basic ingredients for reduced imitation lag periods and greater capacities for these countries to be more innovative with reference to technological endeavors. International technology transfer activities will become more complicated and interwoven as time passes. More importantly, however, the presence of a large technology gap between the technology creators and the technology borrowers is becoming less evident.

More that fifty years ago, the United States developed legislation to prevent the transfer of sensitive technologies to potential adversary states. It believed then, as it still does today, that its national security was dictated by American ability to maintain technological superiority in defense. Examples of legislation include the formation of the Coordinating Committee on Export Control (COCOM) in the late 1940s and the passage of the Export Administration Act (EAA) in the late 1970s. The EAA was later amended in the mid-1980s, requiring the Defense Department to develop and maintain a Militarily Critical Technologies List (MCTL). Around the same time, the Defense Technology Security Administration (DTSA) was established to ensure international transfers of defense-related technology are consistent with U.S. foreign policy and national security objectives. The DTSA continues to serve as the focal point within the Department of Defense for administration of its technology security program. It is responsible for reviewing requests and proposals aligned with the international transfer of defense-related technology goods and services.

The next section of this chapter focuses on these and other contemporary technology transfer issues through a discussion of the

United States' evolutionary political and technical involvements with three separate nations, each of which aspires to acquire and maintain a global leadership position in technological enterprises.

Contemporary Transfer Scenarios

As the world confronts the prospect of establishing and working its way into a new international order, any generalizations made about technology transfer activities or S & T policy for specific nations must remain tentative. The dawning of the 1990s unveiled dramatic events in Eastern Europe, German unification, the Persian Gulf War, and the demise of the Soviet Union. Near the middle of the 1990s, three nations (Mexico, the United States, and Canada) entered into the North American Free Trade Agreement (NAFTA), and during the short time living in the first decade of the twenty-first century, we have already witnessed the Y2K phenomenon, the reluctance of the United States to pursue the Kyoto Protocol (also referred to as the Greenhouse Treaty), and the formation of an international coalition to battle terrorism.

Essentially, the international order is a legal system or network encompassing the principles, ideologies, units of measure, and mechanisms through which international relations are coordinated and common international problems are confronted. The web of technology transfer activities and agreements is, therefore, influenced by the nature of our international order. With the close of the Cold War, an international order based on a few major powers imposing their views on other nations is no longer viable.

The course of establishing a new international order has been and will continue to be uneven. In our minds, the recipe for an emerging international order of coordinated global management should contain the Five Principles of Peaceful Coexistence developed nearly half a century ago at the Bandung conference of Afro-Asian States (1955): sovereignty, equality, peace, justice, and noninterference in internal affairs of states. International acquisition of S & T to ensure subsistence and enhance quality of life will undoubtedly remain an issue of concern throughout this dynamic and ever-evolving period at the dawn of the twenty-first century.

People's Republic of China (PRC)

The People's Republic of China (PRC) has maintained a technology transfer policy in close alignment with its foreign policy. While the PRC was on good terms with the Soviet Union in the 1950s, a substantial amount of Soviet technology was imported, mainly on a turnkey basis. During the early 1950s, China imported more than 150 turnkey plants, mostly in heavy industry, power generation, mining, refining, chemicals, and machine tools. More than four hundred research units were established, primarily focused on reverse engineering Soviet technology. These units ultimately evolved into three groups: Chinese Academies of Science, State Planning Commission, and the State Science and Technology Commission. When Sino-Soviet ties were severed in the early 1960s, China relied more on Japan and Western Europe as sources of technology. The onset of Mao Zedong's Cultural Revolution of the mid-1960s drastically constrained the entire nation's scientific and technological progress.

The Chinese S & T research structure was neglected, which curtailed the careers of a generation of would-be scientists.

The late 1970s hailed the birth of widespread economic reform and an open-door policy. Both movements were construed to be mutually supportive of technology transfer activities. Several aspects of the Deng Xiaoping inspired open-door policy that expanded foreign trade included a reduction of centralized government planning and control; the establishment of special economic zones, regions, and cities with increased decision-making autonomy related to foreign involvement; and approval of joint ventures, cooperative management strategies, wholly owned subsidiaries, and participatory ventures in natural resources surveys.

The PRC's open-door policy significantly modified earlier approaches to technology transfer. The 1950s technology imported from the Soviet Union was accompanied by blueprints, technical documentation, Soviet technical advisers, and training programs. Later agreements with Japan and the West did not involve peripheral technical support. The resultant arms-length turnkey plant transfers resulted in excessive costs and a total lack of the transfer of competencies essential for technological literacy and mastery of the operations. There was an absence of adequate training for the successful absorption of acquired technology throughout the system.

China's modernization efforts throughout the 1980s focused on the infusion of foreign technology to develop export capabilities versus sole dependency on substitution of imports. Technology transfer activities have since emphasized disembodied technology, "know-how," "know why," a complementary mix of hardware and software, and the hiring of foreign consultants and technical training support services. The PRC made great economic strides in the 1980s. Its GNP doubled, and China enjoyed a predominantly amiable relationship with the West. Despite its optimism, this huge nation remains a developing country with an immense population and a weak economic foundation.

Further, after the appalling events of the late 1980s in Tiananmen Square, Chinese Communist party leaders were forced to realize the impossibility of separating economic and political phenomena. The party's dream of being able to reap the harvest of increased production and technological efficiency by liberalizing the economy without slackening its control over the Chinese people was shattered. Their massacre of pro-democracy protesters seriously disrupted relations with the United States. A critical task of Chinese foreign policy for the 1990s was to create an international technology transfer environment allowing countries with different social and cultural systems to coexist.

Beyond the political battles between Washington, D.C. and Beijing, U.S. businesses routinely confront various obstacles and constraints in their dealings with China. U.S. technology suppliers are often willing (at first) to make concessions and defer to the overly bureaucratic Chinese supplier selection process. U.S. firms view technology sales and acquisitions as arrangements that must be mutually beneficial to both sides. From a U.S. technology supplier's perspective, China must continue to revise and reformulate its future technology transfer policies to provide the following:

❑ Greater flexibility in profit remittance to foreign investors.

❑ Satisfactory access to internal Chinese markets.

❑ Nondiscriminatory access to domestic suppliers of raw materials, energy, and labor.

❑ Long-term participatory partnerships between Chinese and American firms. The Chinese disregard for intellectual property rights agreements is also problematic, and piracy is very real.

The Chinese are not without complaints about U.S. trade and investment policies and practices. For nearly two decades, the PRC expressed dismay that it must renew its *most favored nation (MFN)* status each year, instead of it being given in perpetuity as for other developing nations. MFN status is now referred to as permanent normal trade relations (PNTR) and earns its recipient provision of concessional loans, support for World Bank funds, and a Generalized System of Preferences (GSP) treatment. The Chinese also object blatantly to certain U.S. export control laws. The United States still reviews license applications for exports to China quite rigorously, but the PRC will most likely be granted PNTR in the near future.

The Chinese remain anxious to increase their ability for self-reliance through an aggressive reconstruction of their S & T infrastructure. The Chinese Academy of Sciences has 121 research institutes, three universities, and more than four hundred businesses scattered across the country. The state run institutes perform in-house research and also serve as doctoral institutions. China's major universities are considerably ahead of those in the United States when it comes to being entrepreneurial. Many of the top academic institutions oversee subsidiary companies that commercialize technology invented by their skilled professors. Such companies might offer shares on the Chinese stock exchange, but they are still a part of the university structure and report to the university's administration. Unlike what might occur in the United States, these firms do not spin off to become independent operations, and they therefore allow the universities to benefit from a revenue stream.

Another example of contemporary technological prowess in China is portrayed in their gargantuan Three Gorges Dam project being constructed on the Yangtze River. See Figure 7-8. The government announced the project in the early 1990s, and it is slated to be completed in three multiyear stages.

Phase I stretched throughout the 1990s and consisted of building cofferdams within which the river bottom could be excavated and the foundation of the dam begun. A temporary ship lock was also constructed to allow shipping to continue throughout the project. Phase II, charted from 1998 to 2003, involves a good part of the dam itself. At the end of this phase, the projected plan is to fill the reservoir and allow the dam to begin to generate power, thus producing revenue to fund Phase III. The final phase of the project is forecast to culminate near the end of the first decade of the twenty-first century, and it entails completion of the dam across the Yangtze River, including building additional powerhouse units. Once the entire project is finished, Three Gorges Dam will be the world's largest hydroelectric dam, fitted with twenty-six generators rated at seven hundred megawatts each (i.e., equivalent to the output of approximately fifteen of the world's nuclear power plants currently in operation)! Despite the projected benefits of the project, there are severe environmental and social side effects to this transfer of technology; these are discussed briefly in Chapter 9 on environmental issues.

Regardless, this example illustrates China's level of technological advancement, and its plan to become a full partner in world trade and international technology transfer. Amazingly enough, Chairman Mao Zedong, as early as 1953, declared his staunch support of a Three Gorges Dam project for his country. In his mind, this structure would provide a forceful symbol of China's self-sufficiency and ability to

Figure 7-8.
This is the Three Gorges Dam Project on China's Yangtze River.

develop resources without assistance from the West. In many ways, China remains a land of distinctive contrasts, but it has made great strides toward technological modernization. The United States stays optimistic that China will make good progress in coordinating its technology policies and nurturing a compassionate human rights agenda. Both countries are members of the Asia-Pacific Economic Cooperation (APEC) forum, whose member leaders have pledged to meet the goal of free and open trade and investment in the Asia-Pacific region by the year 2020.

India

The Republic of India is also a sprawling land of contrasts and seeming contradictions. Since becoming independent in the late 1940s, the Indian government's basic philosophy of development has been self-reliance. Although India is a developing nation whose economy is heavily dependent on agricultural performance (62 percent of the population depends on agriculture), it is not technologically unsophisticated. The country continues to graduate a high number of science and engineering doctoral graduates each year and also has the world's third largest armed forces regimen. India's industrial base has grown steadily, but slowly. The support structure to bolster its growth is far short of what is needed. Electrical power is in critically short supply. India's communication and transportation networks do not even come close to meeting the needs of an industrialized nation. Its general framework for economic policy has remained fairly stable. India's accomplishments in nuclear power and space studies provide testimonial to its technological efforts. Conversely, a huge percentage of the nation's population remains illiterate and uninterested in the technological artifacts around them.

Technology transfer relationships between India and industrialized nations, specifically the United States, are delicate and complicated. While U.S. foreign policy has been conditioned largely by Cold War politics, India's policies have sought to maintain an independent position outside the Cold War struggle. Over the years, India has befriended America's adversaries, including the Soviet Union, North Vietnam, and Cuba. However, the United States has been the predominant source of military weapons to India's arch rival, Pakistan. India may be a democratic state with political values similar to the United States, but relations between the two have been less than congenial. During much of the four decades preceding 1990, the "world's largest democracy" was, by declaration, neither an ally to nor enemy of the "world's most powerful and technologically advanced democracy."

Throughout this cycle of conflict and cooperation, the Indian government developed a highly elaborate set of policies and practices for screening and regulating foreign direct investments. Many of these mechanisms presented a burdensome nightmare of bureaucracy to American firms interested in technology transfer agreements with India. Through the 1980s, the Indian government emphasized its preference for technology transfer investments in the form of sales or licensing of technology, as opposed to joint ventures. When foreign direct investments were approved, the technology supplier's ownership was generally limited to 40 percent of a company's equity. External investors were required to acquire equipment, components, and raw materials within the country. Imports to sustain the organization were seldom allowed, even when locally available substitutes cost more or were found to be of inferior quality.

Also, much to the frustration of American scientists and technologists, Indian patent laws have afforded significantly less protection than do the intellectual property laws of most industrialized countries. Strict limitations on equity ownership discourage multinational firms from installing state-of-the-art equipment, for fear they will lose control of valuable proprietary information. The lack of patent protection, combined with the government's efforts to immediately diffuse technology know-how to other firms after one local company has obtained it, do not describe an optimal transfer scenario for technology owners who have already invested heavily in R & D.

One final feature of India's industrial policy prior to the 1990s involved insistence that Indian nationals are competent to provide all the people-intensive skills and services to run joint venture facilities. The importance of organizational skills, managerial expertise, operating experiences, marketing plans, and equipment maintenance are often overlooked. The "learn by doing" approach is consistent with the philosophy of self-reliance. Local users are subsequently cut off from valuable training, education, experience, and technical innovation that could be obtained from the foreign companies. This deficiency in the technology transfer process may have contributed to the devastating pesticide plant accident in Bhopal in the early 1980s. To some extent, those Union Carbide workers had received job training based on rote memorization of steps rather than a thorough understanding of the logic behind the technical procedures.

Prior to the mid-1980s, two-thirds of India's defense-related technology acquisitions were from the Soviet Union. Toward the end of the decade, following the signing of the Memorandum of Understanding (between then-President Reagan and the late Prime Minister Rajiv Gandhi), Indo-American technological transfer dialogue involved India's receipt of three main items: advanced aero-engines for the development of a light combat aircraft, satellite and booster rocket technology for India's Space Program, and supercomputers for weather forecasting. These "dual-use" technologies implied the possibility of limited military cooperation between the United States and India. Near the end of the 1980s, our two governments signed an agreement for the establishment of a U.S.-India Fund (USIF) to provide for joint activities such as workshops; exchanges of scientists and experts; and joint research programs in the fields of education, culture, and sciences. USIF was initially set up for ten years, but its life was extended into an eleventh, providing funds for a large number of joint scientific projects.

A series of economic reforms were put into action in the early 1990s. Several of the reforms have enhanced technology transfer flow through various channels. For example, foreign investment and exchange regimes have been liberalized, tariffs and other trade barriers have been reduced, and India's financial sector has seen some modernization. Foreign portfolio and direct investment actions increased steadily through the 1990s. The Foreign Investment Promotion Board reviews proposals for direct foreign investment. In recent years, most receive approval. Automatic approvals are available for investments involving up to 100 percent foreign equity, depending on the kind of industry, which is contrary to the 40 percent ownership rule discussed above. As you might expect, foreign investment is particularly sought after in power generation, telecommunications, ports, roads, petroleum exploration and processing, and mining. Despite this healthy progress, India's economic growth is still constrained by cumbersome bureaucratic procedures, high interest rates, and inadequate infrastructure.

In the late 1990s, the United States was India's largest bilateral trading partner. Principal among U.S. exports to India are aircraft and parts, advanced machinery, and fertilizers. From India, the United States imports textiles and ready-made garments, gems and jewelry, and chemicals. On the downside, also at the close of the 1990s, India's nuclear tests were damaging to Indo-American relations, which resulted in the imposition of sanctions by the Clinton administration. U.S. sanctions on Indian entities linked to the nuclear industry remain a source of friction at the start of the twenty-first century.

Russian Federation[2]

The dissolution of the USSR makes an overview of U.S. technology transfer activities with this region highly speculative. For more than four decades, the controversial aspects of technology transfer activities were most pervasive when the subject of selling technology to the USSR was broached. Prior to the breakup of the Soviet Union, there were some four thousand federally subsidized research institutes located around the vast country—a number

[2]After the December 1991 dissolution of the Soviet Union, the Russian Federation became its largest successor state and inherited its permanent seat on the United Nations Security Council, as well as the bulk of its foreign assets and debt.

incidentally quite similar to the population of federally funded labs found in the United States. Under Soviet rule, an anxiety prevailed among scientists about sharing information in general. This of course was counterproductive to a vibrant technology transfer process even inside the country's borders. All inventions and scientific discoveries became Soviet state property once disclosed, at most bringing the inventor a bonus of a couple hundred dollars.

For several decades, many of the U.S. policy statements concerning technology deliveries outside our borders had been predicated on a bipolar world order. Without question, U.S. government officials had great difficulty authorizing or even tolerating the sale of technology to communist countries. Administrations since World War II have seriously questioned decisions to even allow Soviet scientists access to American universities and research materials when these educational endeavors could ultimately be used against us.

The Soviet Union first began to import technology on a broad scale in the 1920s, but the United States largely curtailed its exports during the Depression. Exports increased substantially during World War II when the United States was in alliance with the USSR against Germany and Japan. The beginning of the Cold War in the 1940s virtually ended these arrangements. Import policies changed again after Krushchev was removed from office in the mid-1960s. Soviet dependence on Western technology artifacts continued to increase into the 1980s.

Toward the end of the 1980s, in return for hard currency and high technology, Soviet research institutes started to sell their S & T back to the West. Much of the discreet technology developed for the Soviet space station went up for sale, as did a broad array of other technologies that had for decades been considered militarily sensitive.

Monsanto Corporation of St. Louis, Missouri formed a research collaboration with four Soviet laboratories in order to secure licensing rights for a process by which diamond films are applied to polymers. Around 1990, the Soviet Association of Biotechnology signed an agreement with Genesis Technology Group of Cambridge, Massachusetts to develop medical products deemed marketable to pharmaceutical companies or midsized biotech firms. It was agreed Soviet scientists would retain patents and receive royalties, but would sell commercial rights in return for future R & D funds.

Shortly after these and many other promising transfer agreements were finalized, the future of S & T in the former Soviet Union became highly uncertain. The fracture of fifteen Soviet republics caused serious concern among the estimated 290 million inhabitants. Twentieth century communism left the Russian Federation and the other republics in a state of economic disaster and environmental chaos. Political leaders were challenged to confront serious problems, several of which included radioactive contamination of massive land areas; aging nuclear reactors too unsafe to operate, but necessary for energy demands to be met; extensive deforestation and topsoil erosion in Siberia; thousands of barrels of oil spills due to faulty construction of oil pipelines; nuclear arms control; and the destruction of numerous chemical and biological munitions stockpiles across the region.

Early in the 1990s, the new democratic Russian government saw it necessary to curtail funding to most research institutes. At the same time, however, these entities were granted both the right and incentive to solicit business from foreign firms in order to generate their own operating incomes. The intellectual foundation underlying Russian innovation in S & T has fewer resources than when the state was involved in technology ownership, but collaboration is now more widely encouraged. Through the 1990s and into the twenty-first century, as the country shifted to a more market-oriented economic structure, technology transfer arrangements and collaborative opportunities flourished. Several examples of the invigorated culture for technology transfer activities within the Russian Federation include establishment by the federal government of a trading company of technologies called The Russia House (www.mnts.msk.su), dedicated patent and licensing offices in most research institutes, and an abundance of foreign and Russian law firms specializing in intellectual property deals.

While these developments sound promising, the intellectual property rights situation in Russia remains unsatisfactory and has been a significant obstacle to foreign investment in sectors like pharmaceuticals. In this field, the cost of investing in the development of new drugs makes companies skeptical about risks regarding their intellectual property rights. The former Soviet technology transfer and licensing monopoly, Licensintorg, lost its grip on transactions in the early 1990s and eventually gave way altogether to private initiatives. Most institutes and research firms now work with their own patent offices, where knowledge of basic intellectual property laws is still deficient. Investors are hopeful this will improve in the short term.

Large U.S. companies like Monsanto and Uniroyal Chemical, which have entered into technology transfer arrangements with firms in the Russian Federation, caution companies to thoroughly investigate potential partners before signing a joint research contract. Otherwise, when the project is finished, it may turn out that the Russians who signed the licensing agreement (or other document) did not have the authority to do so on behalf of their institution. Ultimately, the U.S. company could end up with no rights to the intellectual property resulting from the research, or worse, they could be prosecuted for attempting to steal Russian property.

Two sectors in which the Russian Federation has a large accumulation of human capital and substantial potential for competitive advantage are in the space and aerospace industries. In recent years, major types of collaboration in these fields have led to the following:

❑ Joint ventures aimed at fixing a specific weakness in the capabilities of a Russian industry, such as the agreement between GE Aviation and Rybinsk Motors.

❑ Alliances designed to transfer specific components of Russian hard technology to the Western partner, such as the contract(s) between Pratt & Whitney and Energomask.

❑ Technological alliances, such as those between Boeing, Dassault, DASA, and Airbus with Zhukovskii Central Aerohydrodynamics, relative to specific research projects being carried out by the latter for its Western partners.

A central and valid theme emerging seems to be a recognized value placed on Russian hard technologies. More critical to the formula perhaps is the role of Western soft technology and finance needed to bring Russian S & T to the global market. Without question, it appears the Russian Federation has managed, in less than fifteen years, to articulate an attitude of scientific and technological cooperation and collaboration with its foreign colleagues.

Summary

There are many types of technology transfer agreements occurring on an international scale. Most of these involve some form of horizontal or vertical transfer of a new material or product, design specifications, or capacity for production from one organization or nation to another. Transfer agreements typically consist of a technology donor and a technology recipient who rely on a variety of channels for the linkage to take place. A few of the formal channels for transfer include foreign direct investment, turnkey plants, licensing, and the export of manufactured goods. An assortment of non-commercial and more informal transfer channels also permits the effective exchange of technological information and expertise among world nations.

A comprehensive assessment of the transferability and appropriateness of the technologies being considered should always precede transfer investments. International technology transfer activities are most successful when multinational companies are skillful and able to cope with unfamiliar language and cultural clues, arbitrarily enforced legal protection, and bureaucratic red tape. Current technology transfer interactions between the United States and the People's Republic of China, India, and the Russian Federation illustrate many of these principles. These three nations were selected for review in that they accurately reveal the fact the contemporary international technology transfer network is characterized by numerous political complexities and technical complications, as well as a great many beneficial settlements and transactions.

Discussion Questions

1. Provide a general definition of the term *technology transfer.* Give several contemporary examples to illustrate your explanation of this concept.
2. What are the key elements that can be used to conduct a comparative analysis of the diversified activities associated with international technology transfer?
3. What strategies and procedures are U.S. manufacturers of high-tech products using to prevent or deter the process of reverse engineering?
4. Explain the difference between foreign direct investment and licensing, as they are currently used to transfer technology from one country to another or from the public sector to the private sector. Provide contemporary examples of each transfer channel.
5. Speculate on the causal factors that may be responsible for intermittent declines in the U.S. global market share of high-tech exports.
6. Define the term *appropriate technology* and discuss several examples illustrating its presence or absence in your current living arrangement or daily routine.
7. Select a foreign nation other than those discussed in this chapter and develop a profile of its current involvement with the United States, with reference to technology transfer activities.

Suggested Readings for Further Study

Arora, A., Arunachalam, V., Asundi, J., and Fernandes, R. 2001. The Indian software services industry. *Research Policy* 30 (8): 1267–1287.

Bajpai, K. S. 1992. India in 1991: New beginnings. *Asian Survey* 32 (2): 207–216.

Bzhilianskaya, L. 1999. Foreign direct investment in the science-based industries of Russia. In *Foreign direct investment and technology transfer in the former Soviet Union.* Cheltenham: Edward Elgar, 64–83.

Covault, C. 2000. Russian ventures face tech-transfer gauntlet. *Aviation Week & Space Technology* 152 (12): 34–35.

Dray, J. and J. Menosky. 1983. Computers and a new world order. *Technology Review* 86 (4): 12–16.

Dyker, D. 2001. Technology exchange and the foreign business sector in Russia. *Research Policy* 30 (5): 851–868.

Etzkowitz, H. 2001. The second academic revolution and the rise of entrepreneurial science. *IEEE Technology and Science Magazine* 20 (2): 18–29.

Foreign competitors are challenging U.S. leadership in high-tech trade. 1985. *Business America* 8 (7): 20–21.

Frame, J. D. 1983. *International business and global technology.* Lexington, Mass.: Lexington, D. C. Heath and Company.

Fransman, M. and K. King, eds. 1984. *Technological capability in the Third World.* New York: St. Martin's.

Gee, S. 1981. *Technology transfer, innovation and international competitiveness.* New York: Wiley.

Golden, W. T., ed. 1991. *Worldwide science and technology advice to the highest levels of governments.* New York: Pergamon.

Graham, E. and P. Krugman. 1991. *Foreign direct investment in the United States.* 2d ed. Washington, D.C.: Institute for International Economics.

Graham, L. and E. Skolnikoff. 1992. Soviet science in crisis. *STS News* Special Issue (February): 1–5.

Gwynne, P. 1991. China passes copyright law. *Nature* 352 (29 August): 750.

Heeks, R. 1996. India's software industry: State policy, liberalization, and industrial development. Beverly Hills, Calif.: Sage.

Huo, H. 1992. Patterns of behavior in China's foreign policy: The Gulf crisis and beyond. *Asian Survey* 32 (3): 263–276.

Kiely, T. 1990. Soviet science for sale. *Technology Review* 93 (7): 16, 18.

Kozyrev, A. 1992. Russia: A chance for survival. *Foreign Affairs* 7 (2): 1–16.

LaDou, J. 1991. Deadly migration: Hazardous industries' flight to the Third World. *Technology Review* 94 (5): 47–53.

Lawson, E. K., ed. 1988. *U.S.-China trade: Problems and prospects.* New York: Praeger.

Levitt, T. 1983. The globalization of markets. *Harvard Business Review* 61 (3): 92–102.

Liu, X., and S. White. 2001. Comparing innovation systems: A framework and application to China's transitional context. *Research Policy* 30 (7): 1091–1114.

Mansfield, E. 1982. *Technology transfer, productivity, and economic policy.* New York: Norton.

Marcus, S. J. 1992. After the USSR: Environmental management, market style. *Technology Review* 95 (January): 63–68.

McCulloch, R. 1981. Technology transfer to developing countries: Implications of international regulation. *ANNALS American Academy of Political and Social Science* 458 (November): 110–122.

McDonald, T. D. 1989. *The technological transformation of China.* Washington, D.C.: National Defense Univ. Press.

Miller, A. and F. W. Rushing. 1990. Update China: Technology transfer and trade. *Business* 40 (January–March): 25–33.

Minchener, A. J. 2000. Technology transfer issues and challenges for improved energy efficiency and environmental performance in China. *International Journal of Energy Research* 24 (11): 1011–1027.

Moss, T. 1983. Universities, technology transfer. *Vital Speeches of the Day* 49 (11): 327–329.

National Science Board. 2000. *Science & Engineering Indicators—2000.* Arlington, Va.: National Science Foundation.

Petroski, H. 2001. China journal I. *American Scientist* 89 (3): 198–203.

Ruddy, T. R. (1998). Technology transfer and business development in Russia: A review and status report. *BioTactics in Action.* [Internet]. 1 (6). <http//www.biotactics.com/Newsletter/v1i6/Russia1.htm> [1 November 2001].

Sahovic, M. 1992. The world and China. *Review of International Affairs.* 43 (1002): 26–27.

Shulin, D. 1992. Reform and opening: The pace accelerates. *China Today* 4 (6): 10–13.

Stanglin, D. 1992. Toxic wasteland (former Soviet Union). *U.S. News & World Report* (13 April): 40–46.

Stobaugh, R. and L. T. Wells, Jr., eds. 1984. *Technology crossing borders: The choice, transfer, and management of international technology flows.* Boston: Harvard Business School.

Stoever, W. A. 1989. Foreign collaborations policy in India: A review. *The Journal of Developing Areas* 23 (4): 485–504.

Technology Administration, Department of Commerce. 1990. *Emerging technologies: A survey of technical and economic opportunities.* Washington, D.C.: Department of Commerce.

Teece, D. J. 1981. The market for know-how and the efficient transfer of technology. *ANNALS American Academy of Political and Social Science* 458 (November): 81–96.

Thomas, R. G. C. 1990. U.S. transfers of "dual-use" technologies to India. *Asian Survey* 30 (9): 825–845.

U.S. export control policy hurts manufacturing sales to China. 2000. *Modern Machine Shop* 73 (1): 303.

Vernikov, A. 1991. New entrants in Soviet foreign trade: Behaviour patterns and regulation in the transitional period. *Soviet Studies* 43 (5): 823–836.

Warhurst, A. 1991. Technology transfer and the development of China's offshore oil industry. *World Development* 19 (8): 1055–1073.

Watkins, J. D. 1983. Technology transfer: A costly race with ourselves. *Vital Speeches of the Day* 49 (11): 322–324.

Wendt, E. A. 1989. U.S. stance toward the Soviet Union on trade and technology. *Department of State Bulletin* 89 (2142): 20–23.

Wilkinson, S. 2000. Shopping for R & D in Russia and China. *Chemical & Engineering News* 78 (15): 32–35.

Xusheng, M. 1992. Thoughts on the question of establishing a new international order. *Review of International Affairs* 43 (1001): 33–35.

Zivanov, D. 1991. The disintegration of the Soviet Union. *Review of International Affairs, I.XII.* 42 (998–1000): 17–19.

Technology has many applications that benefit our environment. Here, a biological science technician records water depth measurements made with an electronic caliper during infiltration studies on no-till plots. (USDA Agricultural Research Service)

Chapter 8

Appropriate Technology

Hayward, California

The Internet has been a boon to thousands of Americans who are deaf or hard of hearing. The World Wide Web offers everything from on-line deaf dating services to chat rooms just for people who are deaf.

DeafNation® communications, an Internet Service Provider (ISP) located in Nevada, provides service to over seven thousand subscribers. The Internet allows individuals who are deaf to fully participate in business and social activities without any stigma attached. One businessman who utilizes the Internet is David Smario, a software programmer. David e-mails his resume and proposals to customers like any consultant. A prospective customer does not even know he is deaf until they meet face-to-face....

Introduction

Appropriate technology (A.T.) has been used to cover a wide range of technologies and lifestyles, including sustainable living, alternative fuels, and ethical technology transfers. A technology is considered appropriate if it solves a social problem without many adverse negative effects. In order to understand A.T., first we need to understand the relationship of technology to society. Every new technology has consequences for society. Some of those consequences can be foreseen, and some of the consequences are unforeseen.

The development of the first Apple® computer led to more people using a computer and, therefore, to more people knowing how to use a computer. Since then, we have become so accustomed to newer computers it is difficult to remember when they first exploded on the open market. An unintended consequence of the computer boom has been an increase in hazardous materials. As computers are discarded, they increase the amount of hazardous waste in local landfills. Today, we can ask the question of whether or not the computer itself is an A.T.

Depending on how we define *appropriate,* our answers could be different. If we defined *appropriate* as increasing one's production, while at the same time reducing dependence on natural resources (think, for example, of telecommuting), then we could conclude the computer is an A.T. If, on the other hand, we define the appropriateness of a technology based upon the waste it generates, then we might conclude the computer is not an A.T.

A technology is appropriate when its intended positive consequences outweigh its unintended negative consequences, although some unintended consequences might actually be good, as is the case in our *Dateline* above. *Appropriateness* is not a simple term, either. The decision of whether or not a technology is appropriate can only be made after one investigates a technology, understands the environment or context of that technology, and knows the reason for or purpose of that technology. The situation described in the *Dateline* section of this chapter shows an appropriate use of the Internet among the deaf community. Because of the anonymity provided by this communication medium, deaf consultants, like the person discussed above, can work without any preconceptions about their abilities. Although the Internet was not developed primarily to assist deaf people, this is an example of a positive, unintended consequence of this technology. Most people would agree this is an appropriate use of computer technology.

Often we don't consider the issue of appropriateness when we choose to use a particular technology. In this regard, we are very much like the engineers who design these technologies. A new car or toaster is often designed to meet certain technical and marketing constraints. Experts in various fields develop these constraints. As a product goes through the design and development cycle, often the potential users' needs get lost. Consequently, when the product is on the market, it might only fit the needs for a particular subgroup in society. The traditional automobile industry is a perfect example of this type of thinking. For most of the twentieth century, almost all automobile designers were men, and they designed their vehicles with male drivers in mind. However, in the United States at least, 80 percent of all new car purchases are either influenced or made by women. The U.S. automobile manufacturing companies were confronted with the issue of appropriateness—were the male-designed cars they built appropriate for their female customers?

One solution addressing the issue of appropriateness is to let the product design and development be guided by the potential users. In the 1990s, Ford Motor Company did this by assembling a team of thirty women engineers to redesign the Windstar® van. The team came up with several changes that made the Windstar van more appropriate for women drivers, including a larger gas tank to reduce the need for refueling; a thinner steering wheel; and "sleeping baby" lights that provide indirect, soft lighting inside the vehicle. Although this is an American example, the issues it raises can be applied to other technologies. The end user will not necessarily use a product the way it was designed to be used. Also, a designer can only hypothesize about whether or not a technology will be appropriate. Without the input of the various constituent groups, a new technology may not be successful or appropriate.

The focus of this chapter will be to discuss the issues and emerging technologies falling under the umbrella of A.T.s in the United States, as well as overseas. Some of the questions we will address include the following:

❑ How is smart growth changing the land use patterns across the country?

❑ How can renewable energy sources best be integrated into our existing energy system?

❑ What is the environmental impact of green buildings?

❑ What are the different techniques and technologies used in sustainable agriculture?

❑ How does a company implement International Organization for Standardization (ISO) 14000? How do these standards affect the way a company is managed?

Key Concepts

As you review the material contained in this chapter, you should learn the meaning of each of the following technical terms and phrases.

agroecology
biomass
environmental management
ethnobotany
green building
integrated pest management (IPM)
International Organization for Standardization (ISO) 14000
new urbanism
nongovernmental organization (NGO)
renewable energy
smart growth
sustainable agriculture
technology equity

What Is Appropriate Technology (A.T.)?

There are three ways of evaluating appropriateness: technical, cultural, and economic. A decision about whether or not a technology is technically appropriate can be made after considering the technical knowledge and background of the people who will be using this technology. A decision about cultural appropriateness should analyze the technology in its relationship to the critical systems in the society, including family, religion, labor, and education. A decision about economic appropriateness would be made after looking at a technology's effect on income levels and distribution in a society and disparity between different socioeconomic groups.

A decision about appropriateness is dependent upon the country and situation. Appropriate technology (A.T.) is used both in discussing technology transfers, as well as the use of technology in an individual country. In the case of A.T. in the United States, a general perception is that a technology is appropriate if it has little to no adverse effect on the environment, social structure, and culture of the United States. Some examples of A.T. used in the United States are renewable energy sources, sustainable agriculture, and using recycled materials in the construction of new buildings.

Appropriateness is also used in relation to technology transfers. Here, the decision about whether a technology is appropriate or not can differ depending upon where the technology is being transferred. What one country might consider acceptable, another may not. In many cases in the past, technology has been transferred from a developed country to a developing country without much thought or consideration as to whether or not it is appropriate. Each country has its own constraints that will interact with the technology being transferred. As an example, in general, solar energy is an A.T. because it reduces our dependence on fossil fuels. It would not, however, be appropriate to transfer solar technologies to a country with very little sunlight. Does the country receiving the technology agree that the technology is useful for them? The issue of appropriateness is particularly important when there is a disagreement between the developed and the developing country as to the appropriateness of a technology.

The issue of A.T. is inherently linked to technology transfer. As discussed in the previous chapter, technology transfer is the transfer of a technology from one group to another group, most often from one country to another country. As technologies are being transferred or a decision is being made as to whether or not they should be transferred, an assessment of the appropriateness should be completed. Various factors for this assessment would include the following[1]:

❑ What is the need?
❑ Is there an adequate business environment in place for this technology?
❑ What is the best technical option for the transfer? (Some issues include the requirements for operating the technology, repair facilities for the technology, scope of the technology.)
❑ What are the possible unintended negative effects of the technology?
❑ What are the broader cultural, political, and social effects of the technology?

Many activists argue that the local people in a country understand their individual needs better than people in a country far away. Unfortunately, since many technologies are transferred from one government or industry to another, the needs or desires of potential local users may not be heard and are often overlooked.

---

[1]Everts, S. 1998. *Gender and technology. Empowering women, engendering development.* New York: Zed, 34.

The U.S. government has a mixed record in supporting research into A.T. After the 1973–1974 oil embargo, people in the United States were increasingly sensitized to energy conservation. In the late 1970s, the United States created a National Center for Appropriate Technology (NCAT), which focused on investigating technologies and processes that would be appropriate for low-income communities. NCAT was funded as a federal agency until the early 1980s, when Ronald Reagan became president. During the pro-business 1980s, A.T. was put on the back burner, at least with respect to government funding. However, NCAT found new life as a contractor with the U.S. Department of Energy (DOE). NCAT was hired to evaluate the twenty-two hundred final reports from small grants funded by the DOE and build a database of technical information related to the transfer of technology from the government to civilian applications. Near the end of the 1980s, a new national A.T. center was established and funded by the U.S. Department of Agriculture, called the Appropriate Technology Transfer for Rural Areas (ATTRA). ATTRA, which is managed by NCAT, addresses the issue of A.T. in three areas: sustainable farming, alternative crops and livestock, and innovative marketing.

In the following sections, we will discuss several examples of A.T. in the United States. In the United States, the issue of A.T. is usually linked to a philosophy trying to balance growth with a concern for the environment. Often, these environmentally sensitive technologies are referred to as "green technologies." Whatever we may call them, they are evaluated in a different way than traditional technologies being developed for the marketplace.

Renewable Energy

Renewable sources of energy, by definition, cannot run out. Some energy writers say they are eternal. Generally speaking, they do not blow up, melt down, or pollute the environment. After the first oil crunch of the early 1970s, optimistic forecasters claimed alternative energy sources could eventually meet all of the world's energy needs. Due to the fact that all renewable energy sources have at least two inherent shortcomings, that optimism has faded somewhat over the past few years.

> **RENEWABLE ENERGY**
> Sources of energy that are always available because, in reality, they cannot run out, and they can be replenished.

Renewable sources of energy are diffuse and intermittent. One example of the diffuseness feature is that a one thousand–megawatt solar farm might occupy about five thousand acres of land, while a nuclear power station with the same generating capacity only requires around 150 acres. Since renewable energy is intermittent, it is also unreliable. Although the wind often blows strongest when most power is needed (late afternoon and evening), it cannot be counted on to do so each day. Likewise, the Sun may not always shine when you expect it to—consider the last time you were forced to cancel an outing! For these reasons, any renewable supply must be attached to an efficient means of storing energy for the lean times. Storage systems are becoming more efficient, but commonly waste up to 20 percent of the accumulated energy. Despite these limitations, renewable energy is making a visible contribution to electricity grids in various parts of the world.

Renewable energy industries produce energy-using resources such as sunlight, wind, water current, and organic waste. Eternal forces in nature continually replenish each of these resources. Contemporary renewable energy technologies include solar thermal energy, biomass energy, geothermal energy, solar photovoltaics (PVs), wind energy, small hydroelectric energy, and ocean energy. The first three on the list generate heat energy, while the remaining techniques directly generate electric power. By the late 1990s, renewable energy accounted for 14 percent of the total energy consumption in the world. Two types dominate renewable energy sources: biomass and hydropower. Other forms or sources of renewable energy include geothermal, wind, ocean, PV cells, and thermal solar. All of these are discussed briefly in the following sections.

Biomass

The most common renewable source of energy is organic plant matter. Many developing countries depend on wood and agricultural waste for energy. Almost half of India's energy consumption and nearly 90 percent of the total energy consumption in several small countries in Africa is provided by wood. India obtains another tenth of its energy from animal dung and crop waste, but the use of *biomass* is not limited to developing countries. Sweden has increased its use of biomass dramatically in the last ten years and presently uses fast growing willow trees and other organics to supply 20 percent of its total energy supply.

In many areas of the developing world, trees are being used faster than they are being replaced. Also, when dung is burned for energy, farmers deprive their fields of essential nutrients. As a result, they reduce the amount of crops grown and fed to cattle to produce more waste matter. One solution may be found in biogas. Digesters can use the dung to make methane gas and also yield a residue that is good fertilizer. China is already making use of several thousand digesters across its expansive landmass.

Plants may have a modest future in replacing the oil used as a fuel for vehicular transportation. Brazil presently commands an impressive program of ethanol production from its sugar cane crops. In other parts of the globe, vehicles are running experimentally on coconut oil in the Philippines and gopher weed in Arizona. National Aeronautics and Space Administration (NASA) scientists found that sewage-fed water hyacinths can be converted into biogas. The Japanese are experimenting with microorganisms that eat rice waste to produce hydrogen, which could ideally be used in conjunction with fuel cells.

Hydropower

Hydropower has been used for literally thousands of years, beginning with waterwheels used in mills and for irrigation. Modern large hydropower plants are very expensive to build; however, they have lower operation costs than thermal or nuclear plants. A problem with widespread use of hydropower is that it is not distributed equally around the world. The countries in North America, including the United States, create more hydropower than most of the rest of the world. See Figure 8-1. In the United States alone, about 10 percent of the total electricity is generated from hydropower. Although this number is high, it has dropped since the 1940s, when 40 percent of the electricity in the United States was hydropower.

There are different types of hydroelectric plants. The Tazimina Hydroelectric Project, located around 175 miles southwest of Anchorage, Alaska, is an example of a diversion hydropower plant where the water

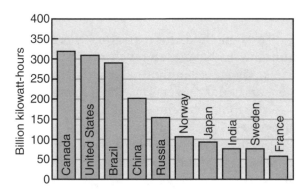

Figure 8-1.
This graph shows the top hydroelectric generating countries. (Energy Information Administration, U.S. Department of Energy)

is channeled through a canal. This way, since there is no dam, there is less disruption of the natural environment. Disruption of the environment is the major reason why there are fewer hydropower plants being built today. Since hydropower plants require water to be channeled away from its natural flow, by design they disrupt the environment. Hydropower has caused several changes in the environment: reduction in both the amount and variety of fish in a river, increase in sediment, decrease in water quality because of decayed material in the flooded area, and human health problems such as malaria. Engineers have been revising hydropower plant designs to try to mitigate these problems. Many new plants include fish ladders allowing fish to bypass the power generation, but even this solution is not appropriate to all countries.

Geothermal Energy

This form of energy is not strictly renewable or eternal because Earth's rocks will eventually cool. Residents of Italy will also argue that, contrary to some publications, it is not a new form of energy. The Italians have been running a power station fueled by hydrothermal power (steam plus hot water) intermittently since the early 1910s. The planet's relatively small number of high quality reserves of subsurface steam and hot water limits the potential of this form of energy. For those countries where active geysers have been found, geothermal power generation is quite effective. Of all the countries in the world, the Philippines has the highest percentage of power generated from geothermal sources; 22 percent of its electricity is generated by geothermal steam. The percentage of geothermal energy is high (at least 10 to 20 percent of the total) in four other countries: Costa Rica, El Salvador, Kenya, and Nicaragua.

There is a great deal of heat trapped inside Earth just waiting to be vented, but larger scale uses of geothermal power are still at least a decade or more in the future. Central America, parts of Southeast Asia, and the western United States have the greatest potential for major reliance on geothermal energy. Promising sites also exist in parts of southern Europe and East Africa. Taking full advantage of this resource will necessitate new technologies to tap into deeper geothermal reservoirs and make use of lower temperature geothermal waters.

Wind

Experts in the field of alternative energy feel wind energy is the most favorable of the renewable resources. Windmills mechanically turn turbines without an intermediate stage of heating water. In the early 1980s, more than eight thousand wind machines were installed in California. One of the largest wind farms is presently found in the rolling, windswept hills of the Altamont Pass, east of San Francisco. Wind farms are clusters of machines, generally located in mountain passes, connected to utility lines.

Attempts to reap economies of scale by building larger windmills capable of generating more than one megawatt of power have been suppressed by technical problems. Capital costs have remained prohibitive. More reliable and less expensive wind machines are being designed and built. One newly approved wind project, consisting of two hundred giant turbines, will be located in the sea off Ireland's east coast. When this wind plant is finished, it will supply 10 percent of Ireland's electricity needs. This large-scale project may stimulate the adoption of this technology in parts of the world where wind conditions are less than ideal. Despite technical problems, world wind generating capacity increased over 450 percent from 1995 to 2001, and this trend does not look likely to stop. See Figure 8-2. Germany leads the world in the use of wind power, with the United States, Spain, and Denmark following.

The Ocean

Three methods for extracting energy from the sea have been reviewed seriously: wave power, ocean thermal energy conversion, and tidal power. The first aims to harness the motion of the waves using a variety of devices. The second seeks to exploit the temperature differences between the warm surface layer and the colder deep waters of the world's oceans. Both could generate sizable amounts of electricity, but neither one has yet been demonstrated much beyond the experimental stage.

The third method is known as tidal power. It is similar to hydroelectric power in the sense that it is severely restricted by geography. It requires long, tapering bays that drive the tide into a large bore as it

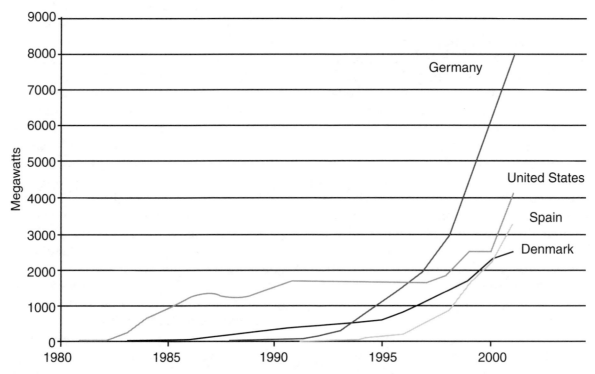

Figure 8-2.
Wind generating capacity has increased dramatically, over 450 percent, since 1995. (Adapted from Earth Policy Institute)

moves along the channel. The incoming tide can then be trapped behind a barrier of some sort and ultimately used to drive turbines on its way out again. There are not very many sites around the world having these specific geographic features. Two suitable locations are found at the Bay of Fundy in Canada and the Severn Estuary in Great Britain, but at this time, generating electricity from the power of the ocean is generally unfeasible. The Severn Estuary plan, which Great Britain decided not to pursue, would have cost the country $12 billion.

Photovoltaic (PV) Cells

Semiconductors have the unique property of being able to turn sunlight directly into electric current. This application is surfacing in a variety of items such as solar powered calculators, refrigerators, and satellites. According to some energy forecasters, solar cells installed on rooftops may allow for a much greater decentralization of electricity than other technologies. Photovoltaics (PVs), however, are still too expensive for large-scale power generation. Developers have not yet been able to make cells cheaper without sacrificing conversion efficiency.

Thermal Solar Power

Thermal solar power technologies and solar ponds are projected to have competitive generating costs by the end of the century. The capital cost for expensive items like polished mirrors to track the path of the Sun is presently exorbitant.

Southern California Energy (SCE), one of the world's leading innovators of alternative energy methods, has used solar energy as part of its production for over twenty years. With the support of a multimillion dollar grant from the U.S. DOE, SCE constructed the country's first commercial scale solar electric facility in the Mojave Desert. Known

officially as Solar One, it was the largest central receiver solar plant in the world. In the Mojave Desert, direct sunlight often produces temperatures of 160 degrees at Earth's surface. Using Solar One's parabolic mirrors, water was converted to steam and used to power a steam turbine. A consortium of the U.S. DOE and a number of electric utilities, led by SCE, revamped Solar One in the mid-1990s and changed it to a more advanced solar technology—molten salt. See Figure 8-3. At Solar Two, solar energy is collected through a field of individually guided mirrors, called heliostats. The sunlight heats salt to 1,050 degrees Fahrenheit, which turns the salt into a liquid (or molten salt). The liquid and hot salt is then piped away, stored, and used to power a steam turbine.

Both Solar One and Solar Two were designed as temporary facilities to test the viability of large-scale solar generation. Solar One produced power only when the Sun was shining. A sharp-edged cloud, large enough to block the Sun for just five minutes, automatically tripped the turbine. System restart took more than an hour. Solar Two solved this problem and proved that solar energy could be produced on a commercial scale. The molten salt technology allowed the energy to be collected during the sunny days and stored until there was demand.

It is unlikely the solar technologies described here or the other alternative energy technologies mentioned will ever totally replace the world's conventional energy options. The most important point to recognize is they most definitely belong in our international energy mix as we continue with our quest toward the creation of a sustainable society. We must maintain our momentum toward the further development and advancement of all potential energy resources in order to make the imminent transition to an oil-independent civilization with the least amount of trauma.

Figure 8-3.
Most of Solar Two's heliostats are large, segmented glass mirrors mounted on frames tracking the Sun throughout the day, continuously directing solar energy onto the central receiver. Solar Two has more than eighteen hundred heliostats, and taken together, they comprise the plant's largest single component. By collecting solar energy during daylight hours and storing it in hot molten salt, power tower technology gives utilities an alternative method for meeting peak loads when demand for electricity is high. For example, during summer afternoons, when air conditioning use is heaviest, the stored energy can be used to generate electricity and supply additional power to the utility grid. (Energy Technology Visuals Collection)

Changing our energy production structure is not easy. It is a long-term process, and it will require commitments from business, government, and individuals. Since most scientists believe that conventional energy resources, including coal and gas, could last another fifty to one hundred years, it is difficult to create a sense of urgency about renewable energy sources. Also, renewable energy, as already discussed, does not mean there will be no environmental damage. The perfect, nonpolluting, widely available energy source has yet to be found. For the near future at least, biomass and wind power have the most potential for use on a broad scale.

Smart Growth

Prior to World War II, most Americans lived either in rural communities or in cities. After the war, there was an explosion in a different type of land use—suburbs. As GIs from World War II got married and had children, they wanted to buy their own houses. These homes were usually single-family houses built on converted agricultural land. Of key importance, these new suburbs were located far from the centers of the older cities. Suburbanization, in turn, led to commuting, and commuting led to more use of personal cars for work and personal travel. This trend of development is the opposite of smart growth.

What do we mean by smart growth? **Smart growth** is development accommodating the needs of a community without sacrificing the environment. Smart growth aims to balance development and environmental protection by creating new developments centered more in the towns and cities; include alternative transit options (trains, bike paths, and safe walkways); and have mixed use development. Mixed use development moves away from the post–World War II ideal of single home only suburbs to a model including housing, commercial, and retail space in the same development. An example of a recent mixed use development is the Paseo Colorado complex in Pasadena, California. The new complex was built in the center of town and includes a two level shopping center with four stories of apartments above the shopping areas.

In 2001, the Environmental Protection Agency (EPA) concluded a comprehensive study of smart growth development.[2] This report found that smart developments can increase air and water quality, preserve open space, and clean up older blighted areas of the city. All of these benefits would increase a community's quality of life. Air quality is increased because people spend less time commuting in their cars, thereby reducing air pollution. Between 1980 and 1997, vehicle miles driven increased annually much faster than the population growth, with an overall increase of 63 percent. Water quality is increased in smart growth developments because there are fewer paved surfaces, and paved surfaces increase polluted runoff. The EPA found that a one-acre parking lot had sixteen times more polluted runoff after a rain than did the same sized meadow. In many cities, there are abandoned or underused industrial facilities—these complexes are called brownfields. There are almost one-half million brownfield sites in the United States today. If a city redevelops a brownfield, it can remove a contaminated site and provide new housing or workplaces in a convenient location.

Chattanooga, Tennessee has become a model of smart growth, and it had one of the worst environmental records in the country when it started. The U.S. Department of Health, Education, and Welfare named Chattanooga as the most polluted city in the country in 1969. After this dubious honor, the city decided to take a dramatic approach to redevelopment. It decided to reclaim its downtown and use this process to provide more jobs, better housing, and a better environment for its people. In just twenty years, the city had a dramatic turnaround. Its air met federal standards, and it had transformed its decaying waterfront into a park. To reduce the amount of air pollution, Chattanooga now has a free electric bus service that shuttles visitors and residents from parking lots located on the outskirts of town.

Chattanooga's overall plan directly addresses brownfield development. The city's largest project aims to transform a former industrial district into an eco-industrial park. This environmentally friendly center is designed to be a zero emissions zone where all waste is converted into fuel.

[2]Environmental Protection Agency. 2001. Our built and natural environments: A technical review of the interactions between land use, transportation, and environmental quality. [Internet]. <http://www.epa.gov/smartgrowth/topics/eb.htm> [1 Feb 2002].

There are direct and indirect effects of development on the environment. See Figure 8-4. Any development modifies the land and changes the existing ecosystem. Development can also directly affect the water quality because buildings, roads, and other structures change the natural flow of water in an area. In the Washington D.C. area, for example, over two hundred thousand acres of open space was lost to development just in the 1980s. The indirect effects are not normally considered. New residential and commercial buildings change the employment opportunities for a community. If a company decides to move from its downtown headquarters to a new office park located in the suburbs, this move affects both the workers and the community. Also, since many suburbs have little to no public transportation, they indirectly increase the amount of solitary driving in a community, thereby increasing the amount of air pollution.

Smart growth means less land can accommodate new development; this development is sometimes called compact development. There are three common techniques to achieve compact development: infill development, brownfields redevelopment, and cluster development. Infill development is development attempting to add additional housing or business facilities inside an existing development. This way, a city can fill unused space in a particular area. Cluster development allows for the same number of dwellings to be built as a "regular" development does; however, the individual lot size for each house is reduced, and room is left for open spaces in the development.

Many cities are encouraging infill development to meet the demands of a growing population. San Jose, California, located at the center of Silicon Valley, has been grappling with a booming population for two decades. Before the digital revolution, San Jose was best known as the "Valley of Heart's Delight" because of the many orchards and walnut trees covering the valley. Over the past three decades, almost all this agricultural land has been sold and developed to become either housing or industrial workplaces. Along with the fruit orchards, San Jose was also home to many fruit canneries. Over seventeen packing plants have been demolished in the last twenty years to make way for new development. One of these former Del Monte canneries, Plant 51, is the site of a brownfield development to convert the cannery into lofts. Most of the historic features of the building will remain, but the inside will be converted to provide living spaces. See Figure 8-5.

The changing attitudes toward home ownership among people in the United States help the reconversion of industrial plants into living spaces. As the U.S. population ages, there is more demand from the so-called "empty nesters" for housing located near restaurants, cultural venues, and shopping. Along with this group, there is a sizable number of young, childless professionals who also are looking for convenient and maintenance free living environments. This movement of middle class people into town living is sometimes called *new urbanism.*

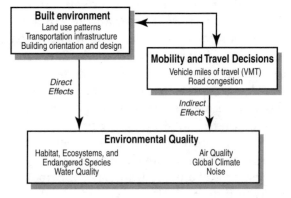

Figure 8-4.
This chart shows the direct and indirect effects of the built environment. (Environmental Protection Agency)

Figure 8-5.
Del Monte's Plant 51 was one of the many fruit packing plants that formerly operated in Santa Clara County, California, when the area was better known as the Valley of Heart's Delight. Now Santa Clara Valley is better known as Silicon Valley. Today, Del Monte Plant 51 in San Jose, California is being renovated and converted into downtown lofts.

Despite the advantages offered by this infill development, most of these new living units are developed for the middle and upper classes of society. The Plant 51 apartments and lofts will range from $300,000 to $1 million each—prices well out of range for all but the wealthy.

The smartest U.S. city, in terms of development is Portland, Oregon. In the mid-1990s, Portland's regional government, Portland Metro, established a fifty-year plan for growth in the region called Region 2040. Portland has dealt with development differently than most other cities in the United States. Instead of expanding outward, Portland established an Urban Growth Boundary (UGB) in the late 1970s. Outside this boundary, there was green space. When the UGB was established, it seemed to allow enough space for future growth; however, the city's growth has occurred more dramatically than expected. Instead of expanding the UGB, Portland decided to embark on a smart growth strategy to accommodate its future needs. Zoning will consist of changes to encourage mixed development, and suburban centers will change from single story to multistory buildings.

In addition to zoning changes, Portland will expand its light-rail network to reduce the amount of driving done by its citizens. Portland is not the only city that has plans to increase public transportation. There is a resurgence of trolleys, light-rail lines, and other commuter train lines across the United States. After decades of dependence on automobiles, cities are discovering the benefits of public transit. Los Angeles, arguably the most car crazy city in the United States, now has a booming subway system that continues to expand, and the District of Columbia is thinking of bringing back its electric streetcars that last ran in the early 1960s.

Smart growth requires us to reconsider the American ideal of a single-family dwelling located in the suburbs. In order for the United States to reduce both urban sprawl and air pollution, there needs to be more effort on reusing the existing developed land and providing alternative public transportation. Without this type of creative planning, the reality will remain longer commutes and increased pollution.

Green Buildings

When we think of pollution sources, we might think of automobiles or factories with smokestacks, but few of us would think of our own homes as sources of pollution. Surprisingly, buildings are a major source of air pollution in the United States. According to the U.S. DOE, buildings emit 49 percent of all sulfur dioxide, 25 percent of all nitrous oxide, 35 percent of carbon dioxide, and 10 percent of particulate emissions. Considering the number of homes and businesses in the United States—over 76 million residential and 5 million commercial buildings at last count—this problem is considerable. Tearing down all polluting buildings is not feasible, but building new ones with green features should reduce the amount of air pollution. By 2010, the DOE estimates that another 38 million buildings will be constructed in the United States. If just some of these new buildings include green features, this would favorably impact the environment.

GREEN BUILDING

A building constructed in a way that minimizes waste and includes recycled, renewable, and reused resources to the maximum extent possible.

There are many different techniques used in the construction of green buildings, including energy efficiency, renewable energy, water conservation, and waste minimization. We will discuss some of these techniques in the following sections.

Designing Energy Efficient Buildings

Energy efficiency is the most important factor in green construction. Since buildings last for decades, even centuries, reducing the energy needs of a building can bring long lasting savings in both energy and money.

Unfortunately, many of these energy saving techniques increase the construction costs.

Green buildings are designed to use higher levels of insulation than conventional buildings. Also, it is important to make buildings as airtight as possible. All houses leak to a certain extent—green houses will lose less of their energy to the environment. However, this trend toward making houses more airtight can lead to increased air pollution inside the structure—an unintended negative consequence of green building technologies.

Another way to make houses more energy efficient is by using renewable energy sources. Solar energy, because it is the energy source with the least environmental problems, is particularly suited to meet the energy demands of green houses. Studies of housing patterns in the United States indicate that 70 percent of new homes could be designed to use solar technologies. To increase the number of houses and businesses using solar energy, the DOE is sponsoring numerous research and development activities. One such program is the Million Solar Roofs Initiative (MSR), which has the goal of installing solar systems on 1 million rooftops by 2010. If this goal is achieved, then the reductions in carbon dioxide emissions will equal the exhaust from 850 thousand cars.

The DOE is studying ways to design zero energy homes. Zero energy homes generate more energy than they use and are not connected to the local power companies. The DOE built a prototype zero energy home in Lakeland, Florida and a traditional home nearby—both homes had the same 2,425 square foot floor plan. They compared the energy consumption of these two houses. The zero energy home used 72 percent less power from air conditioning than did the conventional home. Air conditioning is a severe power drain in the hot and muggy climate of Florida. The builders achieved energy savings through several innovative techniques. Instead of a flat roof, the zero energy house had a three-foot wide overhang; this increased the amount of shade. The builders also used reflective white tiles on the roof and installed solar control windows admitting light to the house, but limiting infrared and ultraviolet light that cause heating. The appliances and lighting in the zero energy house were all high efficiency—this also reduced the amount of heat they produced. Overall, the zero energy house used less energy, while maintaining cooler indoor temperatures.

Solar energy is not the only renewable resource being considered. In certain parts of the country, geothermal energy is also being used to provide power, particularly heat. Although most geothermal reservoirs are located on the West Coast, in Alaska, and in Hawaii, a builder was able to use geothermal energy in a house in Salem, Michigan. The geothermal system in the Kujawski House provides heating, cooling, and hot water for fifty dollars a month. Even in the midst of Michigan's cold winters, the heating bill did not exceed sixty-five dollars a month. Both solar and geothermal energy are limited by demographics, but designing a house with energy efficient appliances is a step that can be applied throughout the United States in building new houses and renovating older ones.

Reducing Material Use in Construction

Smaller is better for the environment; in terms of construction, using less materials is always preferable from an environmental point of view. The trend today, however, is for houses to get larger. This trend inevitably causes more pollution and energy use, but size is not the only constraint. The energy loss from a building is related to its surface area. Small, green design will reduce the outside surface area of a building in order to reduce energy losses.

There are changes inside the house that can always reduce material use. Building America, a partnership promoting green buildings, changed the framing in a new house. They set the studs for the walls twenty-four inches apart instead of the standard sixteen inches. This allowed the builders to put in thicker insulation and use less wood in its construction. Since an average wood framed house will use the equivalent of over one acre of wood during its construction, any wood savings means less adverse environmental impact.

Using Low Impact Materials during Construction

Many construction and building materials contain toxins. Many types of carpeting, for example, emit gases as they age. Also, carpets catch all the pollutants and toxins that come into our houses on our shoes. Research has found, particularly in tightly sealed houses, that the occupant's exposure to dangerous chemicals and pesticides is much higher inside the house than outside the house. For those living in tightly sealed houses, the air inside is generally much dirtier than the air outside.

Builders are beginning to use recycled or renewable materials during construction. In the last few years, several houses in Arizona have used straw bales rather than wood to construct the frame for the house. The bales are covered with a thick layer of stucco on the inside and out. Since straw is an abundant by-product of agriculture (generally produced when growing cereal grains), this technique is a good example of the use of a renewable material in construction. A side effect of straw bale construction is that the thick walls increase the energy efficiency of the home—this effect lasts over time. Another renewable construction material is rammed earth. This technique uses compacted soil in conjunction with stabilizers for wall construction.

Using recycled materials can also reduce environmental impact. Although these materials might be difficult to find in your local hardware store, recycled alternatives exist for roofs, sidings, floorings, and carpets, to name a few. Reusing parts of older buildings and renovating them to serve new uses also reduces the environmental impact. See Figure 8-6.

Another technique in reducing material use is modularization. Instead of building a house entirely on site, the house or industrial building is built using a modular approach. This type of construction is significantly faster than conventional construction, and there is less waste. Although most people think of

Figure 8-6.
Majestic Slate tiles are a premium recycled roofing product manufactured by EcoStar, a division of Carlisle Syntec. (Bruce Beaton)

mobile homes as modular construction, the reality is much different and more permanent. Today, one can build a multistory building from modules constructed in a factory. After the building is finished, it will be indistinguishable from conventional buildings. In Great Britain, the modular built McDonald's restaurant in Peterborough was open for business just forty-eight hours after construction began. In home construction, a group of Boston developers are working with the DOE to design a house using modular design. Their first two energy efficient homes were 20 percent less expensive to build than traditional houses.

Green building faces some obstacles, particularly with building codes designed for traditional construction, and because they use materials that are unusual, they are sometimes more difficult to maintain. The long-range impact of this type of construction is not certain. It may remain a small portion of the construction business, or it might take an even greater share in the future. Much of this depends on the tax benefits granted by states and the federal government, as well as the individual motivations of the builders and homebuyers.

Sustainable Agriculture

Sustainable agriculture is not a set series of practices; rather, it is a philosophical and practical approach to farming and agriculture. Sustainability is built upon three broad goals: farm profitability, improvement of the environment, and increased quality of life for farmers and their communities. Sustainable agriculture is future-oriented—the principle is that we should meet our food needs for the present without limiting the ability of future generations to meet their food needs. Sustainable agriculture is an approach to land management and agriculture.

Agriculture, as a field, has changed dramatically since World War II. Major changes that have impacted agriculture include mechanization, increased use of chemicals and fertilizers, crop specialization, and government policies favoring maximizing production. Agricultural productivity increased throughout the twentieth century; output in U.S. agriculture rose an average of 1.9 percent each year from the late 1940s to the mid-1990s. At the same time, the cost of food has dropped. Although productivity is up, it will need to increase more to meet the needs of the future world population, which is estimated to reach 10 to 11 billion people by 2040. Also, soil continues to erode, and supplies of groundwater are being depleted. In some parts of the United States, California for instance, there is a limit to the water supply for farmers. Sustainable farming in California includes the use of drought resistant crops and the improvement of water conservation.

There are four dominant practices used in sustainable agriculture: integrated pest management (IPM), conservation tillage, enhanced nutrient management, and precision agriculture. Each of these four techniques is used to a varying extent in the United States. IPM has been used for over twenty years, while conservation tillage has a lower adoption rate among farmers.

Integrated pest management (IPM) is a system for managing pests (including insects, diseases, and weeds) to keep them at levels at which they cause minimal damage to crops. It differs from conventional pest management in that IPM attempts to use ecologically sound, non-pesticide methods to reduce and manage pests. IPM does not attempt to eliminate all pests; rather it manages them to keep pests at an acceptable level. Although some farmers use pesticides in IPM, the goal is to use pesticides targeting a specific pest instead of generic pesticides that commonly kill a range of pests. IPM farmers also rely on biological controls, such as using other insects that feed on the unwanted pest. For some agricultural products (particularly in the fruit industry), IPM use is already high. Researchers predict that IPM use in the United States will reach 75 percent in the next thirty years.

There are many different conservation tillage techniques, but overall this refers to any plowing system leaving at least 30 percent of the soil surface covered with residue from the year's plantings. This is done so there will be enough soil coverage to decrease soil erosion. This technique toward sustainability has been driven in part by government policies. The 1990 Farm Bill required farmers to develop strategies to stop erosion on their cropland. This law motivated more farmers to adopt nonconventional tilling of the land to prepare it for new crops. Many farmers adopted a no-till method that leaves the organic waste from last year's crop on the field. See Figure 8-7. Conservation

Figure 8-7.
Rows of soybean plants emerge from a field covered with old cornstalks from the previous harvest. These soybeans were planted in narrower (fifteen-inch) rows because, as they mature, their big leaves will quickly shade the ground, making it harder for the Sun to warm weed seeds that may lie between the rows. This natural canopy from the growing soybean plants can help farmers reduce the need for herbicides. (University of Missouri)

tillage also reduces the amount of fuel a farmer has to use on the field, thereby reducing both air pollution and costs incurred by the farmer.

Enhanced nutrient management includes testing of the soil before using any fertilizer. The goal of nutrient management is to minimize unused nutrients. Farmers do this by conducting soil analyses. After the soil is analyzed, they can apply the fertilizers at the point when they are needed. Crop rotation, when crops on a particular field are changed each year, is another technique used for nutrient management. Crop rotation can reduce the need for fertilizer because certain crops (alfalfa and legumes) replace nitrogen into the soil.

Precision agriculture is the newest and the most technology intensive technique in sustainable agriculture. Precision agriculture uses information technologies, including global positioning systems (GPS) and remote sensing, to achieve optimal farming outputs. This technique is based on the belief that if a farmer matches the pesticide and fertilizer use to smaller areas in the field, then there will be less waste and less materials that could damage the environment. To collect data at specific locations in the field, a farmer can use a GPS to collect a wide range of data about the field conditions. GPS farming systems can create maps of a field, thereby allowing the farmer to change his planting to fit the soil conditions. Using a handheld GPS, a farmer can see features such as weed spots, plant variety areas, or hazards. Another technology used by farmers involves the use of probes. Probes can be placed in the field at various points and can continuously monitor the characteristics of the soil, including its moisture level and soil nutrients. Then, a farmer knows exactly how much water or fertilizer to place on this portion of the field.

Precision agriculture further extends the computer explosion onto our contemporary

farms. It provides the farmer with detailed information about the conditions in the field so the farmer can make more informed choices. Although precision agriculture has been available since the early 1990s, few farmers use this technology. The information being gathered is voluminous and complex and has to be interpreted by the farmer. Also, not all aspects of precision farming are readily adapted. Most precision agriculture farmers use a GPS guidance system; only 2 percent were using telemetry to send information from the field to the home office. Until more farmers can see that precision agriculture gives them a good return on their investment, the use of these information-rich technologies will remain small.

In addition to the 1990 Farm Bill, there are many other laws affecting farmers in their work. Currently, U.S. policies on crop-support systems encourage large farms producing a single type of product. This crop-support system works against the principles of sustainable agriculture. If U.S. agricultural policies could be changed to encourage diversified production, sustainable agriculture would become more prevalent.

Economy versus Ecology

Economy versus ecology—unfortunately, this is often the way the relationship between these two is portrayed. In the United States, even the regulations promoting sustainability seem to take an antagonistic approach to businesses and the economy. The most critical environmental initiatives—the Clean Water Act, the National Environmental Protection Act, and the Endangered Species Act—are top-down regulatory approaches. The advantage of regulations is that they force changes to occur at a faster rate than they would occur in a free market environment. The automobile industry in the United States was changed because of the regulations related to air quality and safety, despite the resistance of the auto manufacturers. The unintended, but positive, results of these changes were enormous innovations in automobiles, including airbags, seat belts, new and lighter metals and composite materials, and catalytic converters.

Sustainability cannot be regulated, as it is an approach to the use and development of technology. However, sustainable technologies generally are those minimizing ecological damage, and companies are realizing that protecting the environment can help sell a product. Recycled paper is one example of a sustainable product. Today, the EPA requires that federal agencies and contractors use paper containing a minimum of 30 percent post-consumer content for uncoated paper and 10 percent for coated papers.

Along with the press toward sustainability in the world, there has been an extensive amount of education on the effects of technology on the environment. All children in U.S. elementary schools learn the environmental 3 Rs—reduce, reuse, and recycle. Reuse is the least used of the environmental ABCs in the United States today, and interestedly enough, reuse in previous generations was much higher. Before the 1970s, soda bottles were made of glass, and most of them were reusable. After you drank the soda, the bottles were returned to the store; the storeowner returned them to the bottler; and the bottler sterilized the bottles and refilled them with soft drinks. This type of process was used for many other liquid household products and is still used in parts of the United States and in some European countries for milk.

The Energy Star program is a good example of the reduce principle. Computers following the Energy Star guidelines must operate on low power when not being used, and they must enter a standby mode when not in use. The Energy Star feature on computers can reduce their consumption of electricity by 50 to 75 percent. The marketplace is demanding greener products. Supermarket aisles are filled with products promoting their environmental friendliness—from compact fluorescent lightbulbs to phosphate free detergent to toilet paper manufactured with recycled paper.

Several companies have created new programs related to sustainability. The New England Power Company opened a new coal ash recycling plant and is selling the recycled ash as a cement substitute. The recycled ash makes the concrete stronger and less expensive to produce. Dow Chemical has a program called WRAP (Waste Reduction Always Pays) that, in addition to cutting Dow's emissions in half, has resulted in annual savings of $10 million.

The amount of recycling has burgeoned in the United States over the past twenty years; the number of tons recycled increased from 34 million tons in 1990 to 64 million tons in 1999. This equates to over one-quarter of the nation's waste. See Figure 8-8. Recycling, however, only considers the end result in the product life cycle. New efforts are addressing the issue of sustainability at the design stage. Computers are some of the most powerful tools in our society; they also produce a tremendous amount of waste. In 1998, more than 20 million personal computers were put out of service in the United States, and fewer than 11 percent were recycled. If more computers were recycled, there would be a positive effect on the waste stream. A new initiative, the National Electronics Product Stewardship Initiative was started in San Francisco in 2001 to address this problem in the electronics industry. Representatives from manufacturers, government agencies, and environmental groups met together to begin working on a plan to reuse and recycle electronics products.

Another approach has been taken by the European Union. A new directive requires

Figure 8-8.
Coca-Cola Amatil Beverage Packaging makes its polyethylene terephthalate (PET) soft drink bottles with 25 percent recycled content. (Charlie Woolford/Coca-Cola Amatil Beverage Packaging)

electronics manufacturers to take physical or financial responsibility for their products when they are at the end of their useful lives. The motivation for this directive is to make the producers of electrical and electronic equipment responsible for the pollution. Since the manufacturers will be responsible for the disposal of their devices, they would be more likely to design and manufacture their products so these products will have longer lives and be easier to recycle or reuse. This directive is expected to become European law in the very near future. The long-term effects of this law on multinational computer companies are hard to foresee.

The issue of sustainability and the implications of sustainable technologies for business will continue to be debated. Although there is increased awareness in the United States about protecting the environment, we have a tendency to consider our resources as without limit. Implementing the sustainable principles and behaviors into our lifestyles as well as the design of new technologies will require a shift in the attitudes of the American people. As with other issues we face, the tendency is to avoid uncomfortable issues until they become pressing. At the same time, the longer we wait to address sustainability, the more remediation will be necessary. Every new technology has consequences for society. Some of those consequences can be foreseen, and some are unforeseen. Considering the issue of sustainability in the deliberations about technology forces us to consider these consequences in a broader perspective.

International Sustainability Efforts

A sustainable society, over the long term, uses resources at the same or lower rate at which these resources are replenished. The production of waste materials in a sustainable society also does not exceed the rate at which these wastes are reabsorbed by the environment. There are three main trends working against sustainability in the world today: human overpopulation, overconsumption, and underconservation.

The population of Earth is increasing geometrically. The world's population in 1960 was 3 billion; in 1975, it reached 4 billion; and by 1999, it hit 6 billion people. There is a limit to the number of people Earth can sustain—what this limit is we can only guess. Overconsumption, however, is a characteristic of developed countries and is driven by rampant consumerism in these countries, including the United States. The third main problem, underconservation, magnifies the problem of overconsumption. We use too many of Earth's natural resources, and we do not reuse enough products. All of these affect sustainability. The goal is to develop technologies that promote sustainability and do not add to the negative influences on the world.

The widely published Brundtland Commission report on sustainable development titled *Our Common Future* forecasted a future world based on trends of today. This world in 2050 would have technology that was twice as efficient, a world population of 10 billion people, and a world economy five times larger than in the late 1980s. If the developed countries continue to consume at the rate they are today, the environmental reality of this future would be terrible.

More than 175 world governments committed in the early 1990s to Agenda 21, a United Nations–sponsored program with the goal of achieving sustainable development. The United Nations (UN) is planning summits every ten years to assess the progress toward sustainable development and to make recommendations for the future. The major goals for Agenda 21 are clean water; the protection of oceans, forests, and biodiversity; the elimination of poverty; sustainable development and technology transfer; and food security.

Sustainable development efforts are only successful for all parties when they are infused with information from both the donor and the receiver. The top-down and externally driven development model has inherent limitations. For development to be successful, providers must listen to the local users. For after development efforts have ended, it is the local users who must sustain any new technologies.

Traditionally, development has been equated with economic growth. A common belief was that positive development occurred when it directly increased a country's economic structure. Along with this emphasis on economic progress, there has been a tendency to focus on the following initiatives: privatizing government industries, such as the telephone or electric industries; reducing the regulations in a marketplace by removing trade barriers, tariffs, or import restrictions; and formalizing intellectual property rights through patents. This type of development is rooted in the capitalistic perspective of the United States and most other Western countries. These types of development activities assume the entire world would benefit from living under a capitalistic system.

Since we in the United States do live under a capitalistic system, it is difficult to imagine residing in a country without these principles. The goal of development under capitalism is to create a world with a single borderless economy. In this economy, corporations would have unlimited access to markets and could sell their products globally. Corporations would also use the cheapest sources of natural resources, labor, and supplies as they build their products.

This type of development assumes any problems with the environment or social structure would be solved through the normal processes imbedded in a marketplace. Unfortunately, the reality shows this has not been the case. For several decades, the United States and other Western manufacturers have been moving their assembly plants offshore to developing countries. The Maquiladora Program in Mexico is a perfect example of this type of offshore activity.

The Border Industrialization Program, known as the Maquiladora Program, began in the mid-1960s. The Mexican government gives financial incentives to foreign companies, most of which are from the United States, to open factories in certain areas in Mexico. Before the Maquiladora Program, Tijuana, Mexico was a sleepy border town of two hundred thousand people, mostly visited by tourists on day trips from San Diego. Today, it is an industrialized complex of light and heavy manufacturing and assembly plants. Because of the increased number of jobs in Tijuana (and other maquiladora cities), there was an increase in population. Maquiladora cities tend to draw internal immigrants from rural areas because of the greater number of jobs available to them. As the number of immigrants increased, the area became crowded with makeshift housing. Today, Tijuana has five times the population it had thirty years ago—almost 1 million people.

Business in the Maquiladora Program has boomed. Almost 1 million Mexican workers are employed in these plants; this equates to more than $33 billion in finished products being shipped out of Mexico. The maquiladora plants range from low-tech woodworking shops to high-tech electronics firms. While the business is booming, the local environment is not. The level of industrial pollutants in the maquiladora plants is very high.

The typical worker in maquiladora plants is a woman aged from the late teens to mid-twenties. See Figure 8-9. This type of worker is popular in the maquiladora industry because the manufacturers consider women more pliable and easier to manage. The problem is that most of these women are also married with children. They are forced to work long hours at the maquiladora

Figure 8-9.
This typical maquiladora worker is forced to work long hours. (James J. Biles)

plants and then come home and care for their families. Considering the paternalistic nature of the traditional Mexican family, these women are both the breadwinners and the caregivers. Indirectly, the maquiladora plants undermine the traditional gender roles in Mexican society.

The Maquiladora Program is a successful development program, but is it an A.T. transfer? From the workers' points of view, they have employment, but at very low wages. However, the alternative would be no employment at all or having to cross the U.S.-Mexico border (possibly illegally) to go to work. From the environmental point of view, this development is not appropriate because both the Mexican workers and the people who live around the plants have to contend with increased pollution.

This is not a good model for appropriate (or sustainable) technology from the perspective of the Mexican workers because it does not help the community; it pollutes the land and exploits the worker. From the Mexican government's and the industrial firms' perspectives, this technology transfer is very appropriate because these maquiladora plants provide jobs, increase the tax base, and allow Mexico to become more industrialized. However, when we judge the appropriateness of the technology transfer from the perspective of the workers, this assessment would most likely change.

Helping less-developed nations deal with the enormous environmental and social problems they face is no longer a simple matter of morality or altruism. Environmental degradation in Third World countries has very direct implications for the United States. Specifically, ecological or environmental instability contributes greatly to political instability in many areas of the world.

Unmanaged population growth outstrips the capacity of less-developed countries to support their people. Uncontrolled soil erosion undermines a community's capability to provide enough food for its residents. Unmonitored forest destruction deprives the local population of raw materials needed for fuel and shelter. Citizens who must confront these problems on a daily basis represent a potential source of revolutionary discontent. It ultimately becomes difficult, if not impossible, to address the question of environmental protection when your own survival has no assurance.

Haitian immigrants to the United States are not only refugees from politics; they also left behind a country whose land has been virtually destroyed. In many areas, 50 percent of the land surface has eroded to bare rock, and there are no forests left on the island. Jamaica provides another example. Massive emigration from this country is due in part to the deterioration of their natural resources.

In South America, the high Andean mountain regions are suffering perilous soil erosion since traditional systems of agriculture have all but disappeared. Developmental aid in the 1950s and 1960s was devoted to the introduction of cash crops for export in order to attain the foreign currency needed to develop industrial centers. Many of these crops were not suitable to the soil. Likewise, the undermining of traditional farming practices in the Sahel region of West Africa has caused severe environmental problems. In contrast to Western-driven development models, *agroecology* attempts to understand how traditional farming systems in developing countries have achieved sustainability. This field of agricultural development, instead of prescribing what is needed first, investigates what practices exist in local agriculture and how these practices relate to the local social, economic, and technological systems in place.

Some years before armed conflict broke out in El Salvador, scientific and technical observers noted problems of deforestation, soil erosion, stream siltation, and the harsh effects of fuelwood harvesting on natural vegetative recovery. Air and water pollution were not monitored, and the birthrate ultimately exceeded the capacities of existing health, education, and welfare programs. Social and ecological decline eventually prepared the ground upon which subsequent social wars have been fought over the years.

Despite these worsening problems and the underlying threat on international stability, governments seem to be unable to effectively come to terms with them. The tendency has often been to delay the implementation of long-term solutions in exchange for improvisational decision-making. This form of stopgap measure is best described as crisis management. Many development efforts evolve because of crisis situations. It is much easier to throw money at a problem than investigate the best way to prevent it. The nature of development efforts also creates a gap between the donor and the receiver. The donor has the money and perceives it has the correct technology, while the receiver has few choices, except to accept or reject the technology. This situation of power inequity makes the development experience even trickier. Added to this equation are cultural and social differences. It is no wonder successful development is hard to achieve for the long term.

In the last few years, some development agencies have moved away from large-scale development efforts to promoting smaller scale, locally conceived projects. An example of a successful development effort is the biogas project in Vietnam. Funded by the United Nations Development Programme (UNDP), this project promotes biogas (gas produced from the decomposition of organic materials) as a renewable energy source. The biogas project has helped families reduce household chores by 40 percent and has provided a way to use animal waste. See Figure 8-10.

This project is especially interesting because it shows a new trend in development activities: a focus on bringing the benefits of technology to women worldwide. The UN has a development program especially targeted to women, the United Nations Development Fund for Women (UNIFEM). UNIFEM focuses on three areas: strengthening the economic capacity of women, enabling women to have more of a voice in decision-making processes affecting their lives, and promoting women's rights and transforming the development process into one that is sustainable. The technology needs of women in the developing world are different than those of men. However, most development programs assume technology will be applicable to users from both genders. UNIFEM has found through its research that women are not receiving equal benefits from technologies, and more interestingly, technologies that would benefit women are often not investigated.

Since technological development is not spontaneous, ignoring a critical constituent group can radically change whether a technology is even developed. An example of this can be seen in the production and processing of sorghum in Botswana. The traditional method of processing sorghum, in addition to being tedious and time-consuming, is considered to be a woman's job. Because of the time sorghum processing takes, more people in Botswana choose to eat maize imported from South Africa rather than spend the time processing sorghum. Because this was considered to be a woman's task, the development of a technology to assist in the processing was not at first considered. A local *nongovernmental organization (NGO)* decided to look into this problem and adapted a Canadian cereal mill to process sorghum.

Figure 8-10.
This tubular polyethylene "low cost" biodigester is on the ecological farm of the University of Tropical Agriculture in Cambodia. Farmers in Vietnam have purchased some ten thousand units of this type (materials cost is between thirty and fifty U.S. dollars). The primary benefit of this design is its low capital cost, relative to other designs, without compromising the numerous other benefits of biogas digester technology as an energy supply source. (The University of Tropical Agriculture Foundation)

Whether a transferable technology is destined for women, men, or both, it is critical that the development agency or country listen to the voice of the user. In-depth assessment is needed before any technology transfer takes place, in order to achieve sustainability. Development partnerships are much more successful in the long run than are turnkey development schemes.

Another issue in development and technology transfer is energy. Access to energy varies significantly among the countries of the world and among people in the countries. One-third of the world's population relies on traditional energy sources—these people cannot take advantage of the possibilities of new technologies because they lack the energy to make them work. Appropriate development and A.T. transfers must take into account the inequities in infrastructure in developing countries before they can be successful. Only after a full assessment of a technology and its demands can we assess its appropriateness for transfer.

The A.T. projects described above, including renewable energy and sustainable agriculture, can also be applied to international technology transfer. In the next section, we will focus on three areas directly addressing the issues of appropriateness from an international perspective. Because of the different perspectives on and definitions of *A.T.*, there have been diverse approaches to applying this concept internationally. Business and corporate interests drive International Organization for Standardization (ISO) 14000, while *ethnobotany* has emerged from the scientific field. Each of these initiatives has the goal of encouraging A.T.s in the world.

International Organization for Standardization (ISO) 14000 and Environmental Management

The International Organization for Standardization (ISO) 14000 standards for *environmental management* were a result of the focus on sustainable development emerging from the UN Conference on Environment and Development in the early 1990s. After this conference, the ISO developed ISO 14000 to provide a model for companies to follow "sustainable business development."

The focus of *International Organization for Standardization (ISO) 14000* is managing environmental issues through appropriate business strategies. ISO 14000 is not a series of steps a company has to implement; rather, it is a management tool describing how a company can manage its business in an environmentally friendly manner. ISO 14000 requires that a company implement an environmental management system (EMS). For companies that have implemented ISO 14000, the benefits of an EMS include reductions in the use of energy and materials and, at the same time, a reduction in waste produced. Other benefits of ISO 14000 are less tangible. Since ISO 14000 is an environmental management approach, adherence to these guidelines and certification can improve the public image of a company. For a company, there are six steps to developing an EMS.

- ❑ Environmental policy. The company defines its own environmental policy.
- ❑ Planning. The company develops its own plan for identifying the environmental impact of its manufacturing and service processes.
- ❑ Implementation. The top management in the company assigns resources and people to implement the plan.
- ❑ Corrective action. The company monitors its success in implementing its environmental plan.
- ❑ Management review. The EMS plan is reviewed on a regular basis.
- ❑ Continuous improvement. The EMS is revised to improve the management of environmental issues in the company.

After a company develops its EMS, an external auditor who can certify that the company meets the ISO 14000 standards evaluates it. The company then joins the list of ISO 14000 certified companies worldwide. Since the first ISO 14000 certificate was issued in the early 1990s, the number of companies receiving ISO 14000 certification has exploded. Across the globe, there is an unequal distribution of ISO 14000 certified companies, both by country and industry. The largest ISO 14000 sector consists of companies in the basic metal and fabricated metal industries, and the smallest sector consists of companies in trades and services. See Figure 8-11. There are also differences in ISO 14000 certification by country. ISO 14000 is more popular in Europe and Japan than in the United States. In 2000 alone, six times as many Japanese companies received their ISO 14000 certificates, as compared to American companies. Despite its slow growth in the United States, ISO 14000 will continue to expand across the world, and as the principles of environmental management

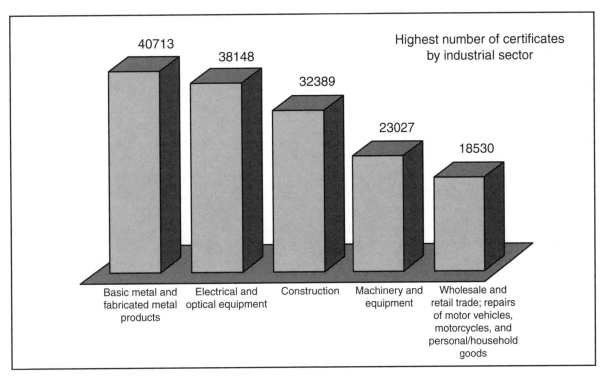

Figure 8-11.
These are International Organization for Standardization (ISO) 14000 certificates by industry, as of December 2000. (Adapted from ISO)

spread, this certification should become more acceptable to U.S. companies.

Fair Trade Movement

Fair or alternative trade provides a contrast to the conventional free market approach to international trade. Instead of putting the consumers first, free trade puts the producers first. Directly, free trade means that any goods produced, whether they are commodities like coffee or tea or manufactured products like textiles or crafts, must bring a good price to the producer of the good. Beyond the issue of a good price, the free trade movement attempts to blend the market-based economy with social justice and environmental concerns.

The largest segment in free trade is food products, with coffee having the most prominent role. Free trade coffee has 2 to 5 percent of the market share in developed countries, but more importantly, free trade coffee is widely available in these developed countries. More than 90 percent of free trade coffee is distributed in conventional supermarkets.

Fair trade is promoted in a variety of ways. Several NGOs, including Oxfam Trading and Traidcraft, act as marketing agents for fair trade producers. In this model, the NGO pays the producer a fair price for their products, then markets the products on behalf of the producers in developed countries. The producers in this model are like subcontractors in a regular business organization. The risk for trading the commodity is borne by the NGO. In addition to trading, NGOs offer technical assistance and support to producers. The International Resources for Fairer Trade (IRFT), affiliated with Traidcraft, provides consultant services in accounting and supply chain management to small Indian businesses. In this way, the NGO helps community-based businesses gain the knowledge and skills that make them able to have greater influence in the marketplace.

NGOs are not the only means through which fair trade is conducted. Other free trade organizations are sponsored by governments or are community-based organizations. No matter how goods are traded, the key issue is fair price. As with many other parts of A.T., there is disagreement as to what constitutes a fair price. Is it a fair price if it is higher than the price offered by other international traders, or is it a fair price only when it is high enough to allow a producer to have a reasonable living standard?

Although the market share for free trade products is small, it is growing, especially in areas with higher income consumers. Switzerland, a wealthy country in Europe, has the highest purchasing rate of free trade coffee. There are over eight hundred fair trade partners in forty-five countries—this represents almost 1 million producer families! As Americans are confronted even more by the disparities in income and living standards across the world, buying products under the free trade banner allows them to personally affect people from another country. Interestingly, a new e-commerce website is being used to promote and sell free trade products. Fair Trade on-line is a new Internet e-commerce site selling fair trade food and crafts from around the world. See Figure 8-12. Although most free trade products are commodities or crafts, in this way, high technology can benefit the producers who are able to sell their products through free trade organizations.

Ethnobotany

Ethnobotany is the field of science investigating how indigenous people use plants in their lives as food, shelter, and medicine. It is in the field of medicine that ethnobotany holds the most promise. For centuries, indigenous people have used their own native plants to cure diseases. Today, the industrialized world is increasingly taking advantage of this information to design modern drugs that might cure a range of diseases. In the United States alone, about half of the twenty-five top selling prescription drugs were developed from natural or plant sources. The remote areas of South America, including its jungles and rain forests, contain a diverse number of plants, some of which have never been seen or used by Western scientists.

The Western world has been developing drugs from non-Western sources for centuries. One of the earliest was quinine, which was extracted from the bark of the cinchona trees found in the Andes by the Spanish in the early 1500s. Other products, derived from plant extracts indigenous to countries in the world's Southern hemisphere include cocaine, curare, and pilocarpine, used for the treatment of glaucoma. Despite the sources for these drugs, the financial benefits were not distributed to the native people who really discovered these medicines. This situation is changing with the new growth of ethnobotany. Most modern ethnobotanists now work with native healers to investigate plants and their properties. There is increasing emphasis on sharing the financial benefits of the new medicines with the people living in the countries where these new medicines are found.

Recently, there was a landmark agreement between the AIDS ReSearch Alliance of America and the country of Samoa. An

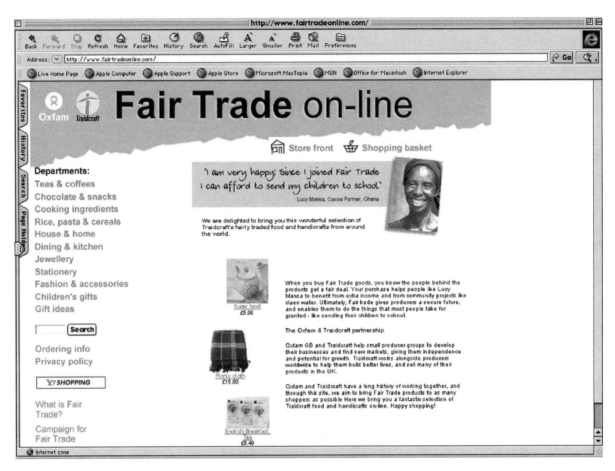

Figure 8-12.
Two nongovernmental organizations, Traidcraft and Oxfam, along with Yahoo, established Fair Trade on-line to provide easy access to fair trade products.

ethnobotanist working in Samoa had found that Samoan healers used the bark of a plant to treat hepatitis. The chemical compound from this plant, called prostratin, was synthesized by the National Cancer Institute and is being tested as an anti-HIV drug. The AIDS ReSearch Alliance of America agreed to return 20 percent of any profits to Samoa; part of this money would go to the Samoan government, while a portion would also go to the village where the drug was found and to the healers who found the drug.

The exploration of wild plants and animals and their uses is called bioprospecting. In contrast to traditional prospecting for gold and other minerals, bioprospecting implies minimal disturbance of the natural environment, and since Western companies have tested only a few of the thousands of traditional medicines, there is potential for profits that would exceed those in farming and ranching. Since most of the tropical forest is lost to farming and ranching, bioprospecting could slow down the rate of loss of the tropical rain forest. Since the early 1990s, Merck, a Western pharmaceutical company, has worked with the Costa Rican National Institute of Biodiversity to search for new drugs in this country.

As more Western companies exploit the resources of the developing world, there should be financial benefits as well for these developing countries. At the same time, the use of plants and other organics from the rural areas can increase their statures, and thereby ensure their survival. In Chapter 9, titled "Environmental Issues," we note that logging and grazing are depleting the rain forest. If, however, developing countries are able to view their rain forests as sources of wealth, there might very well be a stronger motivation to maintain them. Ethnobotany and the medicines derived from it can lead to a more positive environment for these countries.

Technology Equity

As technologies are developed and distributed, they generally are not distributed equally to all different social and income classes in our society. Positive technologies, including the computer and the Internet, are more available to the wealthier segments of our population. On the flip side, negative technologies like waste recycling plants and power plants are found more frequently in areas where poorer people live. The issue in both of these cases is equity. *Technology equity* implies all technologies are available to and used by all socioeconomic segments in a society. How can the United States, never mind the world, guarantee equal access to technology for all people?

From the information-rich vantage point of the United States, it is difficult for us to understand that most of the people in the world do not have access to a telephone, much less the Internet or other advanced telecommunications. In the United States itself, about 143 million people are on-line. This number is high, but it represents only about half of the U.S. population. Computer and Internet use are not evenly distributed by racial group. At all income levels, fewer African-Americans and Hispanic Americans use either a computer or the Internet. Americans living in low-income households have the lowest levels of computer and Internet usage. This is sometimes referred to as the digital divide.

The digital divide exists globally as well. Although Internet use is rising worldwide, it is not equitably divided among all the countries in the world. Experts estimate that 72 percent of the people in the top eight developed countries will be on-line by 2005, as compared with approximately 4 percent in eleven of the poorest countries in the Asia-Pacific region. A lot of this has to do with infrastructure. Many of these developing countries do not have extensive telephone systems. In Bangladesh, for example, for every one hundred people, there is one telephone line, and providing access to a computer does not decrease the digital divide if a large portion of the population cannot read and write. This is the case in India, where approximately 50 percent of the population cannot read. This situation in India motivated a team of Indian scientists and engineers to develop a small, inexpensive computer—the Simputer—that reads out the text on web pages in several Indian languages. Although the cost of the device is still higher than what most Indians can afford, the engineers are investigating adapting it with a smart card so several users can share one Simputer.

To solve the problem of technology inequity worldwide, we need to take a look at the technologies themselves. More access should mean better access and should bring better opportunities for people both in the world and in the United States. We are optimistic that in a few decades the digital divide will be a fact of history. However, the only way this will happen is if the United

States and other developed countries share their knowledge with developing countries in the world.

Summary

The development and use of appropriate technology (A.T.) require us to consider every technology artifact as more than a device. We must consider how it will be used, affect the environment, and impact social and gender roles in society. Many different ways of looking at A.T. were proposed in this chapter. What do they have in common? A.T. is a nontraditional way of evaluating technology. The technology artifact is considered in the context of how it will be used in society. Technology is seen as a social force interwoven into the lives of its users. If we view technology as multifaceted, we can begin to look at it with a more critical eye.

Since humankind has existed, we have been changing this planet, sometimes for the good and sometimes not. In the future, new unimagined technologies will emerge in our society. It is using these technologies responsibly that can make them A.T.s.

The issue of appropriateness is linked to the social and political environment of a particular country. All countries and all people have basic needs—food, shelter, and security. A country must balance the needs of its people with their desires for new and better technologies. Since we in the Western world are only beginning to deal with and acknowledge the issue of appropriateness, it is simplistic to assume we can provide all the answers for the rest of the world. Indeed, appropriateness assumes a level of self-reliance and self-governance about technology, its benefits, and its risks. As responsible citizens of the world, our desire should be to use all our technologies so we minimize the damage we cause to our environment and ourselves.

Discussion Questions

1. Take a trip to your local hardware store. Find at least three different "green" products you could use in building or renovating your home. For each product, describe its advantages and disadvantages.
2. Research an International Organization for Standardization (ISO) 14000 company either in the telephone book or on the World Wide Web. How do these standards affect the way a company is managed?
3. Considering where you live, which renewable energy sources would work best in your region? Why?
4. Describe two different techniques used in sustainable agriculture.
5. Research the availability of free trade products in your local supermarket. What types of products are sold? Are they widely available in your area?

Suggested Readings for Further Study

Alternative Farming Systems Information Center. *Sustainable agriculture resources.* [Internet]. <http://www.nal.usda.gov/afsic/agnic/agnic.htm>.

Börjesson, Pål. 1999. Environmental effects of energy crop cultivation in Sweden—I: Identification and quantification, and II: Economic evaluation. *Biomass and Bioenergy* 16 (2): 137–70.

Boyle, Godfrey, ed. 1996. *Renewable energy: Power for a sustainable future.* Oxford: Oxford Univ. Press.

Brown, Lester. (2002). World wind generating capacity jumps 31 percent in 2001. *Earth Policy Institute.* [Internet]. <http://www.earth-policy.org/Updates/Update5.htm> [8 Apr 2002].

Brundtland Commission [The World Commission on Environment and Development]. 1987. *Our common future.* Oxford: Oxford Univ. Press.

Carr, Marilyn. (1997). Gender and technology: Is there a problem? Paper prepared for TOOL/TOOL Consult Conference on Technology and Development: Strategies for the Integration of Gender, Amsterdam. [Internet]. <http://www.unifem.undp.org/g&tech.htm> [5 Feb 2002].

Cascio, Joseph, ed. 1996. *The ISO 14000 handbook.* Milwaukee: ASQ Quality.

Designing cars for moms—and everyone else. 2001. *Prism* (February): 11.

Eisenberg, David, Robert Done, and Loretta Ishida. (2002). *Breaking down the barriers: Challenges and solutions to code approval of green building.* [Internet]. <http://www.dcat.net/Codes/codes.html> [2 Feb 2002].

Environmental Protection Agency. (2001). *Our built and natural environments: A technical review of the interactions between land use, transportation, and environmental quality.* [Internet]. <http://www.epa.gov/smartgrowth/eb.htm> [1 Feb 2002].

Everts, Saskia. 1998. *Gender and technology. Empowering women, engendering development.* New York: Zed.

Gold, Mary V. (1999). *Sustainable agriculture: Definitions and terms.* [Internet]. <http://www.nal.usda.gov/afsic/AFSIC_pubs/srb9902.htm> [2 Feb 2002].

Goldemberg, José, ed. 2001. *World energy assessment: Energy and the challenge of sustainability.* New York: United Nations Development Programme, UN Department of Economic and Social Affairs, and World Energy Council.

Ha, K. Oanh. 1999. Let's hear it for the Internet. *San Jose Mercury News* E1 (8 August): 6.

Hazeltine, Barrett, Christopher Bull, and Lars Wanhammar. 1998. *Appropriate technology: Tools, choices, & implications.* Academic.

Hrubovcak, James, Utpal Vasavada, and Joseph Aldy. (1999). *Green technologies for a more sustainable agriculture.* [Internet]. <http://www.ers.usda.gov/publications/AIB752/> [2 Feb 2002].

International Organization for Standardization. (2001). *The ISO survey of ISO 9000 and ISO 14000 certificates.* [Internet]. Portable Document Format. <http://www.iso.ch/iso/en/iso9000-14000/pdf/survey10thcycle.pdf> [11 Feb 2002].

[Internet]. (1995). Establishing priorities with green building. *Environmental Building News.* 4 (5). <http://www.buildinggreen.com/features/4-5/priorities.html> [2 Feb 2002].

[Internet]. (1999). Prairie crossing homes. Portable Document Format. <http://www.nrel.gov/docs/fy99osti/26261.pdf> [2 Feb 2002].

[Internet]. (2000). Cannery to become housing. *Silicon Valley San Jose Business Journal.* <http://sanjose.bcentral.com/sanjose/stories/2000/10/02/daily28.html> [2 Feb 2002].

[Internet]. (2001). *On the path to zero energy homes.* Portable Document Format. <http://www.nrel.gov/docs/fy01osti/29915.pdf> [2 Feb 2002].

[Internet]. *Appropriate Technology Transfer for Rural Areas (ATTRA).* <http://www.attra.org/>.

[Internet]. *Center for International Ethnomedicinal Education and Research (CIEER).* <http://www.cieer.org/>.

[Internet]. *Center for Renewable Energy and Sustainable Technology (CREST).* <http://www.crest.org/index.html>.

[Internet]. *Development Center for Appropriate Technology (DCAT).* <http://www.dcat.net/home.html>.

[Internet]. *Environmental health coalition border environmental justice campaign.* <http://www.environmentalhealth.org/border.html> [31 Jan 2002].

[Internet]. *International Organization for Standardization (ISO) 9000 and ISO 14000.* <http:www/iso.ch/iso/en/iso9000-14000/index.html>.

[Internet]. *National Association for Appropriate Technology.* <http://www.ncat.org/>.

[Internet]. *Traidcraft.* <http://www.traidcraft.co.uk/>.

Jackson, Suzan L. 1997. *The ISO 14001 implementation guide: Creating an integrated management system.* John Wiley & Sons.

Jefferson, M. 2001. Heads we win, tails we win. *New Statesman*, 130 (4542): xix.

Layton, Lyndsey. (2002). D.C. transit may go retro. Streetcar revival considered 40 years after departure. *Washington Post.* [Internet]. <http://www.washingtonpost.com/wp-dyn/articles/A11986-2002Jan20.html> [2 Feb 2002].

Metro Oregon Regional Government. (1997). Regional framework plan. *Vision 2040.* [Internet]. Portable Document Format. <http://www.metro-region.org/pdf/frame.pdf> [1 Feb 2002].

Oskamp, Stuart. 2000. Psychological contributions to achieving an ecologically sustainable future for humanity. *Journal of Social Issues* 56 (3): 373.

Parker, Danny S., James P. Dunlop, John R. Sherwin, Stephen F. Barkaszi, Jr., Michael P. Anello, Steve Durand, Donard Metzger, and Jeffrey K. Sonne. (1998). *Field evaluation of efficient building technology with photovoltaic power production in new Florida residential housing.* [Internet]. <http://www.fsec.ucf.edu/~bdac/pubs/CR1044/LAKELAND1.htm> [2 Feb 2002].

Quinn, James B. and James F. Quinn. 2000. Forging environmental markets. *Issues in Science and Technology* 16 (3): 45–52.

Roberts, Michael and Ronald Begley. 1993. Working toward globally sustainable agriculture. *Chemical Week* 153 (19): 42–43.

Rudick, Roger. 2001. Commercial real estate; Rail access can enhance property values; Angelenos' skepticism aside, many cities report that homes near transit stations often command hefty premiums. *The Los Angeles Times* (4 December): C8.

Sale, M. J., G. F. Cada, T. J. Carlson, D. D. Dauble, R. T. Hunt, G. L. Sommers, B. N. Rinehart, J. V. Flynn, and P. A. Brookshier. (2002). *DOE hydropower program annual report for FY 2000.* [Internet]. Portable Document Format. <http://hydropower.inel.gov/transfer/DOEID-10992-FY01.pdf>.

Tiffen, Pauline and S. Zadek. (1998). Dealing with and in the global economy: Fairer trade in

Latin America. In *Mediating Sustainability,* edited by Jutta Blauert and Simon Zadek, ch. 6. [Internet]. <http://www.globalexchange.org/economy/coffee/tiffen98.html> [16 Feb 2002].

U.S. Department of Energy (DOE). *Energy Efficiency and Renewable Energy Network.* [Internet]. <http://www.sustainable.doe.gov/>.

Veilleuxm Connie and Steven R. King. (n.d.). An introduction to ethnobotany. *Access Excellence Resource Center.* [Internet]. <http://www.accessexcellence.org/RC/Ethnobotany/page2.html> [18 Feb 2002].

World Health Organization. (2001). *Legal status of traditional medicine and complementary/alternative medicine: A worldwide review.* [Internet]. Portable Document Format. <http://www.who.int/medicines/library/trm/who-edm-trm-2001-2/legalstatus.pdf> [18 Feb 2002].

Chapter 9

Environmental Issues

DATELINE

Inverness, Scotland

The world's first commercial wave power station was just put on-line, off the Scottish island of Islay. The station generates enough power to run about four hundred homes on the island. The plant, called Limpet, works well off the coast of Scotland because of the strong waves buffeting the coast. The incoming waves power turbine engines that then drive a power generator—this produces the electricity.

Although the cost of the power generated by Limpet is higher than conventional sources, its source (waves) is definitely renewable. Waves push air through a turbine as they advance and recede from the shore. Through the turbine, the wave energy is converted into electrical energy....

Introduction

When we think of the relationship between technology and the environment, we tend to view technology as having a negative impact on our environment. In fact, the evolution of the phrase *environmental impact* as a modifier for such words as *assessment*, *report*, and *statement* is directly related to this mode of discussion and debate. Protection of the natural environment has become a key issue in almost all dialogue surrounding the international advancement of technology.

Although the issue of environmental protection has a relatively short history when compared with the history of technology, it has received a great deal of media attention throughout its existence. Participation in these often-dramatic cover stories has not been limited to the scientific, technological, and political arenas. The general public has continually maintained an active voice at the grassroots level, which has stimulated much of the regulatory legislation currently in effect.

Widespread concern about the long-term effects of technology and its many by-products on the natural system of the biosphere began during the mid-1960s. The United Nations Conference on the Human Environment, held in Stockholm in the early 1970s, placed a series of environmental issues on the official agendas of national governments around the world. Through the final decades of the twentieth century, the environment gradually became an item of greater priority on the prospectus of international affairs.

An attempt to generate a comprehensive list of the critical areas of environmental concern would be challenging. It would also be directly influenced by the degree of technical and scientific background possessed by the respondents surveyed. Without question, some of the more serious matters dictating a sense of urgency include air quality, climatic stability, moist tropical forests, arable lands, water potability, and management of hazardous materials. At mid-twentieth century, none of these topics were political discussion topics, let alone issues of international magnitude.

The intent of this chapter is to familiarize you with several environmental conditions that can no longer be overlooked or ignored if future generations are to be assured of a natural, hospitable environment. The extent to which these conditions are either a direct or indirect result of the advancement of technology in society will most likely remain a debatable issue. This chapter provides a continuing dialogue that can ultimately respond to the following questions:

- ❑ What does it mean to be labeled as an *environmentalist* in today's social circles?
- ❑ Has there been any significant progress made in international environmental cooperation?
- ❑ What effect has governmental intervention had on the quality of air we breathe in this country?
- ❑ What strategies are currently being proposed or implemented to deter the degradation of the world's remaining tropical forests?
- ❑ How can the dangerous waste materials generated by advanced technological societies be managed most effectively?

Key Concepts

As you review this chapter, you should develop an understanding of these concepts:

acid rain
agroforestry
Air Quality Index (AQI)
common space
environmental impact analysis
environmental release
evapotranspiration
global warming
greenhouse effect
modern environmentalism
Pollutant Standards Index (PSI)
Superfund

The Contemporary Environmentalist Profile

Not long ago, being referred to as an environmentalist in American society signified radical behavior and an antiestablishment perspective on most issues. This perception is being replaced with a whole new array of characteristics. Across the United States, Americans' concerns about their environment are increasingly being expressed through the all-inclusive, yet somewhat elusive, phrase *quality of living*. The authors give a simple definition of this term in Chapter 11 of this textbook.

The quality of living concept carries slightly different meanings from one person to another or from one geographic region to another. Consistency among these definitions, as related to the environment, can generally be found within the universal values of clean air, pure water, and an appreciation of the diversity of the natural environment. It seems many individuals have come to expect these items as part of their birthrights. They also realize these values are being threatened by technology.

Daily newspapers continually bring revelations of a toxic spill, a hazardous material incident, or a natural resource in jeopardy to our doorsteps each morning. The disheartening nature of these reports, coupled with the desire to live in a quality environment, has prompted an eagerness on the part of the general public to correct these situations. Being called an environmentalist today no longer connotes countercultural viewpoints.

Environmentalism

Modern environmentalism is a term used to describe an information-based perspective supporting the need for harmony between industrial objectives and environmental goals. For most of the twentieth century, people accepted pollution as an expected by-product of technological progress. In the

1960s, however, this perspective was challenged (along with other societal norms). In the early 1960s, Rachel Carson published her book, *Silent Spring,* which describes the effect the pesticide dichloro diphenyl trichloroethane (DDT) has on birds. News stories about pollution in America's rivers added to the impact of this book, particularly the fire on the Cuyahoga River in Cleveland, Ohio in the late 1960s. The river was so full of industrial waste it literally caught on fire.

For the last thirty years, environmentalism has become a strong movement in the United States. First of all, the modern environmentalist recognizes that environmental goals cannot be pursued without regard to their consequences elsewhere. For example, effective energy policy cannot be based solely on environmental values. A clean environment is only one of the many results we want in the development of a national energy policy. A clean environment is not the central driving force. The second component of the modern environmentalist perspective involves the realization that it is both necessary and possible to minimize the destructive environmental side effects of technological progress through careful assessment and planning.

The key phrase in the above description is *information-based.* Environmentalists who suffer from chronic narrowmindedness today seem to be in the minority. Environmental rhetoric is gradually being replaced with scientific and technical data accumulation, resulting in reflective and systematic decision-making. In other words, a brainstorming session between upper management and the production workers in a manufacturing firm regarding its safe and effective storage and disposal of toxic solvents is more indicative of contemporary environmentalism than the bumper sticker reading "You Choose: Cancer or Clean Air!"

Industrialists and environmentalists alike have mellowed with age. Neither faction in this struggle has a monopoly on virtue. Both groups create opportunities to enhance our standard of living. Both groups also create problems challenging our ingenuity and willingness to confront changes. Together, these individuals have begun to consider options to stimulate technological development and simultaneously strive to preserve the natural environment.

Public Support of the Movement

In many countries, the viability of public programs, political platforms, and social movements is very dependent on a strong measure of public support. One of the more remarkable aspects of the environmental movement in the United States has been the durability of public support for its programs, projects, and proposed regulatory legislation. Despite the fact many sociologists and politicians predicted public interest in ecology would fade as soon as some other cause surfaced on the scene, the public's support for environmental programs is still strong. Widespread dissatisfaction with the social costs of careless technology moved environmentalism from a concern of a few individuals to an accepted truth among the populace.

Regardless of support, most people do not identify pollution, tropical forest resource management, and other environmental issues as the most serious tasks at hand for our nation's administration. World peace and economic concerns are more commonly found to overshadow environmental protection as national priorities. This tendency does not necessarily detract from the strength of the environmental movement or the extent to which people demonstrate personal interest in its programs. It merely connotes that the environmentalist movement has gone through a process of maturation and achieved a credible position on our country's political agenda. As compared to many other countries, environmentalism in the United States operates within the existing political structure rather than being a separate political unit.

We can compare environmentalism in the United States with Germany, for example. In Germany, there is a separate political party, Die Grünen (The Greens), that has increased its membership significantly since its foundation in the late 1970s. As a political entity, it is so powerful that it became part of the coalition government of Germany under Chancellor Gerhard Schröder in recent years. In Germany, as in most other European countries, the environmentalism movement is a political movement exerting considerable influence over the policies of the government. Currently, because of the influence of the Greens, Germany has begun to shut down its nuclear plants and increase the taxes on fossil fuels. In the United States, in contrast, the environmentalism movement tends to work within existing political boundaries.

Another measure of public support for the environmental movement can be gleaned from the membership rosters of several influential environmental organizations. Whereas the number of members who contribute to the organizations remained somewhat stable throughout the environmental decade of the 1970s, the Sierra Club and the Wilderness Society have dramatically increased their memberships since then. The Wilderness Society now has over 200 thousand members, and the Sierra Club now has more than 700 thousand members (from 114 thousand in 1970). Active participation in smaller organizations, such as Friends of the Earth and Defenders of Wildlife, has also increased significantly in recent years. Furthermore, these organizations have begun to receive strong verbal backing from many American citizens who are not even actual members. See Figure 9-1.

On a less optimistic note, public interest in environmental policies may have waned over the past few years due to the complexity both of the issues and the ways these issues are presented to the public. The National Environmental Policy Act (1969) established a set of guidelines federal agencies must follow before they undertake any significant action. Subsequently, a network of federal laws, supported by amazing volumes of regulatory materials, has been created to protect the air, water, land, and seas—at least on paper. Almost every new construction project built in the United States must include an *environmental impact analysis.* The analysis includes an in-depth study of the potential impact of the construction on the environment. These voluminous and complex reports are difficult for the layperson to comprehend. This legalistic character of modern environmentalism has brought steadiness, resources, and power to the movement, but it may also be responsible for disrupting a certain degree of public interest in its principles.

A major problem confronting the leaders of this movement may center on how to recapture the energy of the American public in a new commitment toward improving environmental quality. The fact that government has taken such an active role in environmental affairs has possibly created the assumption that the federal government is adequately handling these issues. Many people may feel their votes for an environmental initiative (such as clean water and hazardous waste sites) on the ballot is the largest, and often final, measure of support needed to get things going in the right direction.

Recognize that it is not enough to be sensitive to environmental issues, as some public opinion surveys seem to surmise. Support for the environmental protection movement must be personified through

Figure 9-1.
People stop the logging truck from removing the killed old-growth trees from the Siskiyou National Forest. Other people chain themselves to the truck's axle and on top of the load. (Francis Eatherington)

actual changes in behavior. This change is often difficult for many of us. A case in point to illustrate this brings to mind a small screen-printing shop during the mid-1980s. The proprietors expressed verbal support for the safe use of cleaning solvents by their employees. However, they did not seem overly concerned about their casual disposal of methyl ethyl ketone (MEK) onto the company's parking lot.

In all likelihood, this isolated situation was corrected years ago. Many of the global issues discussed in the following section may very well have originated at such a seemingly innocent and localized site. If the present environmental movement expects to make any sort of dent in devising solutions for these problems, then active public support must be reignited at the grassroots level throughout the world.

Key Issues on a Global Scale

The United States is confronted with numerous international environmental issues essentially falling into three general categories:

❑ Handling domestically important issues requiring the cooperation of other nations to solve.

❑ Dealing with areas of our world falling under some sort of international jurisdiction.

❑ Assisting developing countries to contend with their own seemingly localized environmental problems.

The extent to which progress is expected to occur in any of these areas is still questionable, but we remain cautiously optimistic.

International Cooperation

There is an increasing number of technological by-products posing a threat to the stability and quality of the U.S. environment that can only be dealt with on an international level. Potential menaces to the global climate, such as ozone depletion and carbon dioxide (CO_2) buildup, are the clearest examples. Strategies enacted by the United States alone will be insufficient to deal with these problems. The magnitude of these dismal trends requires a concerted, cooperative action by a number of national governments.

The track record of accomplishments over the past couple of decades justifies optimism, albeit qualified, regarding the abilities of nations to cooperate on international environmental issues. Joint efforts have occurred in marine navigation, air quality, public health, and agricultural practices. Specifically, a large number of international treaties of major environmental significance have been initiated. Many of these have only recently begun to seriously take effect: Bonn Convention on Conservation of Migratory Species of Wild Animals (1979); Convention on Wetlands of International Importance (1971); International Treaty on Long-Range Transboundary Air Pollution (1983); Vienna Convention for the Protection of the Ozone Layer (1985); Montreal Protocol on Substances that Deplete the Ozone Layer (1987); and Kyoto Protocol to the United Nations Framework Convention on Climate Change (1997).

Besides these recent agreements, older international treaties have been given renewed significance due to the growth of concern for the environment. The Boundary Waters Treaty that was signed in 1909 between the United States and Canada has been applied to a growing number of present-day environmental issues. Although much still remains to be achieved in order to improve the quality of water in the Great Lakes, some observable progress has been made. On another border, water quality agreements between Mexico and the United States have updated a 1945 treaty via the more recent construction of a large desalinization plant on the lower Colorado River.

The most notable progress in international cooperation appears to have been made through multinational scientific investigations. Scientific programs organized through specialized agencies in the United Nations (UN) have been implemented cooperatively with various international scientific and technical associations. The International Council of Scientific Unions (ICSU) has been most effective in launching these ventures.

If one reviews the direction of international cooperation rather than the actual results of the above described projects, there is more reason for optimism. Some governments have been willing to sign treaties or adhere to international programs only in which very little follow-up action was expected or required. Other countries have refused to sign international agreements related to the environment. In 2001, for example, the United States was the only industrialized country refusing to sign the Kyoto Protocol on Climate Change to reduce the emissions of CO_2 and other greenhouse gases. Furthermore, in those environmental agreements in which decisive activities were accepted among the signatories, implementation has not always followed. In many instances, the inertia of human behavior has been detrimental to full-scale progress in solving global environmental problems.

It is atypical for people, as well as political entities, to plan for living more than a few years into the future. On the other hand, we must begin to act with a view toward future consequences in order to solve today's environmental problems. Some researchers project it is already too late to bring serious and often dangerous situations under our direct control.

Acid precipitation is a major environmental concern in over a dozen industrial nations, but very little action has been taken to alleviate the problem. International business in agriculture and forestry is altering the ecosystem of Earth quite rapidly. Unfortunately, affected nations have been unable to control these changes in ways that will allow the advantages of technology to be retained without destroying the riches of the natural environment. Effective environmental policy (as well as international cooperation) requires a type of foresight and benevolence for future generations often rare in society.

International Jurisdiction of Common Spaces

Several regions of the world are regarded as common spaces and are, either in whole or in part, considered international territory. These areas are becoming increasingly significant as more is learned about their potentials as sources of strategic raw materials. Areas presently beyond national jurisdiction include the open ocean outside the territorial sea limits of a country, the extraterrestrial space surrounding Earth, and the continent of Antarctica. Despite the existence of international common spaces, there is disagreement among countries as to the extent of these spaces. The sea claimed by coastal countries directly affects the size of the common space in the ocean.

> **COMMON SPACE**
> A region of the world entirely or partially considered to be international territory. These areas are beyond national jurisdiction.

Balancing environmental concerns with the economic advantages of technological development and exploitation continues to be a major issue for these areas. Until recently, these areas were customarily regarded as belonging to no one, and the common availability of their usage was not debated. Advancements in technology have subsequently created a scenario in which the natural resources indigenous to these areas can be more readily pursued. Even still, no general agreement for the governance of these regions has been reached. Perhaps this is due either to the fact that the environmental problems and opportunities these areas present are very unfamiliar or to the fact that the abilities of nations to explore, occupy, or exploit these regions differ a great deal. Only a few technically advanced countries presently have both the economic and technological capacities to conduct research activities on the ocean floor, in Antarctica, or in outer space.

The *Law of the Sea Treaty* was sponsored by the UN and negotiated over a fourteen-year period commencing in the late 1950s. This agreement provides for an international authority to license and regulate deep-sea mining and to allocate a share of any of the potential profits among less advantaged countries. Several governments, including the U.S. government, have expressed dissatisfaction with this arrangement. Although there is no general governance in place for the high seas, treaties to prevent ocean dumping and other forms of marine pollution have been ratified. Proposals regarding the protection of whales and other marine life species have also been negotiated.

The fact that not all nations possessing maritime capabilities are party to these treaties creates the potential for oceanic resource depletion and degradation. Even though some measure of progress has been made toward international responsibility for the oceanic environment, its future viability remains vulnerable to pollution and exploitation. Nations placing their immediate economic interests ahead of a marine environment characterized by richness and diversity pose the greatest danger in the upcoming years.

Similarly, *outer space* has no general governing body. It is unlikely one will be established in the near future. The penetration of outer space by artificial satellites and manned vehicles after the close of the 1950s opened a new domain for possible national jurisdiction.

A body of proposed "space law" rapidly appeared on the scene. Treaties concerning telecommunications, remote sensing, and military uses governed specific aspects of outer space. The UN sponsored a general treaty in the late 1960s favoring the peaceful use of outer space. Conversely, present military proposals among leading industrial nations make future agreements regarding the integrity of the outer space environment seem improbable.

Antarctica presents a case in which the time to choose between international cooperation and conflict has most definitely arrived. The continent has been treated as a scientific reserve since the Antarctic Treaty of 1959. It has been maintained in its natural state without permanent occupancy, resource development, or pollution. A consultative committee of nations has administered the terms of this treaty. At the time of the initial signing of the Antarctic Treaty, technological capabilities were insufficient to justify economic interest in this unique environment.

Today, due to the prospect of future demands for minerals, continuance of Antarctica as a protected state is dubious. The United States and several other nations have begun to reinvestigate Antarctica's resource potential. It seems that the fate of the continent, together with the marine life of the surrounding seas, may be at stake before the end of the century.

Unlike the governments of other nations, the U.S. government recognizes no national claims to Antarctic territory and makes no claims of its own. Not surprisingly, it also possesses the greatest fund of knowledge, experience, and technological prowess essential for Antarctic resource development. Therefore, it seems the United States is in the proverbial driver's seat with reference to Antarctica's future status in the international community. The directions for development here will be toward either a reserve for scientific cooperation or an arena for international conflict.

Localized Environmental Problems

At first glance, environmental problems tend to fall into two categories—local and global. However, the line between these two categories is often blurred. In the case of pesticides, a local problem with pesticides could be a polluted field or stream. Since an international corporation sells the pesticide, however, the problem could then be considered global.

Many environmental issues have both local and global aspects. Here, we will focus on three current environmental problems—pesticides, water scarcity, and climate change. Each of these is a local problem for selected countries, but they also are global problems affecting multiple states and countries.

Pesticides

Pesticides have become a critical part of the food cycle, not only in the United States, but globally as well. Over the last four decades, the major increases in the efficiency of crop production can be linked to the use of pesticides in farming. A pesticide is any substance used for the prevention or destruction of any pest, including insects, animals, weeds, or microorganisms, such as viruses or bacteria. Most chemical pesticides create some risk to humans or the environment because their purpose is elimination of pests. The problem has always been to try to choose the pesticide causing the most good with the least harm.

The Environmental Protection Agency (EPA) is responsible for the regulation of pesticides in the United States. However, not all countries around the world have the same level of pesticide protection. DDT is a good example of a pesticide that has caused problems throughout the world. DDT, discovered in the late 1930s, was used widely in the United States and globally from after World War II to the 1970s because it was particularly effective in killing the mosquitoes spreading malaria. After it was used for many decades, however, problems emerged with DDT. DDT is very toxic to fish, and it builds up in the fatty tissue of animals over time. These problems brought DDT into the food chain and exposed many people to unacceptable levels of DDT. Also, DDT lasts a long time after being spread on a field (from two to fifteen years). Because of these problems, the use of DDT was banned in the United States in the early 1970s.

The key word is *use*. A U.S. company could no longer use DDT; however, a U.S. company could manufacture DDT, and that is what happened. Despite its ban in the United States, American chemical companies still manufactured DDT and other banned pesticides for export into other countries. See Figure 9-2. In addition to educating the developing world about hazardous pesticides, there is the problem of the storage of these pesticides. According to the Food and Agriculture Organization of the United Nations, developing countries have gained possession of more than one hundred thousand tons of obsolete pesticides, including DDT.

This problem is close to being solved on a global scale. In 2001, 127 countries adopted the Stockholm Convention on Persistent Organic Pollutants. After the treaty is ratified by at least fifty countries, the production and use of DDT and eleven other chemicals will be banned, but this ban will cover only the twelve pesticides. New pesticides are being developed at a fast rate each year. How long will it take for us to know the impact of these new pesticides on the environment? And, then, how long will it take for us to ban the undesirable ones?

Water Availability

In the United States, we take water for granted. We get up in the morning, walk to our kitchens, and make our morning coffee by turning on the water faucet. We never think of where the water comes from and how clean it is—at least, not very often. Water, however, is not as limitless as we suspect. True, Earth is 71 percent water; however, only 3 percent of that water is freshwater. In addition, most of the freshwater is inaccessible to humans—it is present as ice and snow at the two poles. Only about 1/10,000 of Earth's water is readily available for human needs.

Water, like other natural resources, is not evenly distributed around the world, and much of the problems about water relate to its distribution rather than to its global availability. The world's population is rising by about 86 million people a year. This population growth also is not evenly distributed—much of the growth will occur in three

Figure 9-2.
Middle Awash State Farm is massively oversupplied with pesticides now leaking into an environmentally sensitive flood plain. An international expert task force visited Ethiopia in the late 1990s to assess what needed to be done to rid the country of huge quantities of widely dispersed obsolete pesticides that have accumulated over the past three to four decades. (Pesticide Action Network, UK)

regions: Africa, Asia, and South America. At the same time, these regions have severe water shortages (or increasing water pollution). In the 1990s, there were 143 droughts in the world; these droughts impacted 185 million people. Even if a country has sufficient rainfall, often the people are not located where the rain does fall. In Central America, half of the population lives on the Pacific seaboard. Unfortunately, the water for these people comes via heavily polluted rivers from the interior of their countries.

Even when water exists in developing countries, it is more likely to be polluted. In most developing countries, wastewater goes from the home directly into rivers and streams without any treatment. Experts estimate that, in sub-Saharan Africa, two-thirds of rural people and one-quarter of urban people are without safe drinking water. In just one developing country, Bangladesh, between one-fourth and one-half of the population drinks water that has unacceptable levels of arsenic. Disputes about water are common in areas where rainfall is low. Currently, there are several water sources being disputed by one or more countries, including the Euphrates River (Turkey, Syria, and Iraq), the Sea of Galilee (Israel and Syria), the Nile River (Egypt, Sudan,

Ethiopia, and others), and the Senegal River (Senegal and Mauritania).

In the United States, despite the Clean Water Act of 1977, the cleanliness of your water depends on where you live. Americans living in rural areas generally have poorer water quality than those living in cities. At least 2 million Americans living in rural areas either lack adequate water or have unclean water. Also, certain states are more vigilant than others in addressing the issues of toxic water. The National Wildlife Federation (NWF) report found that one-fifth of all drinking water in the United States is unhealthy. Also, certain parts of the country have other water problems. Because of the large amounts of fertilizer and animal waste runoff from farms, there is a "dead zone" at the mouth of the Mississippi River about seventy-five hundred square miles in size (about the size of the state of New Jersey). Mercury contamination of water is increasing, and more alarmingly, an estimated nine hundred people are killed and another nine hundred thousand get sick each year as a result of drinking contaminated water.

As population increases throughout the world and more stress is put on existing water supplies, water will increasingly become an issue. Unlike other resources such as oil or coal, there is no substitute for water. We all need water to live, and in the future, countries will have to do a better job managing their water supplies so there will be enough good water to drink.

Climate Change and the Global Warming Debate

Some of the more potentially serious air pollution problems are not even occurring near Earth's surface. The upper atmosphere has been selected as a laboratory, where a dramatic technological experiment is taking place. For over a century, scientists have been debating about what will occur as a result of loading a large amount of CO_2 and several other gaseous elements into Earth's upper atmosphere. Some degree of agreement among them has evolved over the years, and trends toward *global warming* seem quite probable and possibly irreversible.

CO_2 in the atmosphere has unquestionably increased over the past several decades. This increase is the direct result of human activities largely related to industrialization and the advancement of technology. Specific examples of the types of activities that have contributed to the CO_2 buildup include the combustion of a greater number of fossil fuels and tropical deforestation (discussed further in the following section). If the production and use of fossil fuels continues to increase at the same rate, projections indicate CO_2 concentrations in the atmosphere will double preindustrial levels by the mid-twenty-first century and will lead ultimately to severe changes in global climate.

While it is not totally clear how an increase in the mean global temperature will affect weather patterns, some scientific analyses have been made. Various studies have surmised the buildup of CO_2 could eventually melt the ice cap at the North Pole or South Pole, leading to a rise in sea level that would submerge many coastal cities. Between the late 1950s and the late 1970s, the depth of the Arctic Ocean's ice cap was reduced by an annual average of 4.3 feet. A small change in the mean global temperature could cause large shifts in weather patterns creating a higher frequency of droughts, hurricanes, freezes, and floods. A decrease in annual rainfall within the world's prime agricultural belts, coupled with other losses of water from rivers and foliage, could virtually ruin the productivity of some of the most fertile acres of land on Earth.

If warming brings about climate changes, they will most likely start occurring during the lifetimes of roughly half of the people alive today. Although it is easy to envision the dates 2035 or even 2050 as excellent movie titles, they do have a certain unrealistic quality to them. Even something closer at hand—perhaps increasing numbers of hurricanes for the next twenty to thirty years—commands less attention than the dollar's robustness in the international money market. What should the government's policy response to CO_2 be and what sort of research in this environmental area needs the greatest financial backing?

What is the *greenhouse effect?* Energy from the Sun heats Earth's surface. See Figure 9-3. The surface then radiates the heat back into space; however, some of this heat is trapped in our atmosphere and warms the planet. It is called the *greenhouse effect* because it acts somewhat like the glass panels in a conventional greenhouse. Although many people do not know this, the greenhouse effect is critical for life on Earth. Without the natural greenhouse effect, temperatures would be much lower.

The problem is not the natural greenhouse effect; it is the increasing temperatures because of the use of fossil fuels. Since the Industrial Revolution, there has been an increasing

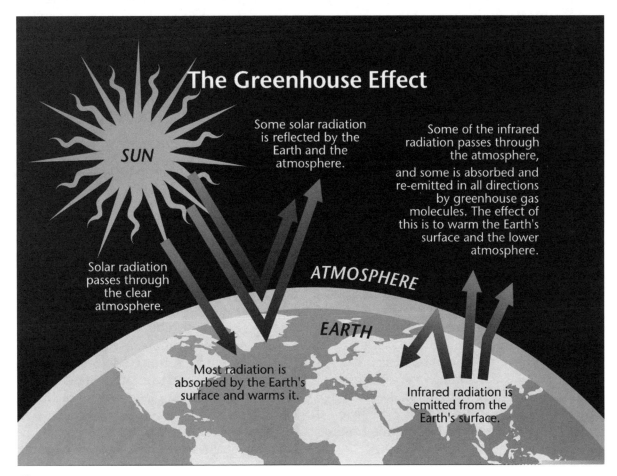

Figure 9-3.
This visual depiction of the greenhouse effect was developed by the U.S. Global Change Research Program (USGCRP), which has initiated a national assessment on the potential consequences of climate variability and change for the nation. (USGCRP)

dependence on fossil fuels to create power. In the United States alone, the burning of fossil fuels is responsible for about 98 percent of U.S. CO_2 emissions, 24 percent of methane emissions, and 18 percent of nitrous oxide emissions. The United States is the biggest contributor, by far, to greenhouse gases in the world—about one-fifth of all greenhouse emissions are from the United States. Individually, the average American adds over fifteen hundred pounds of greenhouse gases each year to the environment.

Few people debate that Earth has been warming; however, many debate how to solve this warming problem. Overall, global temperatures have increased more than one degree this past century, and the hottest decade ever recorded was from 1990–2000. The UN-sponsored Intergovernmental Panel on Climate Change projects that the temperature will rise between two and ten degrees by 2100. There are other changes related to global warming. Sea levels have risen approximately six to eight inches in the last century, precipitation has increased by 1 percent, and glaciers are beginning to melt. If warming continues, as it will most likely do, these changes will be amplified in the next one hundred years.

An issue like the greenhouse effect of global warming requires Congress to look decades ahead. This legislative branch of our government is typically not interested in solving long-term problems. Many lawmakers seem to be unable to think ahead more than a couple of years, and because the United States is the number one greenhouse gases polluter in the world, it makes the situation even more difficult from a political point of view. One of the reasons the United States decided not to sign the Kyoto Protocol (to reduce fossil fuels emissions) was that the U.S. economy would be hurt. Another reason was that the Kyoto Protocol applies only to developed countries.

Whether one believes global warming is important or not, the problem is that the climatic changes will occur long after an individual legislator or president leaves office. Politically, this makes any serious reduction in fossil fuels emissions a difficult prescription to take in the United States.

At the very least, it is imperative to embark on an energy conservation program that would also encourage the development of other energy sources less likely to increase atmospheric CO_2 levels. There should be more research into the climate dependencies of various staple crops, and farmers should be encouraged to use water more efficiently. Irrigation projects and large-scale water supply networks should take into account the possibility of future shifts in rainfall patterns. International reforestation programs are essential, and countries like Brazil and Malaysia should be admonished to discontinue the destruction of their rain forests.

Air Quality Standards

The central goal of the monumental Clean Air Act and its subsequent amendments has been to manage air quality to protect human health. It was the first of the major environmental regulations adopted in the United States. This legislative decision required the U.S. Environmental Protection Agency (EPA) to establish strict limits on several major pollutants existent in industrial centers, including sulfur oxides, nitrogen oxides, and particulate matter. These limits were set at the federal level during the early 1970s. As a result, this environmental law has been the most costly one ever enacted, requiring businesses to bring their production facilities into compliance. During the first twenty years after the establishment of the Clean Air Act, the EPA found that the actual costs were over $500 billion.

In terms of the pollution problems caused by the types of pollutants regulated in this federal act, air quality in the United States continues to improve. Throughout the 1970s, the number of days when air quality in densely populated areas reached harmful levels declined by at least one-third. Since many areas of the country still continue to experience periods when air pollution levels create human health risks, the struggle is far from over. These tough areas of urban pollution (such as Los Angeles and Houston) may ultimately prove to be too expensive and difficult to clean up—however, optimists hope this is not the case.

The World Health Organization (WHO) has ranked deaths from air pollution as one of the top ten causes of disability. In the late 1990s, WHO estimated that each year seven hundred thousand deaths are related to different types of air pollution, from both indoor and outdoor sources. Present control efforts do not effectively address these hazardous and other unconventional air pollutants. Furthermore, existing sulfur oxide controls appear to be grossly inadequate to protect lakes and forests against the problems caused by acid rain. The relentless buildup of CO_2 in the atmosphere may spell disaster for global climatic stability. Finally, even though most Americans spend a majority of their time indoors, there are minimal (if any) efforts being made to control air pollution levels in these areas.

Procedures used to monitor air quality are not as reliable as some of the people reporting cleaner air trends would have us believe. We do not have a national system for monitoring air quality. Instead, area monitors are operated by different states and localities across the country. The locations of these monitoring stations change periodically, records are often incomplete, and quality control practices are *not* always satisfactory.

On the surface, data may suggest air quality has improved significantly over a specified period of time. Recognize, however, that year-to-year improvements may occur for a variety of reasons not directly indicative of the success of pollution control efforts. For example, a major source of pollution may have closed down and moved to another site where pollution levels are increasing; temporary economic slumps may have caused an equally short-lived reduction in industrial emissions; or more favorable climatic conditions may have developed with steady breezes dispersing the pollutants elsewhere. Since monitoring data is not usually adjusted in consideration of these variables, caution should be exercised in the overly optimistic interpretation of air quality changes on a yearly basis. Generally speaking, it takes more than a year or two to substantiate a trend in any field of study.

Ambient Air Quality

National efforts to control air pollution have commonly been focused on improving ambient air quality in high-tech and urban centers. The six conventional pollutants toward which most of the attention has been directed are total suspended particulates (TSPs), sulfur oxides, nitrogen oxides, carbon monoxide, lead, and ozone. Numerous industrial emitters of these pollutants have subsequently switched to cleaner fuels and have adopted control technologies, such as electrostatic precipitators. The EPA estimates that emissions of the conventional pollutants from most of the major sources have declined or held steady since concerted cleanup efforts began in the early 1970s. See Figure 9-4.

In the mid-1970s, the EPA developed the ***Pollutant Standards Index (PSI)*** in cooperation with the Council on Environmental Quality in an attempt to standardize air

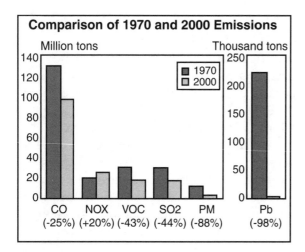

Figure 9-4.
The Environmental Protection Agency estimates that, except for nitrous oxide, the major pollutants have decreased since 1970.

pollution reporting. This index enabled comparisons to be made between different regions of the country. The PSI converted the concentrations of several commonly occurring pollutants to a number between zero and five hundred. In 2000, the EPA updated the PSI and renamed it the *Air Quality Index (AQI)*. The AQI includes data on two additional pollutants (ozone and fine particulate matter) and a new health risk category, "unhealthy for sensitive groups." See Figure 9-5.

The index value of one hundred represents the concentration for each pollutant below which adverse health effects have not been observed. It was selected to correspond to the short-term (twenty-four hours or less) National Ambient Air Quality Standards set by the EPA. The highest AQI value of five hundred corresponds to the concentration of each pollutant that should never be reached. State and local agencies are required to take emergency action to prevent air pollution from ever reaching this level by enforcing such measures as temporary restriction of automobile traffic or manufacturing processes. On the average, ambient air quality in U.S. metropolitan areas, as measured by the AQI, has been improving consistently since the early 1970s. The table presented in Figure 9-6 reveals the extent to which air quality has, for the most part,

Air Quality Index (AQI)							
Category	**Good**	**Moderate**	**Unhealthy for Sensitive Groups**	**Unhealthy**	**Very Unhealthy**	**Hazardous**	
Index Value	0–50	51–100	101–150	151–200	201–300	301–400	401–500
Pollutant	**Concentration Ranges**						
CO	0.0–4.5	4.5–9	9–12	12–15	15–30	30–40	40–50
NO_2	—	—	—	—	—	1.2–1.6	1.6–2.0
O_3 1-Hour	—	—	—	—	.20–.40	.40–.50	.50–.60
O_3 8-Hour	.00–.06	.06–.08	.08–.10	.10–.12	.12–.37	—	—
PM 2.5	0–15	15–40	40–65	65–150	150–250	250–350	350–500
PM 10	0–50	50–150	150–250	250–350	350–420	420–500	500–600
SO_2	0.0–.03	.03–.14	.14–.22	.22–.3	.3–.6	.6–.8	.8–1.0

Figure 9-5.
This Air Quality Index table converts the concentrations of common pollutants to a number from zero to five hundred.

Number of AQI Days Greater Than One Hundred at Selected Cities in the United States									
Metropolitan Area	**1991**	**1992**	**1993**	**1994**	**1995**	**1996**	**1997**	**1998**	**1999**	**2000**
Atlanta	23	20	36	15	35	25	31	50	61	26
Boston	13	9	5	9	11	4	8	8	7	1
Chicago	25	6	3	8	23	7	9	10	14	0
Dallas	2	12	14	27	36	12	20	28	23	20
Denver	6	11	6	2	3	0	0	7	3	2
Detroit	27	7	5	11	14	13	11	17	15	3
Houston	36	32	27	38	65	26	47	38	50	42
Kansas City	11	1	4	10	23	10	17	15	5	10
Los Angeles	168	175	134	139	113	94	60	56	27	48
New York	49	10	19	21	19	15	23	17	24	12
Pittsburgh	21	9	13	19	25	11	21	39	23	4
San Francisco	0	0	0	0	2	0	0	0	0	0
Seattle	4	3	0	3	0	6	1	3	1	1
Washington, D.C.	48	14	52	22	32	18	30	47	39	11

Figure 9-6.
This table shows national air quality trends. Air Quality Index values of more than one hundred are defined by the Environmental Protection Agency as unhealthy.

improved and, in a few cases, worsened in selected cities across the United States since the early 1990s.

The Acid Rain Dilemma

Acid precipitation results from the presence of sulfur oxides and nitrogen oxides in the atmosphere, which then react with moisture to form dilute sulfuric and nitric acids. These resultant aerosols are eventually carried back to Earth as rain, snow, fog, or dry particles. In some areas, the acid is buffered by Earth's natural alkalinity and apparently causes little damage. In other locations, however, the acid reacts with soil, leaches away important nutrients, or drains off the land and accumulates in lakes and ponds. The ecosystems of these bodies of water are then severely disrupted, and many of them die and become unable to support aquatic wildlife of any type.

ACID RAIN

A serious environmental problem resulting from the presence of sulfur oxides and nitrogen oxides in the atmosphere. When these compounds react with moisture, they form dilute sulfuric and nitric acids, which are then carried back to Earth.

In recent years, acid rain has received widespread recognition as a serious environmental problem in many parts of the world, including Canada, Scandinavia, Japan, and the Northeastern United States. There does, however, seem to be a certain amount of scientific controversy regarding the extent to which acid rain is actually the sole culprit in these scenarios. Scientists have been systematically monitoring the evidence about acid rain's environmental effects since the mid-1970s. Scientists often indicate acid rain is only a minor contributor to the environmental damages for which it has been blamed. In other words, claims regarding acid rain damage have been based on circumstantial evidence and are still somewhat misleading.

The U.S. federal government established a National Acid Precipitation Assessment Program (NAPAP) in 1980. It was commissioned to conduct a ten-year research, monitoring, and assessment effort to determine the degree to which acid rain *may* be responsible for certain environmental problems. Earlier research attempts usually *assumed* acid rain was the cause of the damage at hand and proceeded to ascertain how the actual process of degradation worked. NAPAP scientists have designed their research around three main questions regarding the acidification of U.S. lakes:

❑ Has there been a significant increase in lake acidity?

❑ Might something other than acid rain be contributing to lake acidification?

❑ Might something other than acidity be affecting the survivability of the fish?

Retrieving straightforward, reliable answers to these questions has not been easy. There still does not appear to be any scientific consensus on this issue. NAPAP was reauthorized as an open-ended program under the 1990 Clean Air Act. The 1990 Clean Air Act also established the EPA's Acid Rain Program to reduce emissions of sulfur oxides and nitrogen oxides. Phase I of the Acid Rain Program began in the mid-1990s and targeted the largest coal-fired power plants. Phase II began in 2000—this phase set restrictions on small coal, gas, and oil-fired plants, as well as on large coal-fired plants.

Most of the NAPAP analyses dispute the viewpoint that acid rain substantially degrades exposed structures and building materials. They suggest natural weather conditions such as freeze thaw cycles, coupled with a variety of air pollutants, cause the observable damage. More importantly, NAPAP scientists have found most urban pollution comes from local emissions rather than from distant emissions borne by rainfall. It therefore seems likely we should focus on local automobiles and nearby industries in order to reduce the impact of pollutants on urban area structures.

None of this discussion is meant to imply acid rain is not an air quality problem demanding some sort of immediate action. On the contrary, long-range correlational studies are being conducted in an attempt to provide a clearer avenue for action and effective investment of funds. The "scientific jury" seems to be unsettled in presenting its verdict in this dilemma. Despite this uncertainty, the Acid Rain Program has been successful in reducing sulfur dioxide emissions from 16 million tons in 1990 to 11.2 million tons in 2000, and a joint U.S.-Canada survey in 2000 showed that rainfall acidity had been reduced nearly 25 percent, as compared to the 1980s.

Despite the successes seen in North America, acid rain is increasing in other parts of the world. As the emissions levels go down in North America, they are increasing in Asia. China now leads the world in sulfur emissions, and experts predict that, with the large increase in industry in Asia, there will be a tripling in nitrogen oxide emissions. Worldwide, there is no doubt that, unless all governments work on this problem, there will be increasing levels of industrial emissions and, therefore, increasing acid rain.

Indoor Air Quality

It is becoming obvious air pollution is no longer an exclusively outdoor problem. Concentrations of air pollutants such as carbon monoxide, radon, formaldehyde, asbestos, and passive tobacco smoke are often far greater inside buildings than outside. Since many people spend a bulk of their time indoors, these pollutants pose significant health threats.

It may be time to establish a formal indoor air quality-monitoring program similar to the ones devised for outdoor air at the federal, state, and local levels. Several factors indicate indoor air quality may be worsening. Ironically, energy conservation practices in homes and office buildings tend to reduce the escape of indoor air and allow pollutant concentrations to increase. The increasing popularity of burning alternative fuels, such as wood and kerosene, in the home is responsible for the release of numerous hazardous pollutants. Additionally, the use of synthetic chemicals like formaldehyde and volatile organic resins in home construction and furnishings can have toxic effects on humans.

The relationship between concentrations of indoor and outdoor pollutants is further complicated by wind speeds, architectural designs, and ventilation systems. Structures with forced ventilation to the outside or natural ventilation via open doors and windows will have roughly similar pollutant concentrations inside and outside. Many new buildings are designed with reduced air infiltration. Reducing the number of air exchanges between inside and outside lowers heating and cooling costs, but simultaneously increases the concentrations of the indoor pollutants listed above.

Historically, regulatory strategies concerning air quality in the United States have centered on the concentrations of ambient or surrounding pollutants. This approach has involved the measurement of outdoor pollution levels at particular locations. These indices are then assumed to be the primary determinants of exposure for the people residing in that area. It is obvious this procedure is far from comprehensive. Further research into indoor air quality programs is necessary if human health risks associated with air quality are to be accurately assessed.

Tropical Forest Resource Depletion

Scientists and world leaders consider tropical deforestation to be one of Earth's most serious environmental problems. The longevity of their concern combined with the current tropical forest destruction rate provides undeniable validity for this apprehension. The term *tropical forest* refers to forests in the humid, semiarid, and arid regions of the world. Forests ranging from moist (or closed) tropical forests to dry (or open) woodlands are generally included in the description.

The destruction of tropical forests in the world today is so extreme and so devastating that humanity may very shortly lose its richest, most diverse, and most valuable biotic resource. The damage being done may not be entirely visible to those of us who reside in the United States, since the tropics are faraway lands about which we know very little. For many people, the word *tropics* brings to mind a picture of pristine nature, mysterious ceremonies, large mammals, and exquisite orchids and palm trees. The scene is quite reminiscent of old movies and is difficult to authenticate anywhere in the continental United States.

Once thought to be limitless and expendable, a renewable resource so to speak, the rain forests are presently disappearing at an alarming rate. Estimates of exactly how much rain forest is being lost or irretrievably changed tend to range from 20 million to 50 million acres per year. Ten million of this loss is in the Brazilian Amazon alone each year. Moist tropical forestlands encircling Earth's equatorial sector cover approximately 6 percent of its surface area. Just since the end of World War II, more than half of this rain forest has been destroyed in the name of logging, farming, ranching, mining, and road building.

What has been lost already is literally irreplaceable. Contrary to the popular cliche, the jungle does not reclaim its own. Even where a healthy second growth manages to come back, the species it breeds are not the same as those previously in existence. The primal tropical rain forests reach back through evolution. Its brilliant flora and fauna reflect continental drifts and the ebb and flow of ice ages. Scientists project that well over a million species of plants and animals could be extinct early in the twenty-first century as the forests are continually degraded. Furthermore, if the current pace of exploitation remains unchecked, the rain forests will have vanished forever within the next several decades.

Importance of Tropical Forests

For the inhabitants of the world's tropical nations, the forests provide fuel, food, medicines, building materials, wood for lumber and paper, and many other basic needs. Forests help to maintain soil quality, limit the erosion process, stabilize hillsides, modulate seasonal flooding, and effectively protect nearby waterways from accelerated siltation. The tropical forests can also represent an important source of foreign exchange, export revenue, and employment for local residents.

The potential benefits acquired from tropical forests are not limited solely to the tropical nations. World trade in tropical wood is significant to the economic prosperity of both the producing and consuming nations. The United States is one of the largest importers of tropical wood and wood products. Its demand has been growing steadily over the past several decades.

Tropical forests also provide a broad array of non-wood products, including oils, spices, and rattan. These items are valuable for both subsistence and commerce. The annual world trade in rattan and the oils and spices extracted from tropical plants exceeds $2 billion. Some of the more familiar oils and spices found in many American homes include camphor, cassia, cardamom, citronella, and cinnamon. Thus, the export of industrial wood and non-wood forest products earn substantial foreign exchange for nations trading with the United States.

A cursory glance through the contents of the typical American family household's medicine cabinet provides further verification for the importance of the tropical rain forests. The wide array of rain forest plants and animals provide at least one-fourth of the ingredients needed to produce the prescription drugs on the market today. The snakeroot plant of India's monsoon forests yields reserpine, which is an alkaloid providing the base for many tranquilizers. The corkwood tree of eastern Australia provides scopolamine, a drug used in the treatment of schizophrenia. In the Amazon basin, one finds the plant derivative curare, which is an essential agent for treating Parkinson's disease. There are research biologists who estimate that over one thousand plants found in the tropical forests may have potential anticancer properties.

The agricultural stability of the United States is also directly related to the existence of tropical rain forests. These regions provide habitats for many of the world's migratory birds and various endangered avian species. Approximately two-thirds of the birds breeding in the North American continent migrate to Latin America or the Caribbean for the winter season. Some of these birds play a significant role in controlling agricultural pests found to afflict U.S. farmlands. Depletion of their tropical habitats may ultimately spell their extinction.

The tropical forests are also critical to the economic viability of U.S. investments overseas and, as previously mentioned, to the political stability of developing nations. For example, U.S. foreign assistance programs (technical, as well as financial) can be erratically disrupted by flooding, siltation, and other environmental damages caused by deforestation. Inappropriate deforestation practices also reduce food sources and jobs, both of which are necessary to maintain political stability in these countries.

In sum, the tropical rain forests represent the planet's "biological warehouse," where the diversity of plant and animal life found within a single acre plot is simply staggering. Destroying these rain forests means losing 80 percent of the world's vegetation and up to 4 million varieties of life forms, many of whose benefits are still unknown.

Causes and Consequences of Tropical Forest Destruction

The main causes of forest destruction and degradation stem from the following situations: rapidly rising populations, requiring more land for food production; increasing levels of mining and logging operations; primitive methods of agriculture that produce low yields and ruin soil fertility; and the chronic failure of political institutions to execute programs with an eye for long-range prosperity. In parts of Latin America, encroachment on the forests is simply a response to social pressures where a vast majority of the arable land is controlled by a very slim percentage of the population. The availability of land in the rain forests is a tempting solution to this land distribution problem for many countries. Overall, the nonindigenous population of the Brazilian Amazon has increased from 2 million to 20 million since the 1960s; this increased number of people put pressure on the rain forest. Millions of Central American inhabitants do not have enough land to provide for their subsistence. Government officials subsequently view the land currently covered by forests as the answer to overcrowded conditions and extreme levels of poverty among their constituents.

Starting sometime in the mid-1970s, vast tracts of Central American forest were cleared for ranching. One stimulant for this activity, which destroyed some eight thousand square miles annually through the 1980s, came from the U.S. demand for low cost ground beef and pet food. It is doubtful very many Americans consider the environmental consequences of tropical deforestation when they feed their dogs or purchase a deluxe hamburger at their neighborhood fast-food establishment. The most unsettling aspect related to this particular economic activity is found in the short life span of these graze lands. The "slash and burn" methods of cultivation used limit its productivity cycle to a period of three to five years. On the average, pastures become severely degraded and heavily weed infested within five to ten years after they are initially developed. In other words, the tremendous investments required to develop pasture in tropical forest regions are not justified by the short-term yields acquired.

Elsewhere in the Third World, the simple need for wood used as fuel, together with logging by international corporations, is denuding (stripping of covering) tropical rain forests. Nigeria has lost more than 90 percent of its forest cover—45 percent of the country used to be forested, while today only 5 percent contains forests. Traditional timber producers, including Malaysia, Thailand, and the Philippines, may soon face the prospect of becoming timber importers themselves, due to excessive logging ventures.

The potential costs of tropical deforestation discussed in the scientific and technical literature seem to fall into three general categories:

❑ Severe changes in global weather patterns.
❑ Loss of cultural and biotic diversity.
❑ Desertification of the forest lands.

Stated more dramatically, in less than one hundred years, Siberia could be the bread-basket of the world, New York City might be at the bottom of the sea, and one-fourth of all existing wildlife species may be extinct.

Due to the vastness of the tropical rain forest segment of the planet, its disappearance poses an extreme threat to global climatic stability. Scientists estimate the respiration processes of forest trees and plants return at least half of the rainfall dropping on it back into the air. Conversion of the rain forest to pasture drastically reduces the evapotranspiration cycle. *Evapotranspiration* is the water removed from the soil by evaporation from leaf and plant surfaces. For instance, as the South American forest shrinks, the Amazon basin could dry out irreversibly, grow warmer, and shift weather patterns in the United States and Canada, pushing the grain belt northward.

There is also a growing concern about the contribution of deforestation to the greenhouse effect. The dramatic climatic effects related to an atmospheric buildup of CO_2 were discussed in a previous section of this chapter. There appear to be uncertainties, however, about how much carbon is actually released as a result of forest clearing, as well as the amount of CO_2 taken in by healthy forests. Conclusions regarding this particular consequence of deforestation remain tentative.

Destruction of many plant and animal species is another serious consequence of tropical forest destruction. Many species of tropical rain forest trees live at low densities and lack seeds adapted to long periods of stressful environmental conditions. The species of plants and animals native to primary tropical rain forest conditions may be unable to recolonize large clearings or open areas.

In addition to the loss of entire species, we face the morally unforgivable prospect of ethnocide (death of a cultural group) of many small, isolated, and vulnerable groups of indigenous or local people. These individuals hold much of the knowledge and culture necessary to understand and utilize the rain forest's resources. Their obliteration or social displacement cannot be ignored in the midst of technological development.

As mentioned above, the bold razing of the tropical forests under the guise of farming or ranching produces only short-lived profits. The ash released from the burnt logs, initially a rich fertilizer, disappears quickly from the plot of cleared land. Denuded of its protective canopy and deprived of its fertile carpet of decaying plant life, the thin topsoil typical of these rain forests is soon washed away by tropical downpours.

The process continues for a few years until the land can no longer support human life. Farmers and ranchers then cut still deeper into the forest, leaving behind desolate scrubland of near desert conditions. Once the new patch of trees has been cleared, the torrential rains will once again erode the soil and further expand the problem of desertification.

In an era when environmental issues are of overriding importance to the well-being of society, the scientific and technological communities must focus their concern on those deemed most critical. Tropical forest resource management is an issue of great importance. It appears the technologies needed to abate the process of tropical deforestation do exist. Even still, significant inroads into the problem of deforestation and resource depletion have been slow to appear. There have been tremendous political, cultural, and economic constraints to overcome. Many of the actions needed to halt and correct forest resource deterioration can only come from the governments and citizens of the tropical nations themselves.

In light of this reality, the former Congressional Office of Technology Assessment concluded the United States can still play a key role in sustaining tropical forests through its expertise in research, environmental impact analysis, technology development, resource management, and education. It identified a number of existing and emerging technologies that could very well provide solutions to the crisis in the tropical rain forests. Some of these suggested actions include farming techniques combining trees with crops or livestock, *agroforestry;* genetic improvement of the trees; improved charcoal production; more efficient woodstoves; and new approaches to park design and management.

It may be true technologies developed in the United States can be used in the tropics, but it is critical they be adapted to their different climatic and social conditions. For example, while satellite technology can provide very clear imagery of the forests and other vital information that could be used to improve tropical resource management to determine the extent of global deforestation, desertification, and habitat destruction, the primary constraint to developing countries is a lack of adequately trained personnel. Other technologies, such as those for improved

harvesting and handling of timber, can also be modified to be appropriate for use in the tropics.

Technology's Role in Management

As the *Dateline* newsflash indicates, there are more opportunities to create energy without causing any environmental damage. This type of appropriate technological intervention can only occur following comprehensive environmental impact analyses. Water itself can naturally create energy by transmission of its wave power to electric power without causing any adverse ecological effects. Instead of farming or ranching, eco-tourism can replace subsistence farming as a source of income for dwellers of the rain forest. Both of these are examples of technology management of environmental problems. Using modern technology to save natural resources can be a new way to manage the problems caused by older technology on the environment.

Water Management

When water is scarce or unavailable in certain places, governments manage the water supply to provide water for people, farming, and industry. The history of water management shows that each time a water source is diverted or altered, there are unintended consequences. A major reason water is diverted is for agriculture. As populations increase globally, so does the need for food, and unfortunately, modern agriculture requires that water be diverted.

After the crops are harvested, countries are left with the by-products of agriculture. Runoff from farmland increases the sediment in rivers and nutrients drained from fields flow into the water supply. In the United States alone, agriculture accounts for about 65 percent of the total national phosphorus and nitrogen discharges annually. Industrial waste and by-products add to the pollution of the world's waterways. The United States is not alone in dealing with these issues. A by-product of the breakup of the Soviet Union and the Communist bloc was increased industrialization and pollution in these formerly Communist countries.

There are two primary ways countries manage their water supplies: irrigation and dams. Both dams and irrigation have existed for thousands of years. However, dams in particular have had a resurgence in popularity in the last half-century. Since 1950, the number of large dams (at least fifty feet in height) has increased from fifty-seven hundred to forty-one thousand, worldwide. Along with the size, dams are diverting more water. The 610 foot high Three Gorges Dam in China, scheduled for completion within this decade, will stretch for more than one mile across the Yangtze River and will generate hydroelectric power. See Figure 9-7. The technology underlying the Three Gorges Dam is described in detail in Chapter 7 (Technology Transfer) of this book. The water behind this dam is planned to be diverted to provide 10 trillion gallons for agricultural irrigation. Also, this dam requires the movement of between 1 and 2 million people to make way for it. The problem with this dam is the changes in the flow of the Yangtze River will affect more than China. Scientists predict that, if as little as 10 percent of the river is diverted, temperatures in Japan might be raised several degrees.

Figure 9-7.
A boat navigates the Yangtze River through gorges that are expected to flood when the $25 billion Three Gorges Dam is completed in 2003. (China Photos)

What is China to do? It is faced with an increasing population and the need to produce more food to feed this population. The possible long-term effects of this dam seem small in comparison to the starvation of the Chinese people. This example highlights the problems with water management. Often, these strategies are designed to solve the pressing problems of today. Unfortunately, they then cause new problems for tomorrow.

Hazardous Waste Management

At thousands of hazardous waste sites across the United States, toxic chemicals are contaminating the land, air, and groundwater supplies. During the mid-1980s, the EPA estimated dangerous chemicals were leaking out of at least sixteen thousand landfills throughout the countryside. The management of the toxic wastes produced by a progressive technological society has become yet another environmental issue demanding immediate attention.

Toxic waste has become a national problem in this country. It is an issue that has emerged through a confusing web of scientific, legal, community-based, and industry-related perspectives and discoveries. Often out of ignorance and neglect, Americans have carelessly disposed of immeasurable volumes of waste materials since the end of World War II. Manufacturers, farmers, and the general populace have poured synthetic organic chemicals, pesticides, herbicides, and industrial solvents into ponds, streams, storm drains, and poorly prepared ditches, with little regard for geological limitations.

Research has revealed the potential health hazards created as a result of these neglectful activities. They range from minor discomforts (headaches and nausea) to life-threatening ailments (cancer and birth defects). The environmental damages related to this form of contamination may involve loss of vegetation, destruction of animal habitat, and the eventual elimination of certain species of wildlife within the affected geographic area. Public anxiety has stimulated widespread demand for immediate action to somehow rectify the current dismal scenario with reference to handling hazardous wastes. In most cases however, these demands seem to overlook basic questions:

❑ Who should engineer the cleanup?
❑ Who should pay for it?
❑ What should be done?
❑ How clean is clean?
❑ How should it be implemented?

The Comprehensive Environmental Response, Compensation, and Liability Act (CERCLA), commonly referred to as *Superfund,* was enacted to provide for cleanup at sites where hazardous wastes had been abandoned or previous hazardous waste disposal procedures had already contaminated the environment. Under CERCLA, the EPA is responsible for collecting information about sites needing emergency or long-term cleanup. The act essentially required the EPA to generate a National Priorities List of the most critical dumpsites and to develop formal regulations for implementing the Superfund Program. In the late 1990s, there were 1,228 identified Superfund sites in the United States.

Generation of Waste

Efforts to estimate the amount of hazardous waste materials generated each year in this country seem to be almost futile. It is difficult to identify those corporations, small manufacturing firms, or individuals directly responsible for producing these waste materials. A portion of the difficulty stems from an array of inconsistent definitions regarding the specific wastes considered to be hazardous.

For instance, the Resource Conservation and Recovery Act of 1976 (RCRA) defined *hazardous waste* as improperly managed solid waste found to cause an increase in morbidity and mortality or posing a significant threat to human health or the environment. If you find that explanation rather vague, then you are probably not alone.

The EPA adds that substances are considered hazardous if they possess the following characteristics: ignitability, corrosiveness, reactivity, and toxicity. The EPA further includes some, but not all, carcinogens on its list. The EPA feels the testing procedures akin to carcinogenesis are still somewhat unreliable. Unlike the description provided in RCRA, the EPA holds that hazardous wastes occur in any physical form—solids, liquids, semisolids, or gases. On the other hand, the EPA definition does not include releases to sewage treatment plants, treatment in tanks possessing permits under the Clean Water Act, or wastes beneficially used or recycled at the factories where they are generated.

According to the Environmental Technology Council, more than 200 million tons of hazardous waste is generated each year. The process of identifying those establishments generating dangerous substances in the course of their production and manufacturing activities is difficult. Pursuant to a survey conducted by the EPA, some seventy thousand firms reported they were possible generators or handlers of hazardous materials. The petrochemical companies and metal-related industries seem to account for a majority of these wastes; however, even the supposedly "clean" industries are contributing to the problem.

The computer revolution is fueling the production of waste worldwide. Not only do computers generate hazardous waste during their manufacturing, they again generate hazardous waste when they are discarded. Semiconductor plants use arsenic, phosphorous, and boron to improve the conductivity of silicon crystals. Subsequent solutions of nitric and sulfuric acids are used for etching the prepared chips.

To add to the waste, the average computer is now being used for a shorter time. Today, with the reduced costs of new computers, it is frequently cheaper to buy a new computer rather than upgrade or refurbish an old one. Where does the old computer go? Most go into a basement, in a garage, or to the dump. The National Safety Council estimates that only 6 percent of computers are recycled, when compared to the number of new ones being sold each year, and in the United States, within the next few years, there will be over 315 million obsolete computers. See Figure 9-8.

Treatment, Storage, and Disposal Techniques

Guidelines provided by the RCRA indicate that some form of treatment, storage, or disposal process should manage hazardous wastes. In some instances, waste materials may go through all three stages. It is also possible for a single process, such as surface impoundment, to be used for all three purposes.

Facilities that treat, store, or dispose of hazardous waste are required by law (RCRA) to obtain operating permits from the EPA. These permits require periodic monitoring, record keeping, personnel training, emergency planning, and plant closure procedures. Specific design and operating procedures, such as approved double liners and restraining walls for landfills, are also recognized as components of more recent permits issued by the EPA.

Figure 9-8.
A dumping ground for old computers creates even more waste. (Jack Klasey)

Land disposal remains America's dominant method of managing toxic wastes. Secured landfills, industrial lagoons, and deep well injection systems all rely on a timeworn trust in the resilient nature of soil. Approximately 60 percent of America's toxic wastes are injected into deep wells, with about 30 percent more kept in industrial lagoons and surface storage tanks. Other less widely used management techniques include incineration, waste recycling, and the use of genetically engineered bacteria to destroy these substances, thus neutralizing their toxic effects on the environment. The latter example illustrates the role of biotechnology in environmental protection. It also provides evidence of public concern about an accidental *environmental release* of these powerful bacteria that could have severe unwanted effects on the habitat they were engineered to clean!

A great deal of attention has been directed toward the need to limit the types of substances deposited in landfills and the increased use of protective measures, such as more comprehensive monitoring and more effective containment liners. Nearly one-third of the sites identified on the EPA's National Priorities List developed for Superfund were surface impoundments.

Many of these establishments were found to be negligent in their use of monitoring wells to ensure that nearby aquifers used for drinking water had not been contaminated.

Underground injection wells are often used to dispose of liquid wastes, including acid and alkaline solutions, solutions containing metals, and solvents. These wastes are pumped into underground strata, where they are assumed to be contained by impermeable rocks. The technology of underground injection has been used extensively in the oil and gas industry, but its viability on the toxic waste management agenda remains questionable.

Some waste management experts believe incineration of toxic wastes could spell some form of relief and substantially reduce the amount of dangerous substances going into landfills. Modern technology has introduced incinerators designed to destroy 99.9 percent of the organic waste material they handle. Materials, which escape up the flue, are then scrubbed out of the flue gas with auxiliary equipment. Needless to say, this is an expensive treatment and disposal process. Moreover, communities near incinerators have objected to them, due to their anxieties over potential emissions from the stacks. Ocean incineration has been proposed, but has not yet been implemented on a wide scale as a solution to the treatment and disposal dilemma.

Perhaps the best disposal solution for hazardous waste would be to recycle it or recover valuable materials from the waste. Although this is not yet possible in all situations, some waste solvents used in the electronics industry have been shown to contain compounds of sufficient quality to be used again as virgin materials in other industrial settings. The EPA estimates a small, but still measurable percentage of industrial wastes could be reused in this manner. It has encouraged the establishment of waste clearinghouses to implement this treatment strategy.

Hazardous Waste Cleanup

A truism is that prevention is usually cheaper than cleanup. In the case of hazardous wastes, however, it is obvious the situation has gotten out of hand to the point of crisis in many areas of the country. Unfortunately, cleaning up the mess after the fact is easier said than done. Regardless of the cleanup procedures selected, the work is slow, complex, and costly. A quick fix is next to impossible.

Although the CERCLA (Superfund) provides funding for cleaning up the more critical hazardous waste sites, it does not provide standards for determining the degree of cleanup required. This oft-referenced "how clean is clean?" issue is at the core of one of the most pressing problems confronting the Superfund Program. Unlike earlier federal environmental legislation, Superfund does not include standards for cleanup actions at hazardous waste sites. Some citizens would like to see this legislation modified.

Nevertheless, the EPA has difficulty determining standard levels of cleanup at this point, since its experience in developing remedies for hazardous waste sites is limited. It has also become an accepted fact that each site has unique characteristics meriting individual attention. The agency has also acknowledged that circumstances frequently occur in which there are no clearly applicable standards for determining acceptable levels of hazardous substances in soil and other media. Developing standards for the literally hundreds of substances encountered in cleanup activities could ultimately deter the progress of the procedures presently being used.

Two types of cleanup actions are classified with the Superfund Financing Program—emergency actions and long-term cleanup. Dangers of fire, explosion, or acute physical effects from direct contact are typically considered as candidates for possible emergency action. Longer-term cleanup activities move more slowly and commonly entail waste removal, treatment, or containment; data collection via testing; feasibility studies; and detailed design of remedial measures.

The cleanup of the massive quantities of hazardous materials across the United States will require leadership, initiative, imagination, diligence, and patience from all segments of society. Several studies illustrate the prospect of speeding up the cleanup process, which is ongoing as we begin the twenty-first century.

❑ New Jersey established an aggressive Spill Compensation and a Hazardous Discharge Fund to help pay for cleanup.

❑ California initiated an aggressive program to phase out land disposal of cyanides, strong acids, polychlorinated biphenyl (PCB) liquids, and all halogenated organic substances.

❑ The Florida Department of Environmental Regulation is laying the groundwork for an in-state system of hazardous waste management.

❑ Major firms, including Dow, Kodak, Dupont, and 3M, are recycling and incinerating their own dangerous waste materials.

❑ The Arkansas city dump was cleaned through a partnership between the Kansas Department of Health and Environment and the EPA.

Financial and manpower constraints make it impossible for the federal government to do it all. The hazardous waste problem is generated chiefly by the private sector, which provides jobs and revenue for local governments and communities. Government sponsored scientific research projects seem to be moving in the right direction, toward a clearer definition of the problem, as well as proposed regulatory and technical solutions. The hazardous waste problem is not amenable to a quick fix, but it is an environmental issue that can be handled through effective cooperation among citizens, industries, and government agencies in our society.

Summary

Environmentalism has surfaced in this country, and around the world, as an information-based perspective on the intelligent use of technology to prevent adverse effects on the environment. The modern environmentalist is no longer perceived as an antiestablishment radical in our society. There are several key environmental issues requiring dedicated, international cooperation during the upcoming decades.

This chapter focused on problems associated with the atmospheric buildup of carbon dioxide (CO_2), jurisdiction of common spaces, global issues, air quality standards, the consequences of tropical forest destruction, water management, and the management of hazardous materials. Each of these situations may seem dismal at the present, but none of them are totally insoluble. Progress can be made in all areas to safeguard environmental quality for the generations to follow. Cooperation between the public and private sectors in all nations is necessary to get the momentum underway.

Discussion Questions

1. Describe what the term *environmentalism* means in today's society. How does your personal opinion concerning environmental issues compare to this description?
2. Identify and discuss some examples of successful environmental policies that have emerged as a result of grassroots movements during the past five years.
3. Develop a future scenario depicting at least two prospects for the fate of Antarctica in the world community.
4. With reference to at least three contemporary environmental issues, describe what is meant by the general cliche "Prevention is usually cheaper than the cure."
5. Considering the need for pesticides in agriculture, what would be the best way to regulate them in an individual country? On an international scale?
6. Explain what is meant by the greenhouse effect. Discuss the possible causes and consequences of the concept labeled *global warming*.
7. How would you respond if you discovered that your state planned to establish a hazardous waste dump on a vacant property near your home? What questions would you ask? What safeguards would you expect? How might you determine whether a comprehensive environmental impact analysis had been completed and by whom?

Suggested Readings for Further Study

Ahmad, Khabir. 2001. Report highlights widespread arsenic contamination in Bangladesh. *The Lancet* 358 (9276): 133.

Bird, Maryann. 2001. Dried out: Floods inundate some parts of the world, while others are parched. Managing our water is a 21st century challenge. *Time International* 157 (18): 48.

Bissell, Richard E. 2001. A participatory approach to strategic planning. *Environment* 43 (7): 38.

Bjerklie, David, Robert H. Boyle, and Andrea Dorfman. 2001. Life in the greenhouse. *Time* 157 (14): 24.

Brown, William M. 1986. Hysteria about acid rain. *Fortune* (14 April): 125–126.

Choucri, Nazli. 1991. Multinational corporations and the environment. *Technology Review* 94 (3): 52–59.

Cifuentes, Luis, Victor H. Borja-Aburto, Nelson Gouveia, George Thurston, and Devra Lee David. 2001. Hidden health benefits of greenhouse gas mitigation. *Science* 293 (17 August): 1257–1259.

Fouere, Erwan. 1984. New global environment commission. *Bulletin of the Atomic Scientists* 40 (August/September): 32–34.

Geiser, Ken. 1991.The greening of industry: Making the transition to a sustainable economy. *Technology Review* 94 (6): 64–72.

[Internet]. (2001). Toxic chemicals outlawed. (22 May). <http://www.CNN.com>.

[Internet]. *Clean cities.* <http://www.ccities.doe.gov>.

[Internet]. *Energy star.* <http://www.energystar.gov>.

[Internet]. *Environmental Protection Agency.* <www.epa.gov>.

[Internet]. *United States Global Change Research Program (USGCRP).* <http://www.usgcrp.gov/usgcrp/>.

Johnson, Nels, Carmen Revenge, and Jaime Echeverria. 2001. Managing water for people and nature. *Science* (11 May): 1071–1072.

K'amolo, Joseph. 2001. Africa will soon face a major water crisis. *Africa News Service* (14 September).

Laurence,William F., Mark A. Cochrane, Scott Bergen, Philip M. Fearnside, Patricia Delamonica, Christopher Barber, Sammya D'Angelo, and Tito Fernandes. 2001. The future of the Brazilian Amazon. *Science* 291 (19 January): 438–439.

Lovejoy, Stephen B., John G. Lee, Timothy O. Randhir, and Bernard A. Engel. 1997. Research needs for water quality management in the 21st century: A spatial decision. *Journal of Soil and Water Conversion* 52 (1): 18.

Maranto, Gina. 1985. The creeping poison underground. *Discover* 6 (2): 74–78.

———. 1986. Are we close to the road's end? *Discover* 7 (l): 28–50.

More acid rain in East Asia's future. 2001. *Science News* 159 (24): 381.

Munoz, N. 2001. Environment: Experts warn of water scarcity in Central America, *Environment Bulletin* (27 March).

National Wildlife Federation. (2000). *Pollution paralysis II: Code red for watersheds.* [Internet]. Portable Document Format. <http://www.nwf.org/watersheds/paralysis/pp2_report.pdf>.

1985 Environmental Quality Index: Nagging problems still ahead. 1985. *National Wildlife* 23 (2): 33–40.

Nordhaus, William D. 2001. Global warming economics. *Science* 294 (9 November): 1283–1284.

Office of Technology Assessment. 1984. *Technologies to sustain tropical forest resources.* Washington D.C.: U.S. Congress.

Perkins, S. 2001. Big dam project in China may warm Japan. *Science News* 159 (21 April): 245.

Ruckleshaus, William D. 1983. Science, risk and public policy. *Science* (9 September): 1026–1028.

Sand, Peter H. 1985. Protecting the ozone layer: The Vienna Convention is adopted. *Environment* 27 (5): 19–20, 40–43.

Sen, Samanta. 2001. Environment: Developing world fast running out of water. *Environment Bulletin* (26 March).

Stanglin, Douglas. 1992. Toxic wasteland (former Soviet Union). *U.S. News & World Report* (13 April): 40–46.

State of the environment: An assessment at mid-decade. 1984. Washington, D.C.: The Conservation Foundation.

Studies show diminished acid rain and leveling CO_2 emissions. 2001. *Power Engineering* 105 (1): 8.

Tschinkel, Victoria J. 1986. The transition toward long-term management. *Environment* 28 (3): 19–20, 25–30.

U.S. Environmental Protection Agency, Office of Air Quality Planning and Standards. 1991. *National air quality and emissions trends report, 1990.*

Weiss, Peter. 2001. Oceans of electricity. *Science News* 159 (14 April): 234–236.

We rely on the most up-to-date technologies to keep our nation, and the whole world, as safe as possible. Here, a crew member readies an F-14A Tomcat for flight operations in support of Operation Enduring Freedom, on the flight deck of the USS *Enterprise*. (U.S. Department of Defense)

Chapter 10

International Defense Profile

Republic of the Marshall Islands

The U.S. Department of Defense (DOD) announced it completed its first successful test of a ground-based ballistic defense missile system. An American intercontinental ballistic missile (ICBM) was launched from Vanderberg Air Force Base, and an interceptor missile was launched twenty minutes later from the Ronald Reagan Missile Site on Kwajalein Atoll in the Republic of the Marshall Islands.

The interceptor missile hit the ICBM ten minutes after launch. It used its onboard sensors to locate and track the target. The interceptor missile was successful in hitting its intended target rather than a decoy that followed. DOD officials applauded this test as proof that the ballistic missile defense system will be successful and could be implemented. The DOD is attempting to conduct tests each quarter, with each test becoming more challenging....

Introduction

Standing amidst the social, economic, and political debris created by the dissolution of the Soviet Union, disintegration of the Red Army, collapse of the Berlin Wall, and a general easing of East-West conflicts, our view of the world is somewhat obscure. This murky perspective follows four decades during which many Americans lived in fear of nuclear annihilation, triggered by an altercation between the United States and the former Soviet Union. As we proceed further into a new millennium, we are in the midst of a dramatic shift of power from the bipolar world of the Cold War to the uncertainty of a multipolar new world order. The profile of international defense policy is anything but cut-and-dried.

There are new terrorist threats to the United States, which have demonstrated weaknesses in our defense strategies. For our purposes, we define *terrorism* as the ability and determination to use readily available technologies as tools of mass destruction. For generations, the American people had thought of national defense in terms of war; now, there is a new, previously unperceived terrorist threat. *Bioterrorism* has become a new frightening term Americans are only beginning to comprehend. The nature of unexpected terrorist attacks has changed the way most Americans view both the military and national defense. Because of these fears, many Americans have begun to question their personal safety and security on their home continent.

With the new century underway, Americans have begun to adopt a more global perspective on the potential threats of pollution, terrorism, overpopulation, gross climatic changes, acquired immunodeficiency syndrome (AIDS), nuclear power plants, and nuclear weapons proliferation. While these concerns have quite possibly diluted the disproportionate emphasis placed on the risk of a global conflict, the United States cannot afford to abandon its commitment to military strength and national security. Even as the American public has seemingly begun to dismiss the possibility of a direct nuclear attack from the former Soviet Union, it can do so only on the premise that the United States remains strong.

While the affairs of the Russian Federation continue to waver between confusion and outright chaos, the uncertainty is unsettling. Perhaps the worst scenario imaginable is a civil war among the newly declared independent states of the former Soviet Union in which nuclear weapons are actually used. Of critical concern are an estimated twenty-seven thousand nuclear warheads that are potentially hazardous to the world's health and stability. Just under 50 percent of these bombs and intercontinental ballistic missile (ICBM) warheads are strategic; the rest are shorter-range tactical weapons for the battlefield. Suddenly, the world has inherited four instant de facto nuclear weapons states, where just recently there had been only one.

When the topics of weapons superiority, defense policy, and military paradigms are addressed these days, traditional models of combat are often set aside. Pentagon officials categorize the type of periodic battles the United States is likely to enter into with regional powers as mid-intensity conflicts (MICs). This evolving paradigm joins previous models that assumed U.S. forces should be prepared to fight the former Warsaw Pact on European soil (high-intensity combat) or guerrillas in Central America (low-intensity war).

As with Operation Desert Storm in the Persian Gulf and the war in Afghanistan that began in 2001, these future MICs will be fast-paced and high-tech. Military planners envision unrestrained use of our most sophisticated weapons. Essentially, this scenario suggests the United States will be conscripted to do battle with emerging Third World powers through the use of weapons initially designed for a war with the Soviets. Gone are the days when conflict in developing nations conjured up an image of simple, low-tech warfare that did not require complex weapons systems. In today's global society, Third World nations are armed with First World armaments. Widespread proliferation

of chemical weapons (CWs), increased access to nuclear weapons capacity, and the availability of cruise and ballistic missiles, submarines, and high performance aircraft mean more nations than ever before can wield deadly forces on the battlefield.

As U.S. military advisors and government officials continue to ponder the long-term implications of the Persian Gulf War, the breakup of the Soviet Union, and the current war on terrorism for future defense policy, there is reason for scientists and technologists who design and build our weapons of war to continue their work. The Cold War may be over, but a massive number of nuclear weapons are scattered throughout the world. This reality, combined with contemporary American worries about whether or not the United States can retain its competitive edge in the world economy, seem to preclude a slowing of research and development (R & D) for military technology. The commonly held presumption that a nation's technological edge often goes hand in hand with its international military stature will not disappear in the foreseeable future. It may, however, be overshadowed by the progressively popular conviction that economic competitors pose a greater threat to national security than military adversaries do.

The extent to which scientists and technologists will ever be able to fulfill past President Reagan's vision in his almost legendary *Star Wars* speech of an "impenetrable shield" remains a distant illusion. Regardless, **Strategic Defense Initiative (SDI)** projects, including exotic lasers, hypervelocity projectiles, sensitive radars, and ground-based interceptors, received a significant share of our nation's R & D budget into the early 1990s. SDI's most recent incarnation has been as the Missile Defense Agency (MDA) Program. Currently, the United States plans to have an initial MDA system in place by 2005 that will include ground-based interceptors, early warning radar, and satellites. It appears that the new justification to keep the comprehensive SDI program alive (and funded) is that emergent nuclear powers are not likely to be deterred, as have been the superpowers, by the specter of mutual assured destruction (MAD). The SDI will eventually need to shift its focus to ground-based interceptors for defense against regional missiles whose flight trajectories are not outside the atmosphere.

For many of us, the sophistication level of the technologies associated with military prowess is beyond our wildest imaginations. International defense policy is dynamic and evolutionary; its profile seems to change a little each time we hear about another new sensor device, smart missile, deadly chemical superweapon, genetically engineered biological agent of destruction, or enterprising missile-basing scheme. As you review the topics introduced in this chapter, you may begin to grasp and understand the reason many historians single out war as a primary trigger for technological change. For centuries, the human instinctive response to any sort of threat to survival has been enmeshed in a technology of some sort. Prior to the onslaught of our contemporary nuclear era, every new weapon prompted a defense that could counter it. Even when civilian populations became the target, effective defense strategies appeared on the scene. The bombs that brought an end to World War II changed that chain of development forever. The instant annihilation of Hiroshima shifted the balance between offense and defense to a point where there was no longer an optimal maneuver to protect the civilian population in war. Nuclear weapons are quite simply instruments of mass murder. During the Cold War, the notion that any country could wage a small-scale nuclear war without risk of global suicide was an absurdity. In our contemporary multipolar world, larger

powers may tolerate regional conflicts (perhaps nuclear) rather than risk globalizing the war in the name of ideology.

Military technology projects may bear some resemblance to other technology-laden disciplines, but their ultimate objectives represent a clear point of departure. A competitive and often nationalistic spirit drives military scientists and technologists. They most assuredly struggle with ethical and moral issues as they improvise and craft more lethal weapons in the laboratory. These same emotions are said to stimulate performance among biotechnologists who are paving the future course of genetic engineering, medical technologists who study alternative procreation methods, and manufacturing engineers who contribute to the global market of technological gadgetry. On the other hand, military technology remains in a class by itself, from the standpoint that its primary goal is defined by its capacity to destroy an opponent's capacity to wage war. This pursuit commonly leads to the loss of human life along the way. The following questions will be addressed in this chapter:

❑ Will the superpowers continue to be the "keepers of the keys" of military superiority in the global arena? Or did the end of the Cold War signal the end of superpower prominence?

❑ How does the emerging military stature of developing nations factor into the equation of international defense policy making?

❑ What projections are being made to describe the nature of war on tomorrow's battlefield? What are the probable environmental consequences of these battles?

❑ Is the MDA program a worthwhile military investment? What other Pentagon black budget projects might you suspect are earmarked for funding now that we have entered the twenty-first century?

❏ From the perspective of economic competitiveness, how will the U.S. military budget influence employment patterns and enhance industrial leadership?

❏ If contemporary government leaders profess to be concerned for the sustenance of human life, how should they attempt to reconfigure their national security objectives?

❏ What will be the effects of bioterrorism and cyberterrorism on the quality of life and our perceptions regarding freedom in the United States?

Key Concepts

The study of military policy and technology is full of technical terms and confusing acronyms. This chapter will examine a small portion of them in its discussion of the international defense profile. As you read through the various segments, stay alert for the following concepts:

biological weapon (BW)
bioterrorism
chaff
chemical weapon (CW)
cluster bomb unit (CBU)
command, control, communications, and intelligence (C³I)
command, control, communications, computers, intelligence, surveillance, and reconnaissance (C4ISR)
compellance
Cooperative Threat Reduction (CTR) Program
counter-force targeting
counter-value targeting
cyberterrorism
decoy
deterrence
first strike
low-intensity conflict (LIC)
mid-intensity conflict (MIC)

mutual assured destruction (MAD)
nuclear winter
Single Integrated Operational Plan (SIOP)
smart weapon
Strategic Defense Initiative (SDI)
swaggering
terrorism

Military Goals in the Global Arena

There is no question national security has always been a topic of concern among the general public. National security is commonly used to justify huge public expenditures to maintain global deployment of armed forces, develop and procure new weapons systems, and manufacture armaments. Much of what we know about U.S. national (as well as foreign) security policy appears to us in brief snippets in the evening newspaper or on one of the numerous analytical commentaries or news programs broadcast on national television. Set against a backdrop of uncertainty about the future, news bites and editorials heighten the impression that military planners continually modify the conditions of our security policy to coincide with the latest weapons developments or foreign event. Regardless of these political shifts, there are some regularities and patterns, which allow a few generalizations to be made about the present and future U.S. defense policy.

Over the past six decades, international security objectives have become inextricably linked with the quest for military supremacy. The timeless nature of the national security issue is based on assumptions that the principal threat to one's safety (and perhaps freedom) comes from other nations around the globe. The idea that countries everywhere should be prepared to do battle and defend themselves at all times from an external invasion is relatively recent. Prior to World War II, countries mobilized their troops in times of war instead of maintaining large, permanent military installations. Figure 10-1 contains estimates of global defense expenditures as a share of Gross National Product (GNP) for selected nations. The dollar figures for military strength have grown enormously. Further, the number of people in the United States alone who are in one way or another employed in a defense-related job remains quite high. Among industrialized countries,

Military Expenditures as a Share of GNP for Selected Countries, 1997	
Country	**Share (Percentage)**
Industrialized countries	
Japan	1.0
Canada	1.3
Germany	1.58
France	3.0
United Kingdom	2.75
United States	3.3
Russia	5.77
Middle East	
Egypt	2.84
Iran	2.7
Syria	5.6
Israel	9.7
Saudi Arabia	14.47
Jordan	9.02
Iraq	4.88
Asia	
India	2.79
China	2.21
Sri Lanka	5.1
Pakistan	5.7
Africa	
Nigeria	1.43
South Africa	1.84
Ethiopia	1.9
Libya	NA
Latin America	
Brazil	1.8
Mexico	1.1
Venezuela	2.17
Argentina	1.16
El Salvador	0.9
Chile	3.8
Nicaragua	1.45

Figure 10-1.
This chart shows military expenditures as a share of Gross National Product for selected countries. (U.S. Arms Control and Disarmament Agency)

the United States spends a greater amount of money and a higher percent of its total research money on defense-related research. See Figure 10-2. About 54 percent of the U.S. government's 1998 R & D budget was devoted to this area, in comparison to less than 10 percent in Germany, Canada, and Japan. In addition to high levels of funding for defense-related R & D, the United States spends more than all of the G8 (Group of Eight—a group of the eight most affluent countries in the world, consisting of Canada, France, Germany, Italy, Japan, the United Kingdom, the United States, and Russia) countries combined on its military budget. In 2002, the overall defense budget for the United States was $312 billion.

For years, the primary stimulus behind global militarization has been rooted in the ideological conflicts between the former Soviet Union, with its socialist allies, and the United States, in alliance with the world's industrial democracies. More recently, the emergence of Third World countries as potential threats to global stability confirms the fact that the process of militarization and weapons procurement is independent of economic status, political ideology, or level of technological advancement.

Since the mid-1940s, the dominant influence in world politics has been entrenched in a bipolar rivalry between the Soviet Union and the United States. One of the hallmarks of this geopolitical struggle has been the intense proliferation of nuclear arms. In light of the recent dissolution of the Union of Soviet Socialist Republics (USSR) and the Warsaw Pact, it is conceivable that the core of this bloc leadership model will be challenged. It will, however, be several years before a single nation or regional federation of nations will usurp the global roles and range of dominance exercised by these twentieth century superpowers.

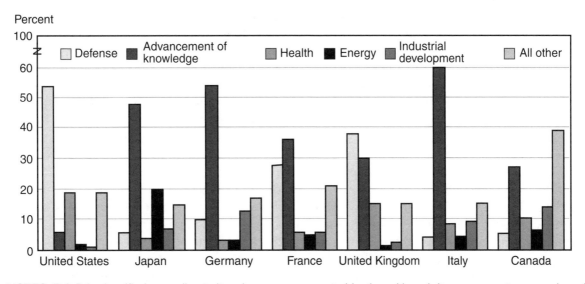

NOTES: R & D is classified according to its primary government objective, although it may support any number of complementary goals. For example, defense R & D with commercial spinoffs is classified as supporting defense, not industrial development. R & D for the advancement of knowledge is not equivalent to basic research.

Figure 10-2.
This graph shows the distribution of government-supported research and development for selected countries. (Adapted from *Science & Engineering Indicators—2000*)

On the other hand, the price for functioning as the world's police officers has reached astronomical levels. Disproportionately high defense spending drains any growth resources critical to a dynamic, industrial society. Much of the initiative for the strategic arms reduction talks (START) in the late 1980s came from the superpowers' realizations that it had become far too expensive to maintain the forces needed to police their worldwide spheres of influence, the nuclear arsenals to guarantee an extended and assured deterrence, and the commitments to preserve their influences in key areas of the Third World.

Bear in mind as you continue to read this chapter, the powerful vision of promoting military security and conquering outer space were instrumental in building U.S. technological thrust and boosting its "ACE complex" to international prominence (ACE includes the aerospace, communications, and electronics companies supporting these objectives). It seems like the perfect time to restructure our economic order so the next round of technological investment will promote an environmentally sound, physically healthy, and politically stable society around the world. U.S. leadership on those fronts should hold at least as much international clout as its long-term quest for military supremacy.

As this new century begins, the American people seem torn about the role of the military in the future. It is too soon to know the long-range effects of the terrorist attacks on the United States in 2001. Will the American people continue to support, even demand, more military spending to increase their perceptions of security? Or will there be a backlash toward the increased militarization of American society? It may be some time before we witness any resolution regarding these events. In any case, we can be sure more lawmakers and citizens will heed the goals and desires of the U.S. defense complex in the new few years.

Rationale for Military Force

Why do governments invest heavily in new weapons, while they simultaneously explore possibilities for arms control negotiations? This is truly the essence of the contemporary arms race between the superpowers. Beginning sometime in the late nineteenth century, the applications of technology in the tools of war were acknowledged for their capacity to provide quantitative and qualitative advantages for their owners.

Unlimited use of armed force as an instrument for attaining meaningful political ends is a well-known agenda. The reasons any one nation decides to mobilize its military forces are far too diverse to enumerate individually. Military historians believe many of these reasons can be loosely categorized into four general rationales, which are discussed in the following sections.[1]

Defense

Defensive use of military force is directed to prevent attacks or reduce damage. It entails the deployment or threatened deployment of sufficient military power to defeat the military force of an adversary or to attack its opponent's territory. Perhaps the most traditional view of the role of the military in our society, the term *defense* encompasses the direct application of military force offensively.

Contemporary nuclear weapons are couched in technological sophistication. *First strike* connotes a nuclear attack so powerful as to leave one's opponent with forces inadequate to inflict substantial damage on the attacker. Second strike is a procedure following a nuclear attack in which weapons can still be used in retaliation against the enemy. This concept implies the ability to survive, if only for a short time, a preemptive first strike.

The defensive use of one's military forces may employ **counter-force targeting**—a

[1]See Stoll, Richard J. 1990. *U.S. national security policy and the Soviet Union.* Columbia, S.C.: Univ. of South Carolina Press.

strategy in which weapons are aimed at the military forces of the opponent, particularly at strategic or theater nuclear forces. Conversely, *counter-value targeting* specifies a program aiming its weapons at "softer" targets, like urban population centers, industrial facilities, and economic enterprises.

Deterrence

In contemporary society, the purpose of deterrence is to protect a country from physical attack by an adversary. *Deterrence* involves the threat of military power to convince an enemy not to undertake certain actions. Essentially, one country deters an opponent by convincing it that the benefits of undertaking certain actions are outweighed by the consequences it will ultimately encounter.

Nuclear deterrence is perhaps the most compelling illustration of a deterrent use of force. Counter-force targeting (offensive first strike) requires extremely accurate missiles in order to be successful against hardened missile silos and obscure military control centers. The Tomahawk missile, a long-range missile that can fly at low altitudes at very high speeds, is a prime example of a U.S. first strike weapon. See Figure 10-3. When both sides have the ability to absorb a first strike and deliver a retaliatory second strike capable of inflicting an unacceptable level of damage, a situation of *mutual assured destruction (MAD)* is said to exist. Retaliatory attacks are generally aimed at counter-value targets such as civilian population centers. Since there is little point in targeting empty missile silos (they would be empty after an offensive first strike had been launched), a counter-value targeting strategy seems likely to have the greatest deterrent effect.

The concept of deterrence has dictated U.S. planning for nuclear war since the late 1940s. Ostensibly, the logic behind this strategy has been the more powerful the nuclear arsenal, the less likely the need for its use.

Figure 10-3.
A Tomahawk cruise missile is launched from the USS *Philippine Sea* (CG 58). (U.S. Department of Defense)

However, this philosophy of deterrence is at odds with the modern types of war, which have been and are expected to be regional conflicts involving both developed and developing countries. The threat of nuclear war does not appear to be a deterrent to many Third World countries, as conflicts continue to develop among them. As the United States and other developed countries evolve their defense strategies to address this changing type of war, we will no doubt have a new definition of *deterrence*.

Compellance

A government's compellant use of military force is often designed to persuade an adversary to change its behavior. *Compellance* involves the deployment of military power in order to get another nation to change its behavior to make it more satisfactory to the aggressor. The selection of targets for compellance in order to either threaten or actually administer punishment is complicated. If, for example, the aggressor (X) wishes for its opponent (Y) to either start doing something or stop doing some other thing, X may compel Y by targeting what Y values most (counter-value targets such as refineries, power stations, communications networks, transportation systems, population centers). If, however, Y is engaged in some type of military operation X finds threatening or objectionable, it may be more effective to target Y's military establishment—its capacity to continue its inexcusable behavior.

Compellance can easily be confused with both deterrence and defense. The motivation for the compellant use of force is not to destroy an opponent's forces (defense) or ensure its continued acceptable behavior (deterrence). The underlying goal of compellance is best envisioned when country Y exhibits a behavior deemed punishable, for whatever reason, by country X.

Swaggering

To swagger, literally, is to walk or conduct oneself with an outward air of boldness or arrogance. In the present context, *swaggering* (also called *posturing*) equates to deployment of military power by a government to enhance prestige, national pride, or the personal ambitions of its rulers. Swaggering military forces are not usually deployed directly against the forces or territory of an opponent, and hard combat does not generally occur.

It is difficult to define the focus or target of swaggering military actions. Two examples indicative of swaggering are general military exercises ("Look at what we can do!") and the procurement of high prestige weapons (the conspicuous consumption phrase, "Look at what we can afford to buy!"). Both examples bring to mind the comely peacock strutting his tail feathers simply to attract attention. Still, government officials find swaggering useful to carve their niches as reputable players in the global defense arena.

Post–Cold War Nuclear Policy

In the confusing and unsettled aftermath of the Soviet Union's breakup, any generalities we make about international nuclear policies will be tentative. In reality, the precepts governing nuclear strategy include decisions regarding war plans, budgets, forces, and deployments. On a practical level, international discourse devoted to the formation of nuclear weapons policies has been molded by the United States and the Russian Federation, each of whose multileveled approach to strategic decision-making has effectively deterred either from a preemptive first strike with nuclear weapons.

A major innovation in U.S. action policy came with the first *Single Integrated Operational Plan (SIOP)* created more than four decades ago. Since that time, all U.S. strategic forces have been part of the same coordinated plan shown in Figure 10-4. The SIOP contains a list of targets that U.S. long-range nuclear forces would attack and assigns weapons to targets, based on established damage criteria, routing, and timing.

Detailed war plans also involve war-gaming (computer simulations and board games) to assure their viability under various conditions. Prior to the breakup of Soviet republics, SIOP specified several basic target categories: Soviet strategic nuclear forces; other military targets; military and government facilities; and economic and industrial structures in

Structure of U.S. Nuclear Strategy

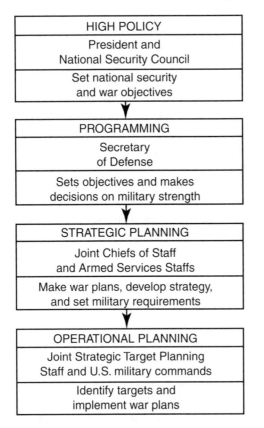

Figure 10-4.
This chart shows the structure of U.S. nuclear strategy, from the standpoint of strategic planning levels in our government. Each of the four levels of strategic planning can head in a direction different from the others, as each responds not just to guidance from levels above, but to separate organizational needs and restraints. These bureaucratic structures help account for contradictions and divergences emerging between the stated policy of the President, the force posture of the Defense Department, the plans of the Joint Chiefs of Staff, and the final implementation of plans by the Joint Strategic Target Planning Staff and the military commands.

urban centers. Attack options have ranged from conducting selective attacks with a few weapons to the engagement in a full-scale nuclear exchange using thousands of

warheads against more than twenty thousand targets inside Russia and the other countries of the former Soviet Union.

While the present likelihood of an all-out conflict between the United States and the Russian Federation seems remote at present, it is far too early to suggest all nuclear weapons be scrapped as relics of the Cold War. International leaders around the globe will watch the unfolding drama between the U.S. and Russian leaders who attempt to downsize nuclear arsenals, rethink nuclear policy, and redefine military plans and force requirements. Regardless of the level or degree of U.S.-Russian cooperation borne out of these negotiations, it is quite probable military planners on both sides will find ways to somehow preserve or even enhance the destructive power of their nuclear arsenals.

The rapid political disintegration of the Soviet Union, coupled with a disjointed, dispirited Red Army, means the country's huge nuclear stockpile is no longer securely under the control of an authoritarian central government. After the breakup, an estimated thirty thousand nuclear warheads remained in the various countries evolving from the former Soviet Union. Independent of political leaders' agreements concerning verification measures and arms reduction activities, the process of moving Soviet weapons from their remote deployment areas to some type of central dismantlement facility is going to be a protractive endeavor lasting for many more years. Along the way, the collapsing Soviet economy may result in the outright sale of weapons materials for desperately needed hard currency.

At the same time nuclear arms reduction is occurring in the former Soviet Union, the United States is undertaking its own reduction. This drop in nuclear weapons is based on the changing approach the military is taking toward war. Overall, the Department of Defense (DOD) estimates the number of nuclear warheads will be reduced from six thousand in 2002 to about two thousand by the early 2010s.

Various names have been ascribed to nuclear policy over the course of the nuclear proliferations era. A few of these include massive retaliation, assured destruction, flexible counterforce, countervailing strategy, and defense dominance. Political rhetoric aside, nations will undoubtedly continue to seek nuclear weapons and also perceive them as a threat. Given this reality, we need a globally accepted policy requiring all countries possessing nuclear weapons to assume a serious posture of minimal nuclear deterrence.

Third World Threats

While nuclear weapons can and must play a diminished role as the Cold War smolders with fewer embers, the specter of what have been adroitly labeled *mid-intensity conflicts (MICs)* remains. Nuclear forces constitute the ultimate defense of every country that has them. The immensity of their destructive power almost makes conventional weapons look like a parody. Nevertheless, military powers have officially devoted a great deal of effort preparing for conventional conflict in a nuclear age in order to prevent a nuclear war from breaking out. Since we always define *special arms* (biological, chemical, and nuclear) to be weapons of mass destruction, does this then suggest tactical surface-to-surface missiles are only weapons of mere destruction? Not really. The profile of contemporary war has already revealed the firepower of modern conventional systems to be quite foreboding. Since the leverage reserved for the superpowers by virtue of their technological superiority has diminished, the line of demarcation between conventional and unconventional warfare may appear somewhat blurred. However, the distinction

between the two is not the weapons used, but the manner in which the war is waged. A conventional war is a conflict between recognized, uniformed armed forces. Unconventional struggles may involve guerrilla warfare, subversion, sabotage, and other terrorist operations of a covert or clandestine genre. When you do not know whom you are fighting (unconventional personnel are commonly out of military uniform) you assume you are fighting everyone—that is the nature of unconventional warfare.

An overwhelming majority of the armed conflicts fought around the world since World War II have been described as *low-intensity conflicts (LICs)*. Some of the principal characteristics of this dominant military motif (border skirmishes, internal battles, police work, civil uprisings, and regional conflicts) include the following:

❏ They tend to occur in less developed areas of the world.

❏ Small-scaled armed conflicts rarely involve regular armies on both sides.

❏ Generally, conscripted forces on one side fight against guerrillas, terrorists, women, or children on the other.

❏ Most do not rely on the use of high-tech weaponry, but they often involve people who willingly engage in suicide missions to inflict damage on their opponents.

By comparison, MICs involve warfare between the large and medium powers of the Third World. Regional powers, who only decades ago were equipped with a few obsolete tanks and subsonic aircraft, are now purchasing a huge quantity of modern aircraft, stealth technology, tanks, lasers, and cruise and ballistic missiles. The subsequent ability of these emerging Third World powers to conduct sustained military operations on a regional or even continental scale is having a profound impact on the world military environment and defense profile.

The perceived risk of MICs will prevail well into the twenty-first century. Nations such as Syria, Libya, Iraq, and Brazil continue to harbor ambitions that could ultimately lead to a collision with the United States. It is a well-known fact that aspiring powers already possess modern arsenals including nuclear, biological, and chemical (NBC) munitions.

For these and many other reasons, it is quite probable that U.S. security policy will align itself with the new military paradigm carved out of the MIC formula. The nations who typically stand out as formidable military powers in the developing sectors of the globe include Argentina, Brazil, Egypt, India, Iran, Iraq, Israel, North Korea, Pakistan, South Africa, South Korea, Syria, Taiwan, and Turkey. Without question, socioeconomic and political stability are the exception to the rule in these nation states. Their overriding commitments to working their own agendas, coupled with the determination of the United States to protect its natural resource and raw material interests, presents a whole new set of challenges for our military forces. An MIC-oriented military posture is evidence in the United States of a de-emphasis of the Soviet threat and preparation for clashes in the Third World.

The nature of battle today is increasingly determined by the technologies being used. The digital revolution did not bypass the military; in contrast, in many ways, the military is exploiting the digital world to a greater extent than the civilian world. Almost every new vehicle and weapon today includes computer-based or digital capabilities.

Arms Control and Nonproliferation

Along with the evolution of war scenarios from large-scale conflicts to MICs and LICs, the United States has added nonproliferation to its defense goals. It is much easier to stop the sale of weapons to a potential unfriendly country than to disarm it. The increasing amount of dangerous weapons worldwide, particularly NBC weapons, have created additional uncertainty for the United States.

One Defense Department initiative is the *Cooperative Threat Reduction (CTR) Program.* The primary goal of this effort is to prevent the widespread acquisition of NBC weapons. Along with the prevention efforts, there is an increased emphasis on arms control treaties. Since the breakup of the Soviet Union, these nonproliferation efforts have assisted three former Soviet republics—Ukraine, Kazakhstan, and Belarus—to become nonnuclear weapon states. The United States has also purchased tons of uranium from dismantled Russian weapons and converted this uranium into reactor fuel.

The CTR Program also aims to improve the transportation and security of nuclear weapons. As part of this effort, it is working with several former Soviet republics—Russia, Ukraine, Kazakhstan, and Belarus—to convert weapons facilities into factories that would produce products for consumers. Also, science and technology centers have employed over seventeen thousand former Soviet weapons scientists and engineers in civilian research.

The United States and the Russian Federation, along with thirty-one other countries, approved a new international agreement on the transportation and sale of weapons, the Wassenaar Agreement. The goal of this agreement is to promote restraint and transparency in the transfer and sale of weapons. This agreement shows the evolution of military planning since the end of the Cold War. Instead of targeting a particular group of countries, the members of the Wassenaar Agreement agreed to prevent the sale of arms to certain countries if they appear to be unstable. Currently, the countries in this category are Iran, Iraq, Libya, and North Korea.

Nonproliferation treaties will only work if they are accepted and followed by all the weapons producers in the world. Unfortunately, each weapon producing country has its own agenda and rationale as to which countries they will and will not sell weapons. Unless there is a new global agreement about weapons control, nonproliferation will remain an uphill battle.

International Peacekeeping

As arguably the only remaining superpower, the United States has taken on an increasingly larger role in law enforcement and peacekeeping exercises worldwide. In the last few years, U.S. troops have been involved in peacekeeping activities in the Balkans, including stationing troops in Bosnia and Kosovo. Most of these peacekeeping activities are coordinated through the North Atlantic Treaty Organization (NATO). About 20 percent of the NATO-led peacekeeping force in Bosnia were U.S. troops.

An unfortunate remnant of modern war is land mines, and a portion of the peacekeeping forces are involved in removing these landmines. See Figure 10-5. The removal of land mines is critical to building the security and peace of a country; however, these products of technology are notoriously difficult to remove. Experts estimate that if we remove or neutralize 75 percent of the existing land mines in twenty-five years, it would cost about $2.25 billion.

The United States and other countries have realized that these peacekeeping activities bring great benefits. The entry of a NATO force into a troubled area allows the development of stability, and since LICs and MICs are most common in troubled developing

Figure 10-5.
An explosive ordinance technician sweeps a path to a suspected anti-vehicle mine at Tuzla Air Base, Bosnia. The suspected mine was determined to be a harmless ammunition casing lid. (U.S. Department of Defense)

countries, peacekeeping can also reduce the number or length of these conflicts. This expansion of the DOD's responsibilities is sometimes at odds with the primary goals of the defense of the United States, but the U.S. military has realized that isolationism is not an effective strategy for maintaining peace in the world.

Nature of War on Today's Battlefield

Real-time exposure to armed conflicts on the other side of the globe is no longer the stuff of science fiction fantasy. Many older Americans grew up with nightly depictions of our soldiers doing battle in the remote jungles of Southeast Asia during the Vietnam War. Younger Americans can remember live network coverage of the Persian Gulf War, Operation Desert Storm, and the peacekeeping forces in Bosnia. The saga of Operation Desert Storm unfolded before our eyes in the dubious comfort and privacy of our living rooms. Even more recently, Americans were mesmerized for hours by live coverage of the horrific terrorist attacks on the World Trade Center towers and the Pentagon. As long as our international communications networks make it possible to beam images of events back and forth across the world's time zones, we, as laypeople, will be privy to the nature of future MIC battlegrounds. Some of the most powerful images that will dominate the television screen are technology-laden weapons. More chilling images depicting the aftermath of an attack using chemical or biological munitions may also be shown, and mixed portrayals of the extent to which the use of highly destructive weapons giving credence to the *nuclear winter* theory may also cross the airwaves.

The world environment has changed. Instead of focusing its planning efforts on one large foe, the Soviet Union, the United States must deal with a future marked by opponents who are not so easily described or determined. It is likely that there will not be another world superpower in the next ten to fifteen years that could amass enough conventional military strength to challenge the United States. However, there will be increasing challenges from smaller countries in a potentially wide range of regional conflicts. In the Middle East, there is a continued state of conflict between Israel and its neighbors as well as tensions between Iran, Iraq, and neighboring states. In Asia, Korea remains split between North and South, and North Korea continues to build its defense infrastructure. Beyond these known areas of conflict, additional ones could easily become problem areas. Whenever the future problems occur, the demand for smaller-scale military operations is likely to remain high in the next ten years.

In the late 1990s, the DOD developed a plan for the Army of today and the future, called Joint Vision 2010. In 2002, they revised this future plan, and it became Joint Vision 2020. The goal of Joint Vision 2020 is to transform war through the use of new information technologies and advanced weapons. By the mid-2010s, the DOD expects to reshape itself so it can more easily respond to threats in the future. Key items on the Pentagon's MIC wish list prescribe mobile firepower; advanced tactical aircraft; command, control, communications, and intelligence (C^3I, defined below); and high-tech weapons, including cluster bombs. Since potential adversaries possess their own cruise and ballistic missiles (notably the Russian-made Scud), U.S. military planners are forced to seek countervailing defense systems. See Figure 10-6.

Command, Control, Communications, and Intelligence (C^3I)

In rapid pace, high-tech MICs, U.S. forces need to quickly detect enemy movements, communicate intelligence data, and channel battle orders through the chain of command. For several years, military planning included R & D related to the use of computers and other communication devices. The C³I systems are vital to survival

---

COMMAND, CONTROL, COMMUNICATION AND INTELLIGENCE (C³I)

Pronounced "see-cubed-eye" or "see-3-eye," this is a military acronym. In the U.S. military, C³I provides information about communications, counterintelligence, security, information surveillance, and reconnaissance to the military forces so they can succeed in every mission.

---

against unfamiliar adversaries in uncharted, forbidding terrain. Sophisticated, highly efficient C³I systems enable commanding officers to initiate lightning offensives and forceful counterattacks and have knowledge of the location and intentions of both friends and foes.

U.S. C³I is heavily dependent on satellites to pinpoint the location of distant missile sites for attack by its bombers. Key links in the network are the Defense Satellite Communications System, the NAVSTAR Global Positioning System (GPS), and CIA/DIA spy satellites. For precise, up-to-the-minute information on enemy movements, the U.S. military has access to Joint

Figure 10-6.
This picture illustrates a Scud missile site used for practice by North Atlantic Treaty Organization forces. (U.S. Department of Defense)

Surveillance Target Attack Radar System (JSTARS). This radar-equipped plane locates tanks, helicopters, and low-altitude aircraft. JSTARS was successfully used in the NATO Kosovo operations. Two JSTARS-equipped planes were used to locate ground targets and proved to be 80 percent effective in this capacity. Recently, the DOD decided to reduce the U.S. JSTARS fleet from nineteen to thirteen aircraft, but supplement that fleet with six of NATO's equivalent aircraft with the Alliance Ground Surveillance (AGS) system. U.S. armed forces use remotely piloted vehicles (RPVs) equipped with video equipment to transmit real-time imagery of the battlefield to tactical commanders, and many special operations aircraft have infrared or night vision equipment to help the soldiers find targets.

Under Joint Vision 2020, the scope of C^{3}I is going to be expanded to encompass more functions. The expanded information plan of the military, called *command, control, communications, computers, intelligence, surveillance, and reconnaissance (C4ISR),* will use improved information technologies to enhance military operations. The five components of C4ISR follow:

- ❏ Multisensory information that can provide information about the battlefield to both commanders and field forces.

- ❏ Implementation of software and hardware to allow battles to be managed faster and more flexibly.

- ❏ Information systems that can interfere with or manipulate an enemy's forces.

- ❏ An enhanced communications grid.

- ❏ An information defense system to protect the DOD's communications from any interference by the enemy.

Information and the effective use of information will be overriding issues for the military of today and tomorrow. Almost all new vehicles and equipment developed for the DOD include information technologies. Much development work has occurred in updating the various planes and tanks the military uses in both its wartime and peacetime operations.

In addition to including information technology on vehicles, the DOD has created a new category of weapons including new information technologies that could be called *smart weapons.* The U.S. Air Force's Air-to-Ground Guided Missile-130 (AGM-130) is an example of a smart weapon. The AGM-130 is an air to surface missile that comes equipped with a television or an infrared camera and data link. After the AGM-130 is launched from a fighter, it allows the pilot to see through the "eyes" of the bomb. The pilot can either lock the bomb onto the target through an automatic guidance system or control it manually from the plane.

The recent versions of the Tomahawk missile have made it a smart weapon also. The current version has three different "smart" enhancements making it more effective. The Tomahawk uses a combination of a GPS (discussed in Chapter 4, Space Exploration), a terrain matching guidance system, and a digital radar image matching system. Unfortunately, this technology does not come cheaply. In 2001, during the war in Afghanistan, the Navy launched about one hundred Tomahawk missiles at a cost of about $1 million each.

In Operation Desert Storm, about 10 percent of the weapons were "smart bombs." In the war in Afghanistan that began in 2001, this figure rose to 60 percent. There is no doubt there will be more use of smart weapons in the wars of the future.

Cluster Bombs

Arguably, some of the technologies listed above suggest an aura of surgical strike capacity on the battlefield; however, conventional bombs and artillery continue to be used in battles. A common workhorse weapon of this genre is the ***cluster bomb unit (CBU).*** Having been gradually refined and improved since the late 1960s, these improved conventional munitions can deliver lethal explosives to personnel over greatly expanded areas. Instead of one large explosion, these bombs contain dozens, hundreds, and even thousands of bomblets called *bomb live units (BLUs).* A series of cluster bombs have been developed since they were first designed in the 1960s as successors to simpler fragmentation or antipersonnel bombs. Today's bomblets can be dropped en masse to disperse themselves aerodynamically, or they can be spewed out by compressed air in streams behind the bomber aircraft. CBUs are deftly equipped with shaped charges intended for use against hardened targets like tanks. Some are designed to disperse smoke, napalm, or chemical agents. CBUs have much wider areas of coverage than standard two thousand–pound bombs developed in the 1950s.

The Rockeye II is a 750-pound unit carrying 717 antitank fragmentation bomblets capable of distinguishing between soft and hard targets. A single F-15E fighter or bomber can haul as many as twenty-two Rockeyes into enemy territory. The BLU-26 Sadeye is a one-pound bomblet that has a cast steel shell with aerodynamic vanes and 0.7 pound of TNT, in which six hundred razor sharp steel shards are imbedded. This weapon can be outfitted with fuses to explode upon impact, several feet above ground, or at some delayed time after landing. It is lethal at distances up to forty feet. The CBU-75 carries eighteen hundred Sadeyes to devastate an area equivalent to 175 football fields.

Air Force representatives have described one 950-pound cluster bomb, the CBU-87, as their weapon of choice during Operation Desert Storm. This ingenious combined effects bomb carries 202 BLU-97/13 bomblets to the targeted area. Each of these 3.4-pound bomblets carries a threefold punch of kill power: a pre-fragmented antipersonnel casing designed to spray deadly shrapnel; a hollow charge antitank warhead; and an added disc of incendiary zirconium to provide the finishing touch of devastation. A single B-52 strategic bomber can carry forty of these CBUs, containing a total of 8,080 bomblets. Using Air Force estimates of each bomblet's radius of kill, this lone B-52 can carpet bomb an area equivalent to 27,500 football fields! These overwhelming paths of destruction bear no resemblance whatsoever to a surgical strike.

Standoff Missiles

These highly accurate missiles are called *standoff* because they can be launched from platforms well beyond the ranges of standard defenses. Ships and helicopters may be used to fire these missiles at hardened targets, such as air bases, command centers, military factories, and tank formations, from distant locations. Many of these weapons were used for the first time in Operation Desert Storm. They depend on sensors and computers to locate, track, and strike their targets. In the future, stealthy cruise missiles may be able to carry half a ton of explosives, essentially unobserved, with almost perfect accuracy, over hundreds of kilometers.

Most conspicuous of the new and experimental standoff missiles is the Navy's Tomahawk sea launched cruise missile. See Figure 10-3. It is used to attack heavily defended command posts, factories, and nuclear reactors. Paveway III, featured in some of the more dramatic televised footage of the Persian Gulf War, is a laser-guided bomb. The High-Speed Anti-Radar Missile (HARM) rides the electronic signal given off by tracking radars, enabling it to hone in on and destroy air defense installations. The Pentagon continues to test and refine progressively more advanced standoff missiles with higher degrees of accuracy.

Anti-Tactical Ballistic Missiles (ATBMs)

A ballistic missile consists of three generic groups of subsystems: propulsion, guidance and control, and the actual warhead. The focal distinction between a ballistic missile and an unguided artillery rocket is the guidance and control system. This segment is obviously the most expensive and technologically most advanced subsystem in a ballistic missile. Additionally, guided missiles are usually multistaged, allowing them to attain considerably longer ranges.

Ballistic missiles are commonly classified as being short-range or tactical (one hundred to one thousand kilometers), medium or intermediate range (also called inter-theater), or intercontinental (ICBMs with ranges greater than three thousand kilometers). Developing countries presently aspiring to obtain or develop their own missiles are motivated by regional tensions, in which short-range or medium-range missiles are far more suitable than ICBMs. Most newly industrialized countries possess missiles that are of the short-range variety, but are extremely inaccurate. Scud missiles widely distributed by the former Soviet Union in the 1970s provide a majority of the missile capabilities of most Third World nations—they have a range of about three hundred kilometers and can be expected to come within about one kilometer of their target.

A key U.S. ground-based antiballistic missile called "the Patriot" was prominently featured and applauded for its performance against Iraqi Scuds during the 1991 Gulf War. Subsequent studies of videotapes taken of the fireworks over targeted cities have since revealed that the Patriot's accomplishments may have been exaggerated. A significant portion of the Scud missiles actually appears to have disintegrated (they were faulty) as they honed in on their targets upon entering the atmosphere and well before the Patriot missiles intercepted them. This observation should not diminish the technical superiority of the Patriot—it merely points out the Scud may not have been a respectable opponent.

Like many other weapons in the DOD's arsenal, the Patriot is continually being enhanced. The most recent version, PAC-3, has a revamped electronic radar system allowing it to be more effective and reliable. In addition, the rocket's altitude control section has 180 small rocket motors allowing the new Patriot to be more agile in its flight. All of the enhancements give the PAC-3 an improved hit-to-kill ratio—it is better at both acquiring targets and eliminating them.

Fuel Air Explosives (FAEs)

Fuel air explosives (FAEs) were engineered to duplicate the blast effects of nuclear weapons without radioactive fallout. Contemporary FAEs form highly gaseous mixtures that, upon detonation, produce a lot more blast than fire. Except for nuclear devices, these weapons provide a much larger blast than any other payload. Widespread use of FAEs occurred first in Vietnam to clear heavily wooded areas, thus creating makeshift helicopter landing sites. A ground-based version of FAE was used to destroy enemy tunnels.

FAEs can be launched from aircraft, helicopters, or ground vehicles. The mass air delivery fuel air explosive (MAD FAE) consists of twelve containers of ethylene oxide or propylene oxide spewed behind utility helicopters. The dirty dozen containers release a cloud of highly volatile vapors that mix with air to cover a huge area with blast overpressures five times greater than TNT. Blast overpressures of three hundred pounds per square inch (psi) have been calculated and shown to flatten everything within a sixty-foot radius and kill any troops nearby, both above and below ground. Just a few pounds overpressure is lethal for human beings.

The BLU-82 (also known as Big Blue 82 or the Daisy Cutter) is a fifteen thousand–pound bomb filled with an aqueous mixture of ammonium nitrate, aluminum powder, and polystyrene soap. Its massive size requires that it be fired from a cargo aircraft by rolling it out the rear door. It then descends by parachute and explodes just above the ground surface. The detonation produces blast overpressures of one thousand psi, thus disintegrating everything within hundreds of yards. In the Gulf War, eleven of these bombs were dropped on Iraq.

None of the weapons mentioned in these last sections—cluster bombs, standoff missiles, ATBMs, or FAEs—suggest an MIC profile that is antiseptic, in which little collateral damage is imposed. These weapons do not conjure up the impression of an impersonal war. They are deadly, dirty, and wreak incredible damage against human beings in and out of military uniform. High technology aside, CBUs and FAEs bring to mind a slaughter.

Bioterrorism

The proliferation of biological weapons (BWs) and chemical weapons (CWs), combined with the capacity to deliver them to unprepared target areas from distant launch sites, is sure to be an important security issue well into the twenty-first century. The rapid advance of technology in the biological field has led to an increased ease with which illicit manufacturing plants for biologically derived chemicals can be concealed. Biochemical technology is entering the weapons field at an alarming rate. Genetic engineering and biotechnology will further refine the art of killing through the possible development of designer concoctions for

which there are no known cures. The subtlety and sophistication of the specter of biochemical weapons may have changed, but the threat itself is not new.

CHEMICAL WEAPON (CW)

A weapon using poisonous, asphyxiating, or other gases affecting humans when they are either inhaled or absorbed through the skin.

In becoming parties to the Geneva Protocol of 1925, more than one hundred nations agreed not to wage biological warfare. This document is an arms control treaty outlawing the use of BWs in war, but does not prohibit their development, production, possession, or transfer. More recently, more than 143 countries became parties to the Bacteriological (Biological) and Toxin Weapons Convention of 1972. This landmark disarmament treaty forbids the very possession of BWs and their associated hardware. Both agreements manage compliance and adherence via shared attitudes, mutual self-interest, and military disutility. Neither one contains adequate provisions for verification, regulatory sanction, nor other forms of punishment. The Biological and Toxin Weapons Convention clearly permits R & D for protection against biological and toxin weapons. The U.S. biological defense research programs scattered throughout the country are open to public scrutiny and environmental legislation. No other nation comes close to this level of openness. In other words, the extent to which treaty signatories adhere fully to these agreements is unknown.

During the past decade, biotechnologists have made remarkable advances in genetic engineering. The ability to more easily manipulate the genetic codes of microorganisms might actually improve the usefulness of the potential BWs already existing or the capacity to create brand new ones. A number of informed individuals have suggested the 1972 Bacteriological and Toxin Weapons Convention is not comprehensive enough to proscribe the expected fruits of genetic engineering. The review conference held in Geneva in the early 1990s concluded it might be an appropriate time to amend this treaty to correct this immediate shortcoming and begin to formulate details for a complete international verification regime. To date, no international agreement has been realized.

BIOLOGICAL WEAPON (BW)

A weapon using a living organism, usually a pathogenic microorganism, for hostile purposes. Biological agents can be bacterial, fungal, viral, rickettsial, or protozoan.

An exhaustive digest of likely biological agents is next to impossible to assemble for several reasons. First, the sheer number of potentially harmful organisms is huge. The potential targets, including many animals and plants, in addition to human beings, span a wide spectrum. The effects of the agent will range from irritating to debilitating to lethal, depending on the immune system of the victim. Further, the scope of potential delivery methods is complex, ranging between overt and covert techniques, as well as direct and indirect inflictions.

Regardless, biological agents share several distinct characteristics. First of all, they are all living pathogens producing diseases, typically lethal, in their target hosts. Second, they have the capability to spread beyond their initial site and precise moment of attack by multiplying and infecting other vulnerable hosts. Finally, since they are groups of living organisms, they will act on groups of other living organisms. Depending on extant field conditions, their initial effectiveness and subsequent spread will vary a great

deal, which makes their behavior all that much more difficult to predict. The following list identifies several known agents clearly having serious and far-reaching effects on all organic life forms. These have been listed by the Center for Disease Control (CDC) as Category A diseases or agents—that is, diseases that pose a risk to our national security.

❏ *Bacillus anthracis.* This is the bacterium causing anthrax, a usually fatal disease. There are three ways to contract anthrax: by contact with the skin, inhalation, or eating contaminated meat. This disease is highly fatal if contracted by inhalation or ingestion of contaminated meat. The symptoms are fever, flu-like symptoms, and sores on the body.

❏ *Clostridium botulinum.* This is the bacterium causing botulism. Botulism symptoms usually occur rapidly, between twelve and thirty-six hours, and can cause severe paralysis. Although most people usually recover, it can takes months of care.

❏ *Versinia pestis.* This is the bacterium causing plague, or black death. Plague usually causes the victim to get a particularly severe type of pneumonia, and it can be fatal if not treated early. Millions of people died during the Middle Ages from plague carried by flea-infested rats.

❏ *Variola major.* This is the virus causing smallpox. The virus is spread from person to person through inhalation. The infected person shows symptoms of severe pain and prostration after about twelve to fourteen days. Then, a rash develops two to three days later. The death rate for smallpox is 30 percent.

❏ *Francisella tularensis.* This is the bacterium causing tularemia, or rabbit fever, characterized by swelling lymph nodes. This bacterium, usually not fatal, can lead to severe weight loss, fever, headaches, and pneumonia.

❑ *Viral hemorrhagic fevers (VHF).* This is a family of viruses that can cause a wide range of illnesses. Two of the most common and deadly are the Ebola and Marburg viruses.

Several of the known viruses remaining high on the list of potential biological agents include rift valley fever, eastern equine encephalitis, smallpox, and yellow fever. Genetic engineers may possess the ominous power of being able to modify normally harmless or relatively benign microorganisms. Consider the possible ramifications of incorporating the genetic code for a component of deadly cobra venom into the influenza virus.

BWs have long been detested and seldom used; however, after the bioterrorist anthrax attacks in the United States in 2001, the specter of BWs is difficult to ignore. The anthrax attacks, although only a few were fatal, proved to the American people that biological organisms could be used to attack this country. Also, overnight, there was more awareness about how difficult it is to detect BWs in the environment. This fact makes a lot of people wonder, "what if?" What if anthrax was delivered on a larger scale? What if other bio-agents, such as smallpox or the bubonic plague, were somehow released in the United States? Would we be able to fight it? How many people would die before we even realized what had happened? The attacks and release of biological agents put a great deal of fear into many people.

Nations do not readily admit to having BWs in their arsenals. At least five nations have possessed BWs in the past: Japan during the 1930s and 1940s, the United Kingdom (UK) during the 1940s, the United States from the 1940s through at least the 1960s, Iraq in the 1980s, and the USSR from the 1970s through the 1990s. Although generally these BWs are safely stored in high security laboratories, accidents have occurred. The United States claimed that an outbreak of anthrax in Sverdlovsk in the late 1970s resulted from an accidental release of *Bacillus anthracis* from a nearby Soviet BW research center. Soviet officials vigorously denied the allegation and attributed the outbreak to the illegal sale of contaminated meat. The world learned only in the early 1990s that the Russians had a significant bio-weapons program, after inspections by joint U.S.-British scientific teams. Today, the CDC estimates that at least two Russian laboratories still store the smallpox virus, but it is the knowledge the ex-USSR scientists have that may cause the greatest concern. With the collapse of the USSR and the resulting economic problems remaining in Russia, some unfriendly countries are recruiting Russian bioscientists to work on BWs. For example, Iran once recruited Russian scientists to work on its BWs program by offering the scientists wages ten times what they received in the Russian Federation.

Unfortunately, as the anthrax attacks in 2001 showed, there is no easy way to determine whether someone has been exposed or infected by the anthrax bacteria. Also, it is difficult, if not impossible, to use a vaccine against every possible biological agent. Several companies are jumping into this void. Tetracore, a company formed by former researchers for the U.S. Navy, developed one product called the BioThreat Alert® kit. BioThreat Alert kits can test for the presence of biological agents in less than fifteen minutes, and this test can be done on-site, rather than in a lab. Cepheid uses another technique for its product, a portable polymerase chain reaction (PCR) machine used to detect biological agents (PCR is discussed in Chapter 2 on Biotechnology). This product is more sensitive and can produce lab results on-site in fewer than thirty minutes.

In addition to new products and technologies identifying biological agents, there is research on other ways to combat *bioterrorism.* Many believe it is more important to prevent a bio-attack than clean up afterwards. So, there is increased emphasis on surveillance and other technologies to make this country more secure. The Federal Aviation Administration (FAA) has installed new high-tech scanners and chemical trace detectors in airports across the United States. Also, the FAA is considering new face recognition technologies to match the faces of air travelers with suspected terrorists on file. This new field, called biometrics, will undoubtedly boom as more products are developed that promise increased security.

Because of the increased uneasiness about bioterrorism, it has received increased attention (and funding) from both the DOD and the U.S. Congress. An additional $2.8 billion was approved in 2001 to deal with bioterrorism. Of this amount, over $600 million went to expand the National Pharmaceutical Stockpile and buy additional supplies of antibiotics and drugs for BWs. Another half million dollars were directed toward speeding the development of a smallpox vaccine.

Biological agents are relatively easy and inexpensive to produce for any nation with a modestly sophisticated pharmaceutical industry. Research, development, production, and even dissemination of BWs could be disguised as civilian activities or easily kept secret. World leaders cannot afford to wage an adverse biological arms race. It is essential that all industrialized, as well as developing, nations reaffirm the treaties signed earlier and amend them as necessary to abate the potential for disaster as soon as possible.

Chemical Weapons (CWs)

Chemical weapons (CWs) killed approximately 1 million people during World War I. The Geneva Protocol of 1925 subsequently banned the use of poisonous, asphyxiating, or other gases; however, it did not ban their development, production, stockpiling, or deployment. Therefore, the United States and other nations continued to produce CWs. Nazi Germany developed a nerve gas much more lethal than those used in World War I. Fortunately, CWs were not used during World War II. During the Nixon administration, the United States reaffirmed its renunciation of the first use of lethal CWs (but excluded tear gas and herbicides) and pledged it would never engage in germ warfare.

Shortly thereafter, the United States and the Soviet Union reached an accord on CWs. This was most likely because their unpredictable nature made both sides hesitant to use them. However, they were unable to come to terms on a ban of the production and stockpiling of CWs, since they could not agree on a method of verification. The United States stockpiled more than thirty thousand tons of chemical warfare agents before production ceased in the late 1960s. Agents included phosgene, hydrogen cyanide, and mustard gas.

CWs, including mustard gas, chlorine, and phosgene were used during World War I, first by the Germans, and later by the Allies. Nerve gases invented in the 1940s included tabun, sarin, and soman. Each is a highly volatile vapor that disperses very quickly. Following World War II, U.S. and British scientists developed VX, which is a more toxic, persistent, and quite deadly nerve agent.

Each of these poisonous gases affects humans when it is either inhaled or absorbed through the skin. They work by inhibiting the action of a key enzyme in the body. By name, acetylcholine esterase is critical to the transmission of chemical messages from nerve cells to virtually all muscles in the body. Muscle paralysis ensues, causing death by asphyxiation. In nonlethal doses, these chemical agents cause partial paralysis, dizziness, vomiting, and diarrhea.

CWs and nuclear weapons are similar from the standpoint that their use directly affects civilian populations. CWs cause death among unsuspecting civilians who have no protective equipment. Nerve gases are so dangerous that any leak, spill, or accidental release can devastate life downwind. The uncontrolled spread of nerve agents by wind beyond a specific military target area could be a particularly serious problem when forces are fighting on their own home territories. Even more frightening, the type of gas masks and protective gear donned by soldiers to ward off mustard gas may be useless against new CWs engineered to penetrate even the most tightly woven fabrics or the most impermeable seals and gaskets. Refer to Figure 10-7.

The broad spectrum of chemicals on the international marketplace offers serious challenges to the verification, control, and monitoring of CW activities. It does not cost a lot to make CWs, and their manufacture requires minimal technological sophistication. Ironically, in almost every case, the chemicals and equipment required to produce chemical agents of war have legitimate industrial applications. Furthermore, they have become more available as the petrochemical, fertilizer, pesticide, and pharmaceutical industries have multiplied. See Figure 10-8.

The issue of CWs was moved to the front burner after Iraq used CWs twice in

Figure 10-7.
Two soldiers adjust their gas masks during a training exercise at Kunsan Air Base, Republic of Korea. (U.S. Department of Defense)

the 1980s, first against Iran from 1980 to 1988, and then against the Kurds from 1987 to 1988. After the Gulf War, United Nations inspectors discovered Iraq had at least five laboratories developing CWs and BWs. This was the final step in pushing the establishment and ratification of the Chemical Weapons Convention (CWC) that took effect in the late 1990s. Over 170 countries have either signed the CWC or accepted it officially.

The different technologies used in the battlefield of today present a horrific variety of choices for a country. In the United States, we have been relatively isolated from the direct effects of war, however, if the future continues in the way it is moving today, we will be faced with more opponents from small splinter groups and renegade countries. These groups are more likely to use both BWs and CWs in a battle. In addition to arming ourselves with high-tech weapons, the United States needs to increasingly use diplomatic means to prevent the outbreak of war.

Dual-Use Chemicals		
Precursors for Both Commercial Products and Chemical Weapon (CW) Agents		
Dual-Use Chemical	**Commercial Product**	**CW Agent**
Thiodiglycol	Plastics	Mustard Gas
Phosphorus Oxychloride	Insecticides	Nerve Agent
Dimethyl Methylphosphonate	Fire Retardant	Nerve Agent
Thionyl Chloride	Pesticides	Mustard Gas
Dimethylamine	Detergents	Nerve Agent
Tris-Ethanolamine	Cosmetics	Mustard Gas
Dimethylamine Hydrochloride	Pharmaceuticals	Nerve Agent
Potassium Bifluoride	Ceramics	Nerve Agent
Sodium Cyanide	Dyes and Pigments	Nerve Agent
Diethyl Phosphite	Paint Solvent	Nerve Agent
Sodium Sulphide	Paper	Mustard Gas

Figure 10-8.
This chart gives a profile of several known dual-use chemicals.

Nature of War on Tomorrow's Battlefield

It appears, from trends seen today, that both wars and warriors will become more computerized and digital in the future. The increasing use of both smart weapons and computers in all areas will mark this futuristic battlefield. Much of what we know and consider to be the realm of video games will be the reality of the future warrior. Infrared goggles and other night vision equipment are used today. The future possibilities seem limited only by the imaginations of the engineers and the pocketbooks of the American public because, as weapons become smarter, they also become more expensive.

As we project the future battlefield, we first must envision our future opponents. Today, our most likely opponents are small to midsized countries over which the United States holds technological superiority. Will there be a new superpower in the future? Both China and the Russian Federation could, in a few decades, rise to superpower status. Then, will the United States be back in a repeat of the Cold War? In planning for the future, military engineers must consider all these possibilities.

Innovative Military Vehicles

It takes the DOD many years, sometimes over a decade, to develop a new vehicle. Right now, there are dozens of airplanes, helicopters, and transport vehicles in various stages of design, development, and testing. As the military is forced by shrinking budgets to be more cost-effective, it is exploring new ways to adapt commercial products for military use. One such adaptation is the SmarTruck® security vehicle, which is built based on a Ford F-350® truck. See Figure 10-9. The SmarTruck security vehicle has many hidden features giving more security, including hidden body armor, high-voltage door handles, detection devices, roof-launched grenades, and a roof-concealed laser. The U.S. Marine Corps is also adapting

Figure 10-9.
Monitor screens inside the SmarTruck security vehicle enable operators to control evasive systems and onboard weapons. (Public Affairs/U.S. Army)

a commercial vehicle for military use. The Marines in Okinawa, Japan are testing an electric car made by Nissan, the Nissan Hyper-Mini Electric Car, for use as a vehicle to travel around the base. Although these two examples are only at the test and experimentation stage at this time, they show the type of changes in equipment development the DOD is undertaking.

Many other vehicles are in the development stage, including the Light Armored Vehicle to be manufactured by General Motors (U.S. Army), the Near-Term Mine Reconnaissance System (U.S. Navy), and the VIRGINIA Class all-electric submarine (U.S. Navy). In this section, we will review two new vehicles at the ends of their development cycles. Both of these are unique in several respects. The Joint Strike Fighter® (JSF) aircraft represents a shift from designing a different plane for each branch of the service to a new plane designed to meet the needs of the three that fly (the Air Force, Marines, and Navy). The Osprey is a hybrid aircraft that can fly like an airplane, while taking off like a helicopter. The Osprey, like the JSF aircraft, is designed to meet the needs of all flying branches of the U.S. military.

Joint Strike Fighter (JSF) Aircraft

The development of the Joint Strike Fighter (JSF) aircraft represents a philosophical change for the DOD. Previously, each arm of the military would fund its own development efforts for new aircraft. The JSF aircraft, in contrast, will be a family of common aircraft for use by all DOD forces. Each flying branch of the military—the Air Force, Navy, and Marine Corps—will configure the JSF aircraft to meet its own particular needs. See Figure 10-10.

There are three different types. The first will be used by the U.S. Air Force and will use standard airfields to take off and land. The second type will by used by the Navy and will be adapted to take off and land on sea-going carriers. The third, designed for the U.S. Marine Corps and the UK's Royal Navy, will allow for short takeoffs and vertical landings. The development costs of the JSF aircraft are distributed between the U.S. Air Force, the U.S. Navy, and several international partners. The UK was an early participant in the development of the JSF aircraft, but several other countries, including Canada, Denmark, Italy, Norway, Turkey,

Figure 10-10.
The Joint Strike Fighter (JSF) X-35C demonstrator aircraft flew near Naval Air Station Patuxent River, Maryland, in 2001. The X-35C demonstrated handling and aerodynamic performance required of a JSF aircraft based on an aircraft carrier. Also successfully tested in 2000 and 2001 were the X-35A (conventional takeoff and landing version) and the X-35B (short takeoff and vertical landing version). The X-35B became the first plane in history to complete a short takeoff, a level supersonic dash, and a vertical landing in a single flight. (Lockheed Martin Aeronautics Co.)

and the Netherlands, have joined this consortium. Together, these new countries are investing over $2 billion in the JSF aircraft.

The JSF aircraft will use stealth technology to avoid being seen on enemy radar. As part of its development of the JSF aircraft, Lockheed used state-of-the-art manufacturing techniques to build its two prototypes. Two of its partners for this plane, Northrop-Grumman and BAE Systems, developed new composites and flexible tooling for the manufacturing of this fighter.

The DOD expects the JSF aircraft will be in full production by 2008. The DOD expects it will buy over twenty-eight hundred planes through 2040, at an estimated cost of $40 to $50 million each.

Osprey

The Osprey uses tilt-rotor technology. It takes off and lands like a helicopter, but flies like a plane. The Osprey will be used to replace the Marine Corps fleet of Vietnam-era helicopters and can be customized to meet the needs of a particular branch of the service. See Figure 10-11. The Marines plan on using the Osprey to transport up to twenty-four troops or a fifteen thousand–pound load at one time. The U.S. Navy plans to use the Osprey for search and rescue missions, air refueling, and medical evaluation. By the early 2010s, the Navy and Air Force are planning on purchasing 50 Ospreys each, and the Marines plan on acquiring 360.

Figure 10-11.
An MV-22 Osprey tilt-rotor aircraft lands at the Pentagon. (U.S. Department of Defense)

Despite the advantages offered by the Osprey, its path to full use in the military has not been smooth. Two crashes in 2000 killed twenty-three marines. In addition, there have been other technical and performance issues related to the design. Right now, the Osprey is back to flight-testing to determine if all the problems have been solved.

Ballistic Missile Defense

As threats from rogue nations seem more likely, there is an increased emphasis on defense. The concept of a nationwide missile defense was first floated by past President Reagan and dubbed "the *Star Wars* Project." The extent to which scientists and technologists will ever be able to fulfill his vision of an "impenetrable shield" remains a distant illusion.

Despite the millions of dollars invested in SDI projects, doubts about the SDI surfaced years ago, shortly after its inception. Scientists, technologists, and laypeople began to realize the impractical nature of critical SDI projects. Various aspects of the perfect space-based nuclear shield either defied the laws of physics or relied on weapons easily neutralized by enemy countermeasures (loads of sand could be hurled onto the surfaces of highly polished mirrors, reducing their ability to deflect powerful laser beams toward targets). Over the years, however, SDI gradually lowered its sights and managed to stay afloat through a quiet, almost unnoticed evolution. Today, the SDI has been transformed into the Missile Defense Agency (MDA) Program. This new program is the result of almost two decades of R & D efforts. In the mid-1990s, then–Secretary of Defense William Perry decided to upgrade the national missile defense efforts from technology R & D to the development of a usable system. After becoming U.S. President in 2000, George W. Bush decided to continue to work on this system and attempt to have a preliminary missile defense system in place by 2005. The MDA has two main goals:

- ❑ To defend the United States from an attack from ballistic missiles.
- ❑ To develop and implement a Ballistic Missile Defense (BMD) system including ways to intercept missiles at all stages of flight.

The *Dateline* section in this chapter describes a test of a BMD system. The interceptor kill vehicle was able to hit the trial ICBM before it reached its target. The test involved the use of both space and ground sensors and radars, in order for the interceptor to find its target. The complexity of this type of activity made this first success even more exciting for DOD researchers. The DOD is planning quarterly tests of the BMD system, each costing about $100 million.

The planned BMD system will be land-based, but will include detection satellites in space. The ground-based interceptor is the missile that would be launched to "kill" an incoming missile. See Figure 10-12. Also, there would be ground-based radar systems and space-based infrared satellites. Tying this all together would be a complex C^3I arrangement that would act as the brains of the system. In addition to building on the technologies developed for SDI, this project is based on the Arrow Program—Israel and the United States have jointly developed the Arrow® interceptor, an experimental ATBM to intercept the Scud as well as other intermediate-range ballistic missiles.

In order to continue development of the BMD system, the United States withdrew from the 1972 Anti-Ballistic Missile (ABM) treaty between the USSR (now the Russian Federation) and the United States. This treaty prohibited the development and deployment of strategic missile defense systems.

Figure 10-12.
A payload launch vehicle carrying a prototype exo-atmospheric interceptor kill vehicle is launched from Meck Island at the Kwajalein Missile Range for a planned intercept of a ballistic missile target over the central Pacific Ocean. The target vehicle, a modified Minuteman intercontinental ballistic missile, will be launched from Vandenberg Air Force Base, California. (U.S. Department of Defense)

Now that the United States is no longer bound to the prohibitions of the ABM treaty, it is moving forward with an even more ambitious plan for missile defense. As defense of the U.S. homeland continues to be a pressing political issue, there will be continuing efforts and money spent in this area.

Directed Energy Weapons (DEWs)

Strictly speaking, lasers are not a science fiction weapon of the future. Operation Desert Storm demonstrated the value of lasers in guidance systems for precision missiles. It is possible lasers and Neutral Particle Beams (NPBs) may soon emerge as precise, deadly weapons in their own rights. The antipersonnel use of contemporary lasers is indeed frightening. Modern tanks are equipped with laser range finders. These range finders are capable of blinding anyone unfortunate enough to view the laser from the downrange side. Soldiers and aviators can be permanently blinded in a period of time measured in nanoseconds by a weapon more than a mile away.

DOD researchers are presently engrossed, with three Directed Energy Weapon (DEW) projects: the Hydrogen-Fluoride Chemical Laser, the space-based Free-Electron Laser (FEL), and the NPB.

DEWs may someday function in a strategic defense system as weapons to be used against missiles in their boost-phases, while modified versions of FELs might be useful against cruise missiles, tactical missiles, and aircraft. Space-based lasers and particle beams are projected to be able to examine a large array of threatening objects and discriminate between warhead reentry vehicles, *decoys, chaff,* antisatellite weapons aimed at themselves, sensor satellites, space mines, and junk. Smart DEWs will be able to identify, attack, and destroy targets without human intervention.

Chemical Laser

The idea of a space-based chemical laser that could shoot down missiles in their boost-phases was one of the first options explored by DOD researchers. Chemical lasers produce power as a high-energy beam of concentrated light, derived from a chemical reaction of the hydrogen fluoride compound (HF) which is used as rocket fuel. HF lasers are the most mature DEWs. They require large mirrors to direct lethal laser power to targets thousands of kilometers away. The Air Force Research Laboratory recently created a new chemical laser combining two gases, nitrogen chloride and atomic iodine; however, this laser is not ready to be used as a weapon, as scientists estimate it will take at least five more years to make it into a weapon.

Free-Electron Laser (FEL)

The word *excimer* is an abbreviation for *excited dimer.* A dimer is a pair of linked atoms. FELs are described in terms of radiation emitted by electrons accelerated in a magnetic field. Short wavelength beams are emitted when the dimer's bonds are broken. An excimer laser or FEL emits beams strong enough to disable an enemy missile in a few seconds. The primary components of the FEL are an accelerator, a device producing a varying magnetic field to change the direction of the electrons (called the *wiggler)*, an optical beam line, and an output mirror. R & D on FELs is still somewhat new by comparison to HF DEWs. Design issues still being ironed out include the complexity of the accelerator and scale-up of the technology from low power to high power devices. Scientists believe FELs, besides being space-based, could also be mounted on ships, on aircraft, or at ground bases to attack cruise missiles, tactical missiles, and enemy aircraft.

Neutral Particle Beam (NPB)

This high-tech weapon uses accelerators developed for experimentation in high-energy physics to produce high current, high-energy beams of neutral hydrogen particles. Neutral Particle Beams (NPBs) can propagate through space without being deflected by Earth's magnetic field. Researchers envision NPB devices to fire a beam of fast moving atomic particles at incoming warheads. The beam shears through the warhead and scrambles its electronics so the nuclear weapon housed inside cannot explode. NPBs would pass right through *decoys,* while the heavier warheads are absorbing them, essentially allowing them to weigh objects in space. At Los Alamos, SDI has a system called the Ground Test Accelerator (GTA) in hand to demonstrate the operation of an integrated NPB system. Developmental research in these and other evolving DEWs is expected to continue into the next few decades.

The Information Warrior

The warriors of the future will not have to depend on brute force to get their jobs done. Today, the DOD is seeking to transform the soldier into an information warrior. The U.S. Army is taking the lead in the integration of computer and other information technologies into the battlefield. They have even established a special division whose job is to be an experimental force for a future battlefield. The Fourth Infantry Division at Fort Hood, Texas has the distinction of being the Army's experimental unit, known as Force 21.

Every fighting vehicle in Force 21 is equipped with a computer linked to the Force 21 Battle Command Brigade and Below system (FBCB2). See Figure 10-13. Inside each vehicle, there is a digital map showing the location of all friendly and enemy forces. It includes a GPS chip, as well as a message system, so the commander can send messages either to one particular vehicle or the entire platoon. If an individual vehicle spots an enemy vehicle, the crew can target it with a laser. Automatically, it will appear on everyone's screen as an enemy.

Another system Force 21 is testing is the Army Airborne Command and Control System, known as A2C2S. A2C2S is a centralized station with five laptop computers integrating a troop's intelligence, fire support, and communication information. This mobile system will allow commanders to continue to monitor and run their troops while they are moving. It allows the company commander to move easily in the battlefield without losing time and information.

The U.S. Navy is also moving toward the digital future. It has developed a new wide area network, named WARnet, to assist them in this transition. WARnet is a wireless network that can provide for ship-to-shore and ship-to-ship communications in a one hundred by two hundred mile area. The nodes in the wireless network are carried by a variety of vehicles, including planes, helicopters, ground vehicles, and unmanned aerial vehicles. (For a discussion of wireless networks, see Chapter 3 on Information Technologies.) All the information from WARnet flows into an offshore headquarters located on a battleship. There, the commander can see what is happening on an immediate basis and make quick decisions about the sea operation.

The Defense Advanced Research Projects Agency (DARPA) is exploring even more fantastic technologies for the future. One DARPA-funded project is to develop an information rich exoskeleton for the soldier

Figure 10-13.
This is a prototype information system for the Army of the future, Force 21 Battle Command Brigade and Below computer system. (U.S. Department of Defense)

of the future. Instead of relying on a vehicle for support, enhanced soldiers would have their heavy weapons, armor, and advanced electronics fastened to their exoskeletons. This exoskeleton is remarkably like the long-term view of the future from the mid-1990s film *Aliens,* in which Sigourney Weaver's character straps herself into an industrial exoskeleton to do battle with the alien. Another DARPA-funded project is a personal flying machine known as Solo Trek XFV.

The battlefield of the future will certainly contain more information and automation, but at the same time, the soldier of the future will have to face foes who will likely use high-tech weapons. As U.S. troops become more dependent on information technologies, they will become more likely to feel the effects of the loss of this information. Just as computer hackers can disrupt a network, a military hacker can cause a threat to the soldier.

This entire issue of cyberterrorism is a new one. *Cyberterrorism* can refer to the disruption of military information, as well as to the disruption of information in our homeland. Imagine the chaos that could result if a foreign terrorist was able to control the stock market electronically or destroy files held in a company's or the government's computers. We have not yet seen these types of attacks from foes, but there is little doubt we will in the future.

The Business and Politics of Defense

The astronomical costs associated with military policy and national security make international defense "big business." Without question, the military industrial complex is an economic giant not to be ignored. This final segment of the chapter takes a very brief look at a few topics influencing future trends in defense spending, including DARPA, black budget projects, the START treaties, and the "greening" of defense contractors.

In the early 1990s, four strategic modernization programs seemed to form the cornerstone of the U.S. defense package. These programs were endorsed for their capacities to provide effective deterrents against missile attacks and give our troops greater capacities to respond to attacks elsewhere around the world. Specifically, these big-ticket items included the MX missile, the Midgetman missile, the B-2 stealth bomber, and SDI. Two short years later, on the heels of the Soviet breakup, the U.S. defense budget called for the curtailment of the B-2 and Seawolf nuclear-powered submarine. The administration capped the Advanced Cruise Missile Program and abandoned production of the W-88 warhead for the Trident II missiles.

As with any large-scale program wishing to keep R & D dollars flowing, the Missile Defense Program has modified and changed its goals as it went along. The contemporary justification for missile defense is that MAD may not deter emerging nuclear powers. We also hear arguments regarding the potential for dual-use benefits (spin-offs) of missile defense technologies. Created by the Pentagon in the late 1950s, DARPA has become something of a technology mission clearinghouse. DARPA's current thrust of operation centers on the two-to-ten-year kind of projects having the potential, with some risks, to have tremendous social impacts. DARPA is often credited as the entity most directly responsible for the whole field of artificial intelligence, interactive computer graphics, computer security technology, time-sharing networks, materials science, and the fields of high definition television and supercomputing.

Less visible to the general public are a host of projects funded through the black budget, the President's secret treasury. This secret source funds all the military activities the President, the Defense Secretary, and the Director of Central Intelligence prefer to keep hidden from public scrutiny. It pays for weapons to keep the Cold War won, as well as to fight World War III—and World War IV. Three such projects released for public view, after spending at least ten years as black budget items, include the Stealth Bomber; the Military Strategic, Tactical, and Relay (MILSTAR) system; and the Advanced Cruise Missile. The Stealth Bomber is probably the most expensive secret weapon since the Manhattan Project's first atomic bomb. The black budget began with the decision to build an atomic bomb, and it remains alive today, concealing many projects bearing innocuous code names. Amidst the throes of a Soviet breakup and Third World tenacity, it is unlikely these black budget items will be dismissed or revealed to us any time soon.

On the brighter side, both the United States and the Russian Federation have agreed to further reductions in intercontinental missiles. In 2000, the Russian parliament approved the START II treaty; this treaty was signed in the early 1990s by former Presidents George Bush Sr. and Boris Yeltsin and ratified by the U.S. Senate in the mid-1990s. The Russian ratification was the culmination of almost twenty years of negotiation and work. The START II treaty requires both the United States and the Russian Federation to reduce their ICBM arsenals to, at most, 4,250 missiles—this is a major reduction from the 10,000 each country had in the early 1980s, when START II negotiations began. The newest nuclear weapons treaty, signed in 2002, between the United States and Russia will cut the existing store of nuclear weapons by about 65 percent by the early 2010s; however, instead of specifying which weapons to cut, this treaty lets each country choose which nuclear weapons it wants to keep. In ten years, each country will only have between 1,700 and 2,200 warheads in its arsenal, a dramatic improvement, but still more than enough to cause massive destruction.

As far as the development of new technologies themselves, we are beginning to witness a slight "greening" of defense contractor industries. Ultimately, the military budget will be pared down—how much and how quickly remains debatable. Political battles are likely to ensue between those congressional leaders who have major weapons programs in their states and others who depend minimally on the military-industrial complex. Concern over increased unemployment and loss of revenue as more weapons programs are curtailed and more armed forces bases are closed has inspired some aerospace companies, like TRW, to direct their research proficiencies toward nondefense-related projects. It has been argued that pollution constitutes another threat to national security. One senator managed to earmark $200 million of the Defense Department's budget for research on environmental cleanup. The Pentagon is reportedly developing a list of projects it might be willing to take on. Let's hope disposal of CW stockpiles and cleanup of nuclear waste sites manage to make their agenda.

Summary

We find ourselves living precariously amidst a vast array of technological conveniences in the shape of personal computers, compact disc players, portable telephones, wonder drugs, nutritionally enriched foodstuffs, and superior optical devices. The same basic research that delivered several of these items to your household is in many ways responsible for the looming specter of our strategic defense technologies. This chapter addressed the topic of national security as it relates to international defense strategies and policy making.

The world has witnessed the end of a Cold War era and the disintegration of the Union of Soviet Socialist Republics. Nuclear weapons have symbolically become the icons of war and peace at the same time. We have lived through policies of massive retaliation, mutual assured destruction, and strategic defense dominance. Perhaps a posture of minimal nuclear deterrence is our only option for a secured global community.

Weapons of the type described in this chapter really do exist—they are not a figment of our imaginations. Government leaders seem to view their military arsenals as necessary forces their constituents have given them the authority and responsibility to discharge if needed. Perhaps we can hope to cut the number of weapons to the absolute minimum and avoid future deployments whenever possible. Minimal deterrence as an international defense posture may not be the ethical, moral, or intellectual choice of all people, but it may be the only choice we have. This, combined with a healthy infusion of cooperative diplomacy, should preserve the peace, making the world a safe and secure place to live.

Discussion Questions

1. Briefly describe each of the four rationales provided when a country decides to use aggressive military force. Make an attempt to provide contemporary incidents as illustrative examples of each scenario.

2. What is the purpose of the Single Integrated Operational Plan (SIOP) in the United States? How might its officials make the distinction between counter-force and counter-value targeting? Provide examples of hard and soft targets in your response.

3. How do you feel contemporary government leaders should go about redefining nuclear policy into the next century? Factor the Soviet Union's disintegration and the waning of the Cold War into your response.

4. List and describe the key characteristics of what the Pentagon refers to as *mid-intensity conflicts (MICs)*.

5. Comment on the ethical and morals issues associated with: [a] Accepting paid employment as a cruise missile technician in the defense industry; [b] Conducting laboratory research experiments to build better bombs; [c] Carrying out battlefield orders calling for the use of biological or chemical munitions.

6. What steps can you individually take to reduce the possibility of a terrorist threat? What should the country be doing to attack terrorism? How does technology fit into your plans?

Suggested Readings for Further Study

Adams, James. 1991. The arms trade: The real lesson of the Gulf War. *The Atlantic* 268 (5): 36, 38, 47, 48, 50.

Agres, Ted. 2001. Biosecurity gets needed attention. *The Scientist* 15 (22): 1, 16–17.

Army without a country, countries without an army. 1992. *The Economist* (25 January): 43–44.

Begley, Sharon. 1992. A safety net full of holes. *Newsweek* (23 March): 56–57, 59.

Bluth, Christoph. 2000. *The nuclear challenge.* Brookfield, Vt.: Ashgate.

Bonvillian, William B. and Kendra V. Sharp. 2001/2002. Homeland security technology. *Issues in Science and Technology* 18 (2): 43–49.

Carnes, Sam Abbott and Annetta Paule Watson. 1989. Disposing of the US chemical weapons stockpile: An approaching reality. *JAMA* 262 (5): 653–659.

Carus, W. Seth. 1991. Missiles in the Third World: The 1991 Gulf War. *ORBIS* 35 (2): 253–257.

Center for Disease Control. (2002). *FAQ's about anthrax.* [Internet]. <http://www.bt.cdc.gov/DocumentsApp/faqanthrax.asp> [25 Jan 2002].

———. *Bioterrorism.* [Internet]. <http://www.bt.cdc.gov/>.

Charles, Dan. 1991. Calls for tougher pact on biological weapons. *New Scientist* 131 (7 September): 20.

Covault, Craig. 1992. Russians forge space pact, but military transition in chaos. *Aviation Week & Space Technology* 136 (13 January): 20–21.

Cowen, R. and J. Raloff. 1990. R & D budget: Civilian gains outpace defense. *Science News* 137 (3 February): 71, 76.

Dao, James. 2001. Panel calls for overhaul of Osprey program, not cancellation. *The New York Times* 19 April, A20 (1).

Department of Defense. (2000). *Chemical and Biological Defense Program.* [Internet]. Portable Document Format. <http://www.defenselink.mil/pubs/chembio02012000.pdf> [25 Jan 2002].

———. (2001a). *Joint Vision 2020.* [Internet]. <http://www.dtic.mil/jv2020/jvpub2.htm> [25 Jan 2002].

———. (2001b). *Quadrennial Defense Review Report.* [Internet]. Portable Document Format. <http://www.defenselink.mil/pubs/qdr2001.pdf> [25 Jan 2002].

Dunn, Michael A. and Frederick R. Sidell. 1989. Progress in medical defense against nerve agents. *JAMA* 262 (5): 649–652.

End of the nuclear threat?, The. 1991. *Nature* 353 (10 October): 483–484.

Enserink, Martin. 2001. Biodefense hampered by inadequate tests. *Science* 294 (9 November): 1266–1267.

Federation of American Scientists. *Military analysis network.* [Internet]. <http://www.fas.org/man/index.html>.

Goldstein, Walter. 1989. Economic growth and military power: Erosion of the superpowers. *Current* 309 (January): 23–31.

Gusterson, Hugh. 1992. Keep building those bombs. *New Scientist* 132 (12 October): 30–33.

Gutman, W. E. 1991. A poison in every cauldron. *Omni* 13 (5): 42, 44–46, 48, 111–113.

Hallin, Daniel. 1991. TV's clean little war. *Bulletin of the Atomic Scientists* 47 (4): 17–19.

Henderson, D. A. 1998. Bioterrorism as a public health threat. *Emerging Infectious Diseases* 4 (3): 488–492.

Hogan, Brian J. 1991. Lasers on tomorrow's battlefield. *Design News* (22 April): 98–102, 104, 106.

Husbands, Jo L. 1990. A buyer's market for arms. *Bulletin of the Atomic Scientists* 46 (4): 14–19.

[Internet]. (2001). Century of biological and chemical weapons. *BBC News.* <http://news.bbc.co.uk/hi/english/world/americas/newsid_1562000/1562534.stm> [25 Jan 2002].

[Internet]. *PBS plague war.* <http://www.pbs.org/wgbh/pages/frontline/shows/plague/>.

Isaacs, John. 1992. Defense spending—workfare for the 90s? *Bulletin of the Atomic Scientists* 48 (3): 3–5.

Klare, Michael T. 1990. Wars in the 1990s: Growing firepower in the Third World. *Bulletin of the Atomic Scientists* 46 (4): 9–13.

———. 1991. Behind Desert Storm: The new military paradigm. *Technology Review* 94 (4): 28–36.

Lancaster, H. Martin. 1992. Don't allow chemical weapons agreement to fail. *Christian Science Monitor* (7 April): 19.

Lepingwell, John W. R. 1992. Towards a post-Soviet army. *ORBIS* 36 (l): 87–103.

Lorber, Azriel. 1992. Tactical missiles: Anyone can play. *Bulletin of the Atomic Scientists* 48 (2): 38–40.

Lumpe, Lora, Lisbeth Gronlund, and David C. Wright. 1992. Third world missiles fall short. *Bulletin of the Atomic Scientists* 48 (March): 30–37.

Markusen, Ann and Joel Yudken. 1992. Building a new economic order. *Technology Review* 95 (3): 22–30.

McGurn, William. 1990. Divvying up the peace dividend. *National Review* (30 April): 28–30.

Meselson, Matthew. 1991. The myth of chemical superweapons. *Bulletin of the Atomic Scientists* 47 (3): 12–15.

Morrison, Philip, and Kosta Tsipis. 1997. New hope in the minefields. *Technology Review* 100 (7): 38–47.

Nolan, Janne E. 1991. Who decides? U.S. nuclear strategy after the Cold War. *Technology Review* 94 (l): 52–57.

Norman, Colin. 1990. Bush budget highlights R & D. *Science* (2 February): 247, 517–519.

Orient, Jane M. 1989. Chemical and biological warfare: Should defenses be researched and deployed? *JAMA* 262 (5): 644–648.

Paine, Christopher and Thomas B. Cochran. 1992. So little time, so many weapons, so much to do. *Bulletin of Atomic Scientists* 48 (l): 13–16.

Pilgrim, Kitty. (2001). *War experts: U.S. campaign cost-effective* [Internet]. <http://www.cnn.com/2001/US/12/14/ret.war.cost/index.html> [25 Jan 2002].

Posey, Carl A. 1992. Nuclear world order. *Omni* 14 (6): 41–42, 44, 76, 80, 83, 86–88.

Postol, Theodore A. 1991. Lessons of the Gulf War experience with Patriot. *International Security* 16 (3): 119–171.

Rathjens, George W. and Marvin M. Miller. 1991. Nuclear proliferation after the Cold War. *Technology Review* 94 (September/October): 24–32.

Rizvi, Gowner. 1991. Has China sold out Third World? Weapons carry political clout. *World Press Review* 38 (12): 12–13.

Rueter, Theodore and Thomas Kalil. 1991. Nuclear strategy and nuclear winter. *World Politics* 43 (July): 587–607.

Scott, William B. 1992. Brilliant pebbles development to change building of spacecraft. *Aviation Week & Space Technology* (4 May): 76–78.

Snidle, Giovanni A. 1990. United States efforts in curbing chemical weapons proliferation. In, United States Arms Control and Disarmament Agency. *World military expenditures and arms transfers—1989*. Washington, D.C.

Stanglin, Douglas and Robin Knight. 1992. Sailing into the sunset: The Soviet military's collapse is dangerous. *U.S. News & World Report* (17 February): 31–32, 37.

Stikeman, Alexandra. 2001. Recognizing the enemy. *Technology Review* 104 (10): 48–49.

Stix, Gary and Philip Yam. 2001. Facing a new menace. *Scientific American* (November): 14–15

Talbot, David. 2001a. DARPA's disruptive technologies. *Technology Review* 104 (8): 40–50.

———. 2001b. Detecting bioterrorism. *Technology Review* 104 (10): 34–37.

Tsipis, Kosta. 1991. A weapon without a purpose. *Technology Review* 94 (8): 52–59.

United States Air Force. (1999). *AGM-130 Missile fact sheet.* [Internet]. <http://www.af.mil/news/factsheets/AGM_130_Missile.html> [25 Jan 2002].

U.S. Arms Control and Disarmament Agency. (1999). *World military expenditures and arms transfers 1997.* [Internet]. <http://dosfan.lib.uic.edu/acda/wmeat97/wmeat97.htm> [26 Jan 2002].

U.S. Department of Defense. *DefenseLINK.* [Internet]. <http://www.defenselink.mil/>.

van Creveld, Martin. 1989. *Technology and war: From 2000 B.C. to the present.* New York: Free.

———. 1991. *The transformation of war.* New York: Free.

Walker, Paul F. and Eric Stambler. 1991. ...And the dirty little weapons. *Bulletin of the Atomic Scientists* 47 (4): 21–24.

Webb, Jeremy. 1991. US to lay bare its biological weapons secrets. *New Scientist* (30 November): 14.

Weiner, Tim. 1990. *Blank check: The Pentagon's black budget.* New York: Warner.

Weiss, Peter. 2001. Dances with robots. *Science News* 159 (30 June): 407–408.

Wilson, Jennifer Fisher. (2001). Biological terrorism. *The Scientist.* [Internet]. 15 (22): 1. <http://www.the-scientist.com/yr2001/nov/wilson_p1_011112.html> [26 Jan 2002].

Chapter 11

Social Response to Technological Change: Breakthroughs and Risks

DATELINE

Baltimore, Maryland

Following a meeting of the National Technological Literacy Conference, many attendees seemed convinced that a large percentage of the American public is technologically illiterate. Many students who are not prepared to grasp the technical content of day-to-day issues are graduating from high schools and universities. Some college graduates are reportedly unable to comprehend the social impacts of technological issues, such as the safety of contraceptive devices and food additives, the practical uses of robotics and gene splicing, and the societal implications of nuclear accidents and space shuttle explosions. Faith in technology, it seems, is not always accompanied by an equal measure of understanding.

Although it is rare to find people who lie awake at night wondering whether or not electricity is mysteriously dripping out of empty wall sockets (a common misconception in the early days of electrification), most people live with some form of high-tech anxiety. There are those "techno-dummies" among us who either believe everything the experts in the white coats tell us or hysterically reject everything the experts tell us, since the products and systems of technology sometimes go awry. Others among us may be labeled "techno-wizards," as the ones who appear to know what they are doing when using technology. The techno-dummies rely on the techno-wizards when their telephones malfunction or their computers break down. Perhaps it is this distinction that provides the basis for technological literacy in our society....

Introduction

Contemporary science and technology have created the amenities of modern civilization. In countless ways, each has increased our capacities to control environmental forces and given us visions of an even more prosperous future. Undeniably, science and technology have also increased both our uncertainties about the future and our dependence on the numerous inventions and innovations they have produced. While we may have increased our abilities to understand, predict, and control the natural environment around us, we may have actually lost the capacity to control the technologies we have created to help us along the way.

The impact of technology on society at large has been a prevalent theme in social, economic, and political thought for decades. Numerous essayists have debated about the extent to which continued technological growth will enhance or hinder the survival of the human race. A diverse collection of writers has drawn our attention to the increasing complexity and rate of technological change in contemporary society. They seem to be continually making projections filled with varying degrees of alarm. Their publications present the specter of an increasingly uncontrollable technology spreading like a cancer among the population. Its consequences cannot always be assessed or accurately predicted. We are led to infer that the human race is on a collision course with ruin and imminent destruction.

On one side of the debate, some social scientists, policy makers, and technologists take the position that society should critically examine the role it wants technology to play and then take action to curtail its destructive tendencies. On the other side, similar groups of professionals assert that more scientific and technological research (knowledge and information) is all that is required to cure all of society's illnesses. In some ways, the debate seems to be perpetuating itself.

While the rhetoric continues, we are living amidst an array of technologies some people contend are simply accidents waiting to happen. When the dark side of technology reveals itself, when disaster hurts innocent bystanders, and when technology causes more problems than it solves, society often points the blame at its creators. When accidents do occur, the general public shakes its head while wondering what is happening to its "technological fix." Regardless, there is little chance the industrialized nations of the world will turn their backs on the promises aligned with new technologies. We have all chosen a lifestyle sustained by high technology, but that does not mean we are not apprehensive about it.

Many people of your grandparents' or even your parents' generations were, for the most part, blissfully unaware of the potentially harmful effects of man-made chemicals on themselves and the environment. They were unconcerned about the finite nature of the raw materials and resources being depleted to fuel technological progress. These days, as our reliance on science and technology intensifies, we have begun to worry about the effects of acid rain on our lakes and forests. We worry about the future consequences of the buildup of carbon dioxide in the atmosphere. We worry about what is happening behind the closed doors of

weapons research laboratories. We are fearful that chemical dumpsites may be leaking into our supplies of drinking water. We also worry about which foods are safe to eat and which drugs are guaranteed to not have harmful side effects on our offspring. It seems a greater amount of information regarding all the technological conveniences we take for granted has most directly created a higher level of social anxiety. Even more so in the aftermath of the terrorist attacks on the United States, people are looking to technological solutions to protect themselves from deadly biological and chemical agents.

Further, as the demand for energy to drive the massive engines of science and technology increases, so too does society's concern about the sources of future supplies and the ramifications of exploiting our current energy technologies. These anxieties range from concern over the disposal of radioactive wastes to worrying about the political and economic consequences of our dependence on foreign oil. (Some of our dependence on Middle Eastern suppliers possibly is the root cause of several of our contemporary military battles.) None of these issues comes with a handbook, a set of instructions, or a glossary of terms to chart a definitive course of action for all of humanity. Technological growth is replete with a series of uphill struggles—but then, it always has been.

The current international proliferation of technologies has led to an exponential widening of risk for all of Earth's inhabitants. The social excuse of being "an innocent bystander who is not responsible for technology" is severely outmoded. The concept of universalism is more evident today than ever before. This characteristic of modern technology refers to both an interdependence among nations and the diminishing number of unaffected cultures. Technological developments and the risks associated with them do not exist in a vacuum. For example, an accident in a nuclear power plant located in one country may indeed have negative consequences for people residing in a neighboring country. The effects of airborne radioactive contaminants do not simply stop at the border.

This chapter is structured to assist you in your evaluation of how technology and its associated risks affect all social groups—regardless of their level of technological development. The various sections present some of the issues pertinent to society's capacity to accept technological change and simultaneously take responsibility for its by-products. It is next to impossible to precisely distinguish a "good" technology from a "bad" one. Yet if society expects to benefit from the fruits of its scientific and technological labor, it must retain the strength and ethical conscience to control and direct the course of its activity. In order to keep science and technology on track, it is necessary to conduct routine risk-benefit analyses, blended with a heavy dose of information, logic, and common sense. Throughout this chapter, the following questions will be addressed:

❑ What are the implications of the accelerating rate of scientific and technological change for society at large?

❑ Are there any particularly effective strategies we can use to cope with the pitfalls associated with social lag?

❑ Is society "at risk," by virtue of its ever-increasing dependency on the fruits of technology and science?

❑ How does the concept of technological illiteracy fit into the threads of our contemporary social fabric?

❑ Is it plausible to trust the experts' abilities to carefully scrutinize the development of new technology in order to ascertain whether or not it should be endorsed or abandoned?

❑ Should these techno-wizards be authorized to decide whether the inevitable risks outweigh any reasonable benefits with reference to unproven technologies?

Key Concepts

Many of the terms and phrases commonly used in the science, technology, and society literature may already be familiar to you. A few of them are listed here and will be mentioned in this chapter. You should be able to explain what is meant by each of them when you finish your review of this material:

busyness syndrome
durational expectancy
information overload
response mode
risk-benefit analysis
social lag
technological literacy
technology accordance
technology trap
technophobia
technostress
universalism

The General Nature of Change

The key word in the title of this chapter is *change*. The next word, based on its level of importance, is *response*. Before attempting to discuss response modes social group members rely on to deal with change, it is essential to understand the meaning of the elusive construct or term *change*. In the simplest terms, change has been the only constant throughout history. In essence, change is constant, progressive, and universal. Change is not always as great or far-reaching as we sometimes imagine it will be. To repeat a timeless phrase, the more things change, the more they stay the same.

If these assertions are valid, why do people fear change? Why do social groups band together to fight against change? Why do some individuals feel inclined to avoid making changes at any cost? Why do some professionals insist on saying things like "I don't mind progress as long as I don't have to change!" The word *change* produces emotional reactions, since it is not a neutral word. To many people, change is threatening. Without any sort of modifier, the word *change* promises no respect for present cultural values. Change may even disrupt the values.

In some instances, the mere mention of a proposed change conjures up visions of revolution, dissatisfaction, and discontent. More inoffensive words referring to the process of change include *education, training, modification, orientation, guidance, indoctrination,* and *therapy*. We are much more amenable to a situation where others *educate* us versus one where others *change* us. We tend to feel less guilty if we view ourselves as *training* other people rather than *changing* them. These substitute words are safer or less emotional, which is why they are more tolerable. At first glance, they carry the implicit assurance that the only changes produced will be good ones, compatible with society's current value system. Unlike the assortment of synonyms presented above, the word *change* is crisp and directional. It is not lost in the translation of a phrase or couched in preconceived expectations.

Take a moment to reflect on *your* personal ability to confront changes head-on and deal with them as they surface in your daily living arrangement. Within a very short period of time, it should become obvious to you that there are two basic forms of change: *theirs* and *ours*. This might provide a basis for the fact that society's feelings about change are often contradictory. On one hand, we welcome change, and on the other hand, we fight against it. We acknowledge *our* individual rights to stimulate change, but oppose *their* rights to force us to accept changes we do not fully sanction. See Figure 11-1.

Social groups display an uncanny ability to cope with adversities. They have a great deal of difficulty, however, coping with uncertainties. The uncertain element of change is frightening and often threatening. Uncertainty suggests change beyond *our* control. This implies a union between *uncertainty* and *them*.

One tenet of social behavior states that if you want people to be supportive of a change, make them believe they had a part in bringing it about. Consider the production manager who makes the decision to move the hourly employees' break period from 10:00 A.M. to 10:45 A.M. Rather than adamantly posting this announcement in an autocratic manner on the bulletin board next Monday morning, this person would be wise to involve all those affected by the change in the decision-making process. The main objective should be to convince the employees the following:

- ❏ The change will be beneficial to them.
- ❏ Their input is valuable.
- ❏ They possess a measure of control with regard to their working environment.

Although it is not always possible to be in total command of changes affecting our lives, it is possible to assume a profile of "readiness." Readiness to make changes implies a willingness to take charge of any number of unanticipated situations. It does not imply "wimpy" behavior. A lack of readiness is generally accompanied by a period of **social lag,** which often varies based on the perceived severity of the changes at hand. We really do have the capacity to make *their* changes *our* own if we keep

Figure 11-1.
Numerous peace demonstrations were organized during the late 1960s and early 1970s. These activities often revealed that personal beliefs are not always sanctioned by the legal sector of society. The original caption for this photo taken in 1970 read "Protestors were herded by police up California Street in San Francisco into the main line of march. An estimated 5,000 persons gathered in front of the Fairmont Hotel to protest the visit of South Vietnam's Vice President, Nguyen Cao Ky. Small outbursts of violence occurred when police attempted to clear various adjacent areas." (*Spartan Daily Newspaper,* San Jose State University)

abreast of developments taking shape each day. The real challenge in the modern world is not so much related to a willingness to accept change, but to the speed with which we are willing to do so.

Pace of Change

If you have ever had the occasion to eavesdrop on a lively discussion in which the participants were reminiscing about the "good old days," the chances are at least one of the members said something about the pace of change. When people reflect on the past, they have a tendency to recall the events from a nostalgic or sentimental perspective. They remember that people

seemed to be less harried and that technical changes appeared to occur at a much slower rate. It is typical to view the changes that took place in science and technology a few decades ago as evolutionary and to describe the ones happening right now as revolutionary.

We, as a society, are bombarded with marketing phrases like "new and improved," "now, with even more power," and "hurry, offer ends Friday" on a daily basis. We have become accustomed to this hectic pace of change, but not everyone is able to tolerate it. Technology in all disciplines is changing so quickly, individuals often believe if they understand certain information today, it must be obsolete. In other words, technological advancement has accelerated to a point such that a span of a few decades can easily contain more developmental breakthroughs than a span of several hundred centuries.

Some futurists tell us the breakneck speed at which science and technology changes will occur within the next couple of decades will make today's stride seem like a snail chasing a turtle. In the 1970s, Alvin Toffler's critically acclaimed *Future Shock* cautioned us about the physiological and psychological maladies people experience when they are forced to deal with too much change too quickly. He introduced us to novel phrases like *information overload, durational expectancy,* and *throw-away society.* Each of these concepts is directly related to society's acceptance of change and its capacity to assimilate change very quickly.

On a personal level, ***information overload*** refers to an individual's inability to think clearly due to an excessive amount of cognitive stimulation. Overloading the human nervous system with too many bits of information (stimuli) at one time causes a breakdown in performance and logic. When this type of system demise is examined on a social level, data seems to imply that society does not respond well to what might be described as an information explosion. Once again, the element of speed surfaces through the choice of the word *explosion* to describe a social condition. As change continues to accelerate in society, it forces a parallel acceleration within each of us. Each time that new information reaches us, we are pressured to revise and update our master image file folder. More comforting images are continuously replaced with newer, less familiar items.

Some of our capacity to tolerate the pace of change in our daily lives has its roots in our earlier childhood years. From the moment of birth, humankind's perception of time is closely aligned with an internal circadian rhythm (a biological cycle recurring in twenty-four hour intervals). The manners in which people respond to time, however, are culturally conditioned. One of the first things a child learns about time involves an understanding of how long things last. This knowledge eventually leads to the formulation of a vast array of what Toffler labels ***durational expectancies.*** Young children are quick to learn that "in a few minutes" means something more immediate than "the day after tomorrow." Parents often explain the concept of *tomorrow* by making reference to the activity of going to bed at night and waking up again the next morning—that is tomorrow. For very young children, the concept of *yesterday* refers to every day they have been alive.

As adults, virtually all of our activities are based on certain assumptions or expectations about durations. We expect a visit to the dentist to last longer than a trip to the post office, but shorter than an excursion to the theater. We make assumptions about the relationships we have with certain individuals—indeed we believe some interpersonal agreements (marriage) are supposed to last a lifetime. Durational expectancies will differ from one culture to another, but they tend to be deeply ingrained into the pace of life. When the pace of change is altered, these durational expectancies are shaken. The resultant social response to either an accelerated or even a decelerated rate of change will range someplace between immediate adaptability and an acute withdrawal from reality.

Those people who adapt quickly to changes in their lives have most probably internalized the principle of acceleration. An individual who anticipates that certain situations will endure for shorter periods as time passes is less likely to be caught off guard than someone whose durational expectancies are intractable. The pace of technological change in today's world is such that many of us wonder if the word *perma-nence* should even be included in the dictionary.

Human beings have apparently begun to resign themselves to the notion that their relationships with certain items are increasingly ephemeral. The fleeting nature of the vast number of technological convenience products in the home is further testimony we have become a throw-away society. See Figure 11-2. Some parents are actually quite reluctant to allow their children to view early morning or after-school television programs, for fear they will discover that an "even more challenging" electronic plaything has supplanted their newest wonder toy. Household consumers invest in new appliances that do everything short of eating their food for them, only to be informed in the following week's news magazine the

particular model they purchased is already obsolete. Some technology skeptics sit back and wait for a newer model, believing it will ultimately cost less and perform more tricks. In the interim, these hesitant consumers observe with envy the conveniences they are missing. Neither of these situations seems very pleasant.

Regardless of your ability to deal with an accelerated rate of information overload in a world where obsolete technologies prevail, you cannot deny the fact that the relentless pace of change gets trying at times. Most of us experience the need for a rest periodically, but seem to rejuvenate ourselves quickly in order to jump back into the fast lane. While many social observers spend their time discussing the ailments associated with erratic change, a considerable segment of modern civilization seems to thrive on it.

Figure 11-2.
A stroll down the paper products aisle of your local supermarket or drugstore provides numerous examples of disposable items taken for granted and purchased as a matter of course in our society. These items provide a measure of convenience, but also reinforce the short-lived nature of many products in our homes. (Jack Klasey)

Increased conveniences afforded by scientific and technological advancements have seemingly produced a culture of quickness.

Instant Gratification and the Busyness Syndrome

Do you remember the timeless cliché that says, "Good things come to those who wait"? How do you think the busy corporate executive on the fast track to success would respond to that suggestion these days? Contemporary American society is caught up in the need to be busy all the time and has no time to wait for anything. In response to the spirit of constant movement, science and technology continually provide numerous artifacts designed to meet our desire to have something called "It All." Having "It All" generally includes a highly respected college degree, an upwardly mobile career, a warm and fulfilling family life, a lean body, and an influential status in one's local community. To be well off on all fronts, one must be outwardly busy at all times. Our fast-paced society is ill bent on instant gratification and terrified of having nothing to do. What a paradox!

The modern technological household in the United States is a resplendent island of convenience. A quick stroll through the high-tech kitchen will reveal digitized dishwashing machines and microwave ovens that talk to you. You might even spot a personal laptop computer, linked to the Internet and perched on top of the butcher block, that will plan meals, provide nutritional information, and guide the user through the necessary calculations to scale down a recipe that serves twelve in order to prepare dinner for two. See Figure 11-3.

If you stop to glance into the broom closet, you might spy a power-driven vacuum cleaner and an iron that shuts itself off automatically. The modern bathroom is

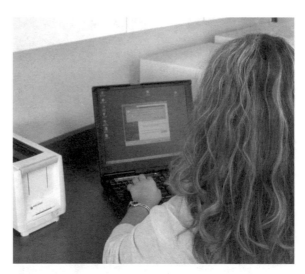

Figure 11-3.
The use of a personal computer in the kitchen is noteworthy.

devoted to satisfying a number of contemporary pleasures. For instance, the bathtub is complete with multiple Jacuzzi jets, a mirror framed with three different types of lighting, a water-resistant radio mounted in the shower for listening to the news rather than reading the newspaper, a microprocessor that controls the scale to announce the owner's weight in a digitally synthesized voice, and a classic English heating unit to warm the residents' towels while they sleep. The medicine cabinet is full of remedies for one ailment or another, and each of them promises "fast" relief. Who would purchase an aspirin-like product that insists it will "slowly remove those aches and pains from your tired body"? Are you kidding? This is the era of instant gratification!

Another island of technological wizardry is encased in what is fondly referred to as the *home entertainment center.* See Figure 11-4. It makes one stare in awe as the compact disc (CD) player, videocassette recorder (VCR), digital videodisc (DVD) player, and the remote controlled dual screen television ensemble reside in harmony along a portion of the

Figure 11-4.
This is a homeowner's entertainment center.
(Sauder Woodworking, Archbold, Ohio)

wall of a normal size room. The high-tech entertainment center certainly entertains its owners. It enables the viewer to watch two programs at once, record a third one, and preview the highlights of a live football game in a small insert found at the lower corner of both screens! That certainly sounds a lot like information overload, but it also provides a great deal of instant gratification.

Along a slightly different vein, Americans may still value such business ethics as courtesy, reliability, and economy, but what really impresses them is speed. We demonstrate this through our affinity for such things as one-hour photo developing services, instant replays, memory redial telephones, and posh suntanning parlors. After all, tanning the old-fashioned way takes too much time! Americans value speed, and a number of U. S. corporate empires were established for all of these people who like to be in a hurry (McDonald's, Federal Express, Domino's Pizza, and Polaroid).

Advertisements for smaller convenience stores criticize full-service grocery stores where customers are required to wait in long lines to purchase a quart of milk. Here again, the need for constant movement is exemplified.

Astute business planners recognize that Americans often refuse to wait in line. Rental car firms still experience their share of bottlenecks, but they have tried, with much publicity over the past few years, to whisk customers into their awaiting vehicles. Airline companies now issue electronic tickets, allowing passengers to use airport vending devices that dispense boarding passes in less than sixty seconds. Most major banks provide automatic teller machines for their extremely busy depositors. Across the nation, supermarkets have made an attempt to improve their slow performances by reserving a few lanes for express check out. Doesn't it anger you when you approach one of these quick service aisles with two items for which you intend to pay cash, only to find the person ahead of you has a cart clearly containing more than the requisite fifteen items?

On the other hand, refusing to wait may mean missing out on an event or losing the opportunity to purchase an item at a reduced price. Although most of us hate to wait for anything, the classic military refrain "Hurry up and wait" has become a fact of modern life. We wait in traffic, wait for buses and trains to arrive, wait in lengthy airport security lines, and wait to be seated in crowded restaurants. We line up to buy tickets for trains that leave late and for movies that do not start on time. Waiting is an unpleasant exercise, and it seems to be on the rise in our society.

As a greater portion of our technology-dependent population commits itself to the doctrine of the busy class, the lines are getting longer. Waiting is a power game. People who hold high-status positions generally do not wait as long as those who are part of the general working class. Similarly, busyness is akin to status in our country. People who always have full calendars are usually the same ones who hold more status in the corporate business world. This is an excellent example of the contradictory nature of the current social buzzwords: *instancy* and *busyness*. Where one goal is attained, the other is partially curtailed (for example, when things get done instantly, people may not seem to be as busy as they feel a need to be!).

The theory of the busy class is based on the notion that keeping a calendar filled proves you are in demand. Interestingly, calendars and schedules have their roots in religion. In the sixth century, the Benedictine monks, who believed it was a sin to waste God's time, popularized calendars and schedules. It is ironic to note that the conspicuous organizers and personal digitized assistants (PDAs) in today's business world have actually taken on a whole new aura of reverence.

A daily organizer is commonly described as a portable set of calendars with an address and telephone book and other peripherals, such as a section in which to record auto mileage expenses and a small tablet of graph paper to satisfy that sudden urge to draw up plans for a new backyard deck. Daily organizers often contain three different types of calendars: daily planner pages to list hour-by-hour obligations up until 9:00 P.M.; the basic year-at-a-glance format; and the six-year planner for the time conscious professional. Anyone who requires three calendars for only one life must indeed be very busy. The diligence with which its owner fills it out, keeps it up-to-date, and retrieves it from a briefcase for consultation almost resembles a religious ritual. Perhaps the organizer is really designed to serve as a stabilizing anchor in a fast-paced, constantly changing society. Some individuals are switching to PDA devices, instead of carrying an organizer. See Figure 11-5.

Another ingredient of the **busyness syndrome** is found in the burgeoning technology of the communications industry. We now have electronic accouterment, such as call forwarding, call holding, memory dialing, speakerphones, pagers, and palm pilots. These features allow people to operate like chief executive officers in the comfort and privacy of their own home offices. Household consumers can put plumbers and gardeners on hold while checking their references with the Better Business Bureau on another line. In order to be in constant communication with important contacts while also in transit, a large number of busy people rely on cellular telephones to stay in constant touch with their colleagues and clients.

Cellular telephones were introduced in the 1980s for prices well over twenty-five hundred dollars. Their price tags have dropped considerably, making them more affordable among ordinary civilians. In fact, many contemporary wireless communications companies offer telephone devices at no cost if people subscribe to their services. Cellular telephones pack a double hit of status in our society. Not only do they provide the veneer of busyness, but they also lend their owners the glamour of being engaged in very important transactions, regardless if the users are in their local grocery store, sitting in the waiting room of their child's orthodontist's office, or sitting in their automobile stuck in a traffic jam.

If some of these examples regarding the pace of life and the rate of technological change seem rather silly, perhaps it is because you are not yet wrapped up in the express track of competition. If they seem all too real and near to your heart, welcome to

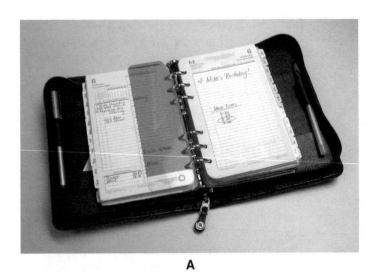
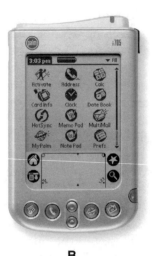

A   B

Figure 11-5.
These are examples of personal time management systems. A—In this traditional daily planner, pages provide space for an appointment schedule with hours identified from 6 A.M. to 9 P.M., a prioritized daily task list, a daily record of events, and a daily expense record. B—Some people have found personal digitized assistants (PDAs) to be more convenient than traditional planners. This handheld PDA, the Palm® i705 computer, provides abilities such as storing address information, calculating, creating appointment calendars, and communicating with others. (Palm, Inc.)

the beginning of the twenty-first century, an era of perpetual motion, and the death of permanence. In either case, your personal attitudes toward the stability of your routine and the relentless pulse of change are a lot less malleable than you might think.

Social Attitudes toward Changing Technology

In the months following the disastrous terrorist attacks on the United States, people lived in a heightened state of anxiety, fear, and helplessness. Many saw this event as an example of our technologies being used against us in a violent and harmful manner. Others remained hopeful that our technological prowess would again enable us to stage a successful war against terrorism in cooperation with several international partners. In a *USA Today/CNN/Gallup* poll conducted

shortly after this tragic event in 2001, 63 percent of Americans surveyed were confident the American way of life would be preserved; 74 percent stated they believed Americans are united and in agreement about the most important values.[1] This suggests U.S. citizens continue to possess an optimistic attitude and want to believe tomorrow will be better than today. It appears that a national conscience has awakened. Much of the credit for this positive trend toward a renewed interest in making the future a better place should actually be given to technology. High-speed printing presses, efficient—often next day—rural delivery service, national public radio programs, and television news broadcasts via satellites have put nearly everyone in this country on the same informational level.

Social attitudes develop over time as one generation provides the fuel of experience and knowledge to drive the next one along.

---

[1][Internet]. (2001). Americans asked opinions on developments after September 11th terrorist attacks. *USA Today/CNN/Gallup Poll Results*. <http://www.usatoday.com/news/attack/2001/11/28/poll-results.htm> [29 Nov 2001].

Attitudes are the product of the capacity to scrutinize events and the freedom to evaluate information. Information is the lifeblood of social progress. American citizens have access to more information now than at any other point in time. The freedom to produce information has led to an almost chaotic abundance of free ideas. Subsequently, not many of us have the time or inclination to stay informed about and keep up with emergent science and technology issues on the public policy agenda. It seems very few of us are even interested in staying abreast of these issues. A study completed in the late 1990s stated the following:

> An analysis of public attentiveness to more than 500 news stories over the last ten years (1987–1997) confirmed that the American public pays relatively little attention to many of the serious news stories of the day. The major exception to this rule are stories dealing with natural and man-made disasters and U.S. military actions.[2]

Information concerning the many benefits we are about to reap from the miracles of science and technology is plentiful indeed. Just as abundant are the reports about technology's faults and the calamities associated with its use. The mass media, both print and electronic, are the basic means through which all parties involved in technological transactions can secure a common database. Social groups are constantly inundated with information and have, at least in this nation, the freedom to decide how to digest it all. The manner in which individuals make these types of decisions is influenced by whatever attitudes they possess with regard to change in general. Overall, it seems safe to conclude, Americans, unlike other national communities, are not a fatalistic people. Most of us have at least a modicum of faith in the idea that everything can change—all

it takes is luck, individualism, freedom, and an energetic drive to move ahead. America is still a place where the play of human nature is allowed the greatest latitude to arbitrate, judge, or develop its tools for progress.

Attitudes and Values

Social scientists define *attitudes* as the general disposition to behave in particular ways toward specific objects or events. They define *values* to be social standards considered worthwhile or desirable. Attitudes and values have a great deal of influence over the directions of a nation's scientific and technological change. The following paragraphs describe a few of the attitudes seeming to be prevalent in contemporary American culture. Each of the examples provides a reflection of a "techno-trend" or "scientific-shift" of sorts.

Any number of social and political adjectives might be chosen to describe the general public's attitudes toward science and technology. A few you may have heard include *cheerful optimism, disenchantment, idealistic support, fearful skepticism, cautious optimism, ambivalence, mild disillusionment,* and *compassionate conservatism.* Although it often seems attitudes are shaped over many years of experience and are not usually all that flexible, the impact of recent events cannot be ignored. Personal perceptions about certain technologies are heavily influenced by contemporary breakthroughs, as well as catastrophes. It takes dedicated interest and knowledge to formulate opinions about science and technology. In studies conducted during the last two decades by the National Science Foundation (NSF), about nine out of every ten U.S. adults reported being very or moderately interested in new scientific discoveries and the applications of new inventions and technologies. Other findings reported by the National

---

[2]Parker, K. and C. Deane. 1997. *Ten years of the Pew News Interest Index: A report for presentation at the 1997 meeting of the American Association for Public Opinion Research.* The Pew Research Center for the People and the Press.

Science Board provide additional insight about the mixed and somewhat contradictory nature of our contemporary culture of science and technology awareness:

❑ The number of adults who feel well informed about science and technology is disappointingly low. In the late 1990s, only 17 percent of adults surveyed described themselves as well informed about new scientific discoveries and the use of new inventions and applied technologies; approximately 30 percent of those surveyed thought they were poorly informed.

❑ About three-quarters of Americans lack a clear understanding of the nature of scientific inquiry and the manner in which scientific investigations are carried out.

❑ Overall, data shows increasing percentages of Americans agreeing that "science and technology are making our lives healthier, easier, and more comfortable" and disagreeing that "we depend too much on science and not enough on faith."

❑ In 1999, only 12 percent of America's adult population were described as the "attentive public" by the National Science Board researchers, which was down from 14 percent two years earlier, and from 25 percent in the mid-1980s. These individuals reported a high level of interest in science or technology developments, felt well informed about them, and indicated that they pursued a regular pattern of relevant information consumption (technical journals, science newsletters, public media, and special documentaries). The National Science Board labeled an amazing 44 percent as being neither interested in nor well informed about science and technology issues.[3]

The expanding scope and global effects of science and technology have stimulated a great deal of attention toward the need for careful government regulation of scientific research. Through the 1980s and 1990s, most government regulating agencies focused on the applications of selected technologies. The advent of disciplines requesting approval for the use of human subjects, experimentation with biotechnology, and the use of nuclear materials has called for the establishment of government policies relating directly to both basic and applied research. The judicious use of this regulatory power requires a populace knowledgeable about the benefits and risks of scientific research and technological advancement.

In broad terms, an extremely impressive majority of U. S. citizens feel that both science and technology have made their lives healthier and easier. Most of these people seem to expect future benefits as a result of ongoing contributions from this nation's great research and development (R & D) laboratories and that, because of science and technology, future generations will be much better off than we are. See Figure 11-6. Statistics also suggest there is widespread recognition among the public that science and technology simultaneously offer the promise of fulfillment and the potential for destruction. Specifically, a minority of individuals (41 percent) presently express reservations about science and believe it makes our way of life change much too quickly.

An analysis of several other contemporary technological initiatives led the National Science Board to conclude that U.S. residents are quite capable of differentiating between those technologies that are acceptable versus those that should be curtailed. The table in Figure 11-7 presents data summarizing this social disposition. Public faith in scientific research in general is significantly higher than support of riskier and more costly technology endeavors like space exploration and genetic engineering.

---

[3]National Science Board. 2000. *Science & Engineering Indicators—2000*. Arlington, Va: National Science Foundation.

Public Perceptions of the Effects of Science and Technology					
1999					
Item	**Strongly Agree**	**Agree**	**Do Not Know**	**Disagree**	**Strongly Disagree**
Promise of science					
Science and technology are making our lives healthier, easier, and more comfortable.	30	60	1	1	8
Most scientists want to work on things that will make life better for the average person.	8	75	2	1	14
With the application of science and technology, work will become more interesting.	7	66	4	1	22
Because of science and technology, there will be more opportunities for the next generation.	12	72	2	1	13
Reservations about science					
We depend too much on science and not enough on faith.	12	38	5	7	38
It is not important for me to know about science in my daily life.	3	13	1	21	62
Science makes our way of life change too fast.	3	38	2	4	53
	B>>H	**B>H**	**B=H**	**H>B**	**H>>B**
Have the benefits of scientific research outweighed the harmful results, or have the harmful results outweighed the benefits?	47	27	11	10	4

B>>H = benefits strongly outweigh the harmful results; B>H = benefits outweigh the harmful results; B=H = benefits equal the harmful results; H>B = harmful results outweigh the benefits; H>>B = harmful results strongly outweigh the benefits.

Figure 11-6.
This chart shows statistics regarding public perceptions of the effects of science and technology. (Adapted from *Science & Engineering Indicators—2000*)

In sum, regardless of how uninformed they report being, an ever-growing proportion of the American adult population is very interested in scientific and technological events. There is an obvious prevalence of science and technology themes in feature films, television programs, and published fiction. Each of these reflects the scientific and technological profile of our current culture. The expanding numbers and rising sales of popular science magazines and home computers seem to reflect the same characteristic. Prudent investors, brokers, and stock market analysts have become wise to keep abreast of science and technology R & D in order to make money for themselves and keep their clients in the black.

Public Assessment of Technological Developments					
1985, 1990, and 1999					
Development Area		Benefits Exceed Costs/Risks	About Equal	Costs/Risks Exceed Benefits	N
Space Exploration	*1985*	54%	7%	39%	2005
	1990	43	9	48	2033
	1999	49	8	43	1882
Nuclear Power	*1985*	50	6	44	2005
	1990	47	12	41	2033
	1999	48	15	37	1882
Genetic Engineering	*1985*	49	12	39	2005
	1990	47	16	37	2033
	1999	44	18	38	1882
Scientific Research in General	*1985*	68	13	19	2005
	1990	72	15	13	2033
	1999	74	11	15	1882

Figure 11-7.
This chart summarizes public evaluation of recent technological endeavors. (Adapted from *Science & Engineering Indicators—2000*)

Despite the **technostress** and the **technophobia** surrounding the lives of many individuals, there is still a dominant mood of optimism, although cautionary, in the United States. Perhaps this trait is deeply rooted in the general American character. As a people, we seem to be strongly supportive of science and technology. We believe much of the prosperity of the past few decades can be most directly attributed to the contributions of science and technology. For the time being, it is safe to conclude that contemporary American social groups will continue to look to science and technology to provide an improved, though always changing, quality of life.

Response Modes

Quality of life is an elusive concept not easy to define. One simple explanation suggests it is a balance between what people need and want, and the extent to which the environment can provide for them. Some of the typical quality of life indicators sociologists and cultural anthropologists use to study this concept include standard of living or per capita income, psychological and mental well-being, environmental stability and quality, general climatic conditions, and opportunities for self-sufficiency. Technological pursuits serving to enhance or improve these general criteria are defined as both beneficial and progressive for society. On the other hand, technologies having the imminent potential to detract from, degrade, or destroy these criteria are defined as hazardous, risky, or detrimental to social well-being.

A number of examples have already been presented to illustrate situations where technology is either applauded for its contribution to the quality of our lives or blamed for the problems we face. The manner in which society responds to both the conveniences afforded and the problems caused by technology are commonly associated with personal bias, preconceived notions or expectations, the amount of information available and the degree to which it is perceived to be accurate, personal interest in the issue, level of understanding and awareness, previous experience, and technology's general track record. Taking all of these factors into consideration, there appears to be a variety of behavioral responses used to confront the equivocal nature of technological change. Also known as ***response modes*** or coping behaviors, six are described here:

❑ Immediate acquiescence. This type of behavioral response to change is one in which the individual verbalizes implicit trust in the experts. Although the "blind faith" attitude is rare, for some people it provides a measure of security with regard to a technology not easily understood. For others, it depicts an ingrained belief that the benefits of technology will always outweigh the costs. In the late 1990s, public confidence in the scientific community was at about 40 percent. These American adults reported they had a great deal of faith and confidence in the scientific community. See Figure 11-8.

❑ Apathy. People who cope with technological development in this manner do not want to be bothered with the "boring" or "gory" details. They typically express concerns about the amount of effort it takes to keep their daily routines in order. They do not have time to worry about whether or not genetic engineering experiments are being properly monitored. Denial is sometimes a part of apathetic behavior as well. For instance, many people refuse to believe we have the technological capacity to destroy ourselves at the push of a button. The visuals presented in *War Games, Star Wars,* and *The X-Files* are fictional portrayals of society, created by Hollywood's tycoons for entertainment, and will not occur in our lifetime.

Public confidence in leadership of selected institutions: 1973–1998

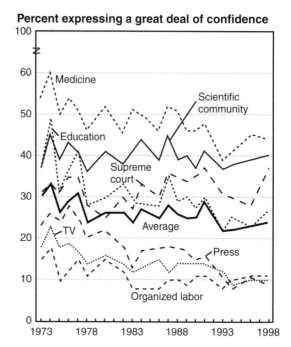

Figure 11-8.
This chart displays public confidence in social institutions that have capacity to influence and create change in our lives. (Adapted from *Science & Engineering Indicators—2000*)

❑ Reluctant acceptance. When people go along with changes for fear of what might happen to them if they resist, they exhibit a response mode called *reluctant acceptance* or *partial approval*. In many cases, this is an interim behavior providing only a slim vote of confidence in the proposed or pending technology change. It is a common attitude held by those people for whom the "scientific jury is still out." This response allows people to cope with the early stages of change by enabling them to sit back and observe the effects of a new technology without going on record with any distinct opinions. While often doubtful of the benefits, this attitude is one of interested ambivalence. Current public perceptions about the future of our costly space program might fall into this category. It is also pertinent to note that there seems to have been significant erosion in the public's confidence in the medical profession. Figure 11-8 shows that our rating for this institution was as high as 60 percent in the 1970s, but has been gradually declining since then.

❑ Avoidance. This response mode best illustrates an individual's inability or unwillingness to cope with changes created by science and technology. Personal emotions such as fear and anxiety often play a role here and may result in avoidance. People who respond with avoidance make some sort of decisive actions resulting in observable, structural changes in their lifestyles. They may decide to change jobs, leave the area, withdraw from a university, or even commit suicide in order to avoid the situation. For this segment of society, the risks associated with technology are perceived to be dominant over any benefits. This response mode may also be predicated on ethics and the moral aspects of pursuing a particular technology. A scientist's decision to leave a nuclear weapons laboratory might fall into this behavioral category.

❑ Reactionist rejection. This response to technological changes is the one that receives the greatest amount of media coverage. People who rely on this response are willing to openly protest the technologies they oppose. Operating on emotional "gut" reaction, they are not afraid of the legal consequences for their actions. They may sabotage the observable artifacts of the technology. These individuals appear to focus primarily on the detrimental side effects of technology, while losing sight of any potential benefits. Many of the rallies and lobbying efforts against nuclear reactors and weapons laboratories are best classified as reactionist rejections of technology.

❑ Accord. The response mode most of us would like to think we use to confront new technologies is labeled *accord*. It literally implies agreement and harmony. However, the behavior patterns used to describe accord go much further. People who either accept and approve, or reject and disapprove a technological proposal in this mode do so with a lot of thought. Their decisions and actions are based on understanding, awareness, accurate data and information, logic, and common sense. It takes more time to reach accord because all the pros and cons are weighed before an attitude is portrayed. The person who takes time to research an issue like toxic waste management before voting on a clean water initiative is using this behavior to cope with change or to actually stimulate it.

Each of us has, at one time or another, used one of these six behaviors or a combination of them to deal with a crisis or cope with changes in our lives. Not one can be identified as the best or most common for all situations. Social response to science and technology is usually a reflection of perceived control, literacy, and independence. Each of these social traits has something to do with society's persistent "love affair" with its technology.

Technology Traps

If we stop and take a good look at the network of technologies sustaining our daily routines, it is hard to deny our dependence on them. This is not to say reliance on technical systems is a bad thing, it is just reality. Some people openly admit their needs to surround themselves with technology's warm blanket of conveniences. Others feel trapped by it.

In a literal sense, the word *trap* refers to confinement or imprisonment where escape is either difficult or impossible. Life in our contemporary society is replete with traps associated with technology. Just about every moving vehicle might be construed as a technology trap of sorts (for example, automobiles, airplanes, bullet trains, submarines, subway trains, and elevator cars). Can you think of others? Each time we step into one of these vehicles and watch the doors close behind us, we take the risk of not being able to escape its confinement. While most people do not stop and think about that possibility when they get into their cars or step onto an elevator, many do feel that way when they board an airplane. In each of these transportation modes, users routinely illustrate their dependence on technology to get them to their final destination quickly and safely.

There are many other situations that can be cited to provide a *literal* translation for the technology trap principle. Modern hospitals and their impressive surgery facilities reveal our reliance on technology to cure illnesses and save lives. A person who is hooked up to a life support system certainly must experience the sense of being confined or trapped. Engineers who work inside government laboratories requiring them to pass through a series of security checks may also feel trapped by technology. Coal miners who descend to work in underground caverns are definitely aware of their dependence on technology to get them back out at the end of their shift. Identifying physical technology traps is not difficult once you start the creative juices flowing.

On a more *abstract* level, a **technology trap** is defined as much more than physical confinement. It is yet another profile characteristic of our modern society. The technology trap concept refers to humankind's dependence on any technological system or device that either is not understood or cannot be controlled. More often than not, we tend to utilize or maintain technology in such a way that we look like we know what we are doing. Behind the facade, quite often lies a vast amount of ignorance, inexperience, confusion, or even frustration regarding the technology being used. Technological systems and devices have advanced to such a state that they are regularly marketed as being "user-friendly." This means a person can use the device or monitor the system without possessing highly technical credentials. Today's automobiles are considerably more complicated than the ones manufactured in the 1930s. They are, however, much easier to operate. Contemporary automobile users do not need to be expert mechanics to drive their vehicles to work. Those who expected to drive the earlier models did not have the same luxury.

Abstract cases of technology traps exist all around you. For instance, unless you are an expert software engineer, the chances are pretty good you do not fully understand why or how your word processing software works the way it does. As long as your term paper is submitted on time, why be concerned with its programming language? We also flip switches and press buttons throughout the day in order to execute a series of tasks. It is rare for most of us to stop and wonder about the location, let alone the internal mechanisms, of the power station providing the electricity we effortlessly use.

We casually pour all sorts of detergents into the water used to wash our clothing. The number of individuals who truly understand the chemical formulation of those products and why they work the way they do is quite small. Consider also the hourly worker at a chemical plant employed to monitor a number of gauges in the company's central control center. Let's say this person was instructed to engage switch "A" if the needle in gauge #201 reaches a certain measurement, or switch "B" if the reading on gauge #417 falls below 43°C. Unless the operator truly comprehends the outcome of these actions or the reasons they are being performed, the control panel may be considered a technology trap.

None of the examples mentioned is intended to make you feel inadequate or hopeless about the manner in which you depend on technology or make use of it without a second thought. They are simply and innocently presented to get you thinking about the kinds of ordinary instances in which every one of us is dependent on the artifacts of modern technology. Some technology educators believe these activities have something to do with one's level of technological literacy. This concept was mentioned briefly in the *Dateline* segment of this chapter.

Technological Literacy

Literacy has been a topic of concern in educational literature for decades. At all levels of instruction, teachers have tried a variety of motivational techniques to enable their students to become literate. In the broadest terms, literacy presupposes the ability to read and write. If an individual is *literate,* he or she is generally thought to be educated and well informed. This generic term has been teamed up with several modifiers in recent years. We hear stories about adult literacy, literacy among minority ethnic groups, computer literacy, technical literacy, and even nutritional literacy. Over the last decade or so, technology educators have espoused the need to design and implement curricular activities to improve our students' performance in the international technological literacy competition. In the late 1990s, the International Technology Education Association published the results of its "Technology for All Americans Project" in a document titled *Standards for Technological Literacy.*

Technological literacy is an educational objective with widespread repercussions throughout the entire academic system. It is also an individual quest toward a meaningful existence in today's society. It is rapidly becoming a desirable managerial trait. The saying "one doesn't need to understand the technology in order to manage it" is finally being disputed.

What makes a person technologically literate? Is there a standardized test one should take to determine his or her level of technological literacy? Responses to these inquiries are obviously contingent on the definition one accepts for this term. Although no one definition for such a comprehensive topic is perfect, the following one is proposed for your consideration. Technological literacy involves an above average level of the following elements:

- ❏ Knowledge of technical and scientific constructs and terminology.
- ❏ Awareness of the impacts (both negative and positive) science and technology can have on society.
- ❏ Confidence and skill in the use of technology's tools, machines, and devices.
- ❏ Control over the operations and functions of technical equipment.
- ❏ Insight into the potential future states of specific technologies in order to project their social impacts.

Wow! This is some definition! The astute reader probably noted the use of the phrase *above average level.* This presents a problem of interpretation. One must ask for clarification as to the meaning of average. What is considered average for one culture or social group may be either above average or below average for another group. Likewise, what is considered average for one scientific or technical discipline may not be true for other areas of study. With these facts on the table, it is safe to say everyone is technologically literate. It is equally truthful to conclude that no one possesses this literacy. This debate could go on forever!

Defining a socially relevant term like *technological literacy* is always a challenge. Measuring the extent to which a person or social group has acquired a prescribed level of technological literacy is impossible. Certain elements of this lengthy definition can be assessed, but the data recovered is heavily influenced by self-examination activities. In a survey conducted by the NSF and reported by the National Science Board, a significant measure of disparity existed between the number of adults who indicated an interest in technology and those who felt well informed (literate) about technology issues. In nearly all cases, the levels of public interest in technological breakthroughs exceed levels of knowledge about these new findings. Interesting to note also, is that men typically express more interest than women in new scientific discoveries and in the applications of new technological inventions. Men also report they are more knowledgeable than women about these issues. See Figure 11-9.

Studies of this nature serve to provide periodic reviews of cultural awareness of science and technology. They do not actually measure technological literacy, but that is not their objective. Even if we cannot quantify the population's scores on a technological literacy scale, it is enough to agree that such a construct exists. Any amount of attention or energy devoted to its enhancement will not be futile, given our dependence on technology.

Indices of public interest in and self-assessed knowledge about scientific and technological issues, by sex and level of education: 1999

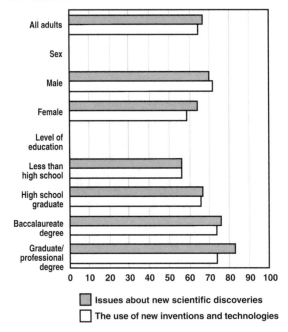

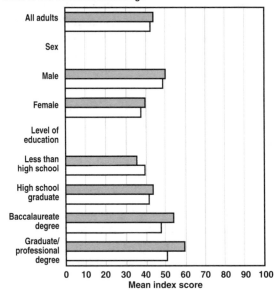

Figure 11-9.
These charts display indices of public interest in and knowledge about scientific and technological issues by gender and level of education. (Adapted from *Science & Engineering Indicators—2000*)

Recognition of Social Dependency on Technology

A series of titles have been proposed to describe our time in contemporary history. It has been labeled the *information age, telecommunications age, postindustrial age, computer age,* and *biotech age.* Regardless of the descriptive term one endorses, we are living in an age of technological enthusiasm, laced with fantasy. Everyone from Hollywood producers to government officials has thrust it upon us. The stunning record of technical accomplishment during the last century lends support to their visions. For many citizens around the world, technology has become an article of almost religious faith. Today's technology is uniquely seductive, since it appears to offer viable solutions to some of this planet's biggest problems. In some ways, it further encourages us to abdicate our personal responsibilities for these problems.

Social glorification of technology took off over a century ago with its cascade of curious inventions, including the steam engine, electric lights, farm machinery, the domestic sewing machine, and even running water. We next witnessed the arrival of the automobile, telephone, radio, television, jet aircraft, miracle drugs, and nuclear energy. Along the way, a manufacturing and marketing infrastructure provided the necessary avenues leading to technology's penetration into all segments of society all around the globe.

Technology was revered during World War II for its potential to become the great equalizer on the battlefield. It helped the United States to purchase victory for its allies via its superiority in industrial production and sophisticated weapons. Following the Soviet space surprise in the late 1950s (*Sputnik*, 1957), the United States adopted science and technology as its chosen vehicles to regain global superiority.

At the beginning of the twenty-first century, we find ourselves inextricably addicted to technology. It underpins every aspect of our existence on this planet. Our affinity for technology and love for the conveniences it offers have led to our taking much of it for granted. Take a moment to review the contemporary businessperson's typical working day. The individual under scrutiny awakes to a radio alarm clock, switches off the electric blanket, dresses in imported clothing made of synthetic fibers, defrosts the frozen orange juice and microwaves some type of breakfast treat with electricity generated by a nuclear power plant, commutes to the office in a highly fashionable automobile, while listening to high-fidelity stereo or making a series of cellular telephone calls, and travels up to the twenty-sixth floor by elevator, just in time to receive an important phone call relayed by satellite from a time zone on the other side of the globe. This is just an example of a few of the machines sustaining our lives. Their influences do not cease with the close of the workday. The same person looks forward to the weekend to play with a host of high-tech toys (DVDs, power tools, video games, ski lifts, luxury yachts, and remote controlled model airplanes). This is the present.

The future outlook for technology is promising, but will inevitably lead to an increased level of "social technohypnosis." In other words, more people will be caught up in the spell of technology. All of us are at least vaguely familiar with implants of electronic devices, such as heart pacemakers. We are witnessing an increasing use of these technologies to drive leg and arm muscles, to serve as implantable substitutes for hearing, and to monitor the metabolic reactions taking place inside the body. Some of the work being done in cryobiology (freezing tissue to store it indefinitely) permits complex organs, such as livers and kidneys, to be retained in a frozen state until needed. This will certainly lead to an expanded use of organ transplants to save more lives in the near future.

The fascinating world of computer electronics promises to continue its trek toward dazzling our imaginations. Electronic maps displayed on a small screen inside a delivery vehicle or your personal automobile reveal details of local streets and landmarks. Their users can also program them to suggest the best route to take to an intended final destination. Amateur photographers can now use very affordable electronic digital cameras that record photographs on magnetic discs instead of film. This feature enables users to preview the photos on a television or computer screen in order to decide which ones to print. Bank credit cards are now actually quite "smart." Some are imbedded with microchips to store checking, savings, and other financial balances, together with personal medical data and important telephone numbers. At present, the Smart Visa® credit card is the newest offering in credit card technology. It enables its user to store cash in the account, permitting use of the card at previously cash only merchants like newsstands and vending machines; have single-click access to account information twenty-four hours a day, seven days a week; request balance transfers, track purchases, and manage accounts on-line; and attach a smart chip reader to a personal computer (PC). See Figure 11-10.

It was not long ago that automated bank teller machines, PCs, high-speed facsimile machines, and on-line databases were figments of the technologist's imagination. When they first hit the market, they were luxuries. Today they are common necessities. In the eyes of some, this type of technology has made the world increasingly incomprehensible. Computers now make loan decisions, diagnose diseases, and represent a vital nerve center for the nation's defense

Figure 11-10.
The APRIVA® x57 smart card reader is the latest in credit card technology. A small 32K microprocessor chip is actually imbedded into the card, giving it much greater information capacity than the magnetic strip. It coexists with other technologies to simplify on-line shopping and preserve security. (APRIVA, Inc. "Wireless Business Solutions")

system. The programs guiding these complex systems are the products of thousands of man-years of diligence. It is rare for one individual to fully comprehend their operations. We have become, in many ways, a world community dealing with abstractions and depending on the interpretation of wiggly marks on paper and digital pulses in machines. In an age of instant gratification, many young people are growing up with the belief that things "just happen," almost magically.

As with any romantic love affair, ours with technology has had its share of euphoric moments interspersed with periodic states of deep depression. The highs and lows are tempered by repeated cycles of reality testing. The ingenious achievements scientists and technologists have made in recent years represent an expression of mankind's finest creative instincts. Society is dependent on the fruits of its technology. Society also has the capacity to direct and control its technology to meet specific goals and objectives. It is the social management of technology that will ultimately determine the extent to which contemporary civilizations can access its benefits without sacrificing their values and cultural ideals.

Living with Technology = Making Trade-Offs

Now and then, physicians have been known to remind their patients "any pill strong enough to help you is also strong enough to hurt you." This warning seems to be applicable to many of the technologies discussed throughout this chapter. Our modern society is further characterized by a concept labeled *universalism,* which refers to a global technology community. For example, a powerful electromagnetic pulse sent through the atmosphere by an adversarial superpower's defense system could knock out this nation's entire electrical power source in just a few minutes. This, of course, would totally immobilize us. Nevertheless, modern civilizations have become hostage to the forces of technological progress. There is no turning back because the benefits, both real and imagined, are enormous. The impending deficits are likewise greater than before and might turn out to be disastrous in the absence of a clear and prudent plan providing for a fallback system of survival.

Risk-Benefit Analysis

Several decades ago, risk analysis was an obscure concept. Beginning in the late 1970s and moving into the twenty-first century, risk analysis has become a major growth industry. Around the world, businesses, utilities, labor unions, insurance companies, hospitals, and government agencies are seeking the services of risk assessors. These institutions are requesting assistance in estimating and controlling risks from toxic chemicals, accidents, and other hazards. Some of these requests are a direct result of regulatory legislation. Regardless of the context, the risk assessor's central concern usually boils down to safety and the question "How safe is safe enough?"

When experts attempt to quantify risk estimates, they engage in a process of prolonged *risk-benefit analysis* studies. They try very hard to define actions that will result in an optimal balance between the benefits from an activity and the risks. In the late 1960s into the early 1970s, an expert in this field distinguished between voluntary activities and involuntary activities to derive formulas related to risk-benefit analyses. Studies revealed that individuals evaluate voluntary activities via their own value systems, whereas an external controlling entity determines the options and criteria for evaluating involuntary activities. Interestingly, the public seems willing to accept voluntary risks roughly one thousand times greater than involuntary risks at a comparable level of benefit.[4]

The proliferation of technologies has expanded the scope of risk at an exponential rate. A century ago, if a maintenance worker failed to attach a carriage wheel properly, the lives of half a dozen people were threatened. In this day and age, when a manufacturer fails to install an O-ring properly in a Boeing 767 aircraft, more than three hundred lives are at stake. The potential for disaster in mass transit, mass production, and mass destruction has grown geometrically. As our world becomes more interactive and

---

[4]Two studies are pertinent to this discussion: Star, C. 1969. Social benefit vs. technological risk. *Science* 165: 1232–1238. and Star, C. 1972. Benefit-cost studies in sociotechnical systems. In *Perspectives on benefit-risk decision making*. Report of the Committee on Public Engineering Policy, 17–42. Washington, D.C.: National Academy of Engineering.

centralized, the easier it is to sabotage its technological life support systems. Our survival on this planet depends on the personnel who tend the military computers and keep them running properly twenty-four hours each day. Somewhat closer to home, your safety is dependent on those people who manage local water treatment plants. At any moment, it seems the lives of thousands are at the mercy of a few who are just as human as we are. What if they make a mistake? Many of us deal with our fears of being powerless by placing our faith in the competence of the people or the technology in control.

We ask what went wrong when presented with the dark side of technologies. A partial list might include the following milestones along our precarious road toward progress: Bhopal, Chernobyl, Love Canal, Three Mile Island, Tylenol, Acquired Immune Deficiency Syndrome (AIDS), toxic shock syndrome (TSS), the *Challenger*, asbestos, human made pesticides, acid rain, polychlorinated biphenyls (PCBs), cancer, Bovine Spongiform Encephalitis (BSE, "mad cow disease"), and Agent Orange. These examples imply we are living in an age of risk, but the human race has always lived at risk. How we react to and deal with risks may have changed, but we must continue to live with the sober truth that there is no such thing as zero risk. Perhaps it is humankind's realization of this caveat motivating us to press forward in the quest for greater safety, while simultaneously creating newer technologies against which we must be protected. While these hallmarks of civilization seem to contradict each other, somehow we manage to keep the motors of social improvement running at full speed.

If one accepts the fact that safety is the absence of risk, but that risk cannot be totally abolished, a method to determine an acceptable threshold level must be derived. Safety, expressed in terms of acceptable risk,

is a social judgment call. Investigating the "how safe is safe" issue necessitates two central considerations. First, from a technical perspective, what are the probabilities of occurrence and the possible severity of harm? Second, from a social perspective, what sorts of cultural and ethical norms are established with respect to the valuation of a single human life, personal property, or the natural environment? Stated in a different way, the technical experts often worry about the quantity of the risk, while the general public worries about the quality of the risk.

This leads into the question of disparities between perceived risk and actual risk. When the experts assess the risk of a technology, they are primarily concerned with how many people die from it in an average year. When laypeople evaluate the risks of technology, they focus on a much broader list of criteria: How well do we understand the risk?, Could it wipe out an entire community?, Will it make a particular area uninhabitable for a long period of time? (For example, only thirty people died of radiation exposure after the 1986 nuclear disaster at Chernobyl, but more than one thousand square miles of land have been rendered unsafe for human life for more than one hundred years.), and Might it affect future generations or some members of society more than others? In answering each of these sensitive questions, experts offer the public a series of numbers and estimates to put technological risks into perspective, while public perceptions warn them against using sheer statistics to assess risks.

We assume the experts know a great deal about the technical aspects of nuclear power, space flight, biotechnology, food additives, and pesticides. This enables them to calculate very narrow definitions of *risk* in terms of factors that can be reduced to numbers and then extrapolated to make probability statements. For instance, one Air Force report estimated that the chances of a fatal accident on the space shuttle were 1 in 35—compared with 1 in 10,000 for a nuclear disaster or a dam failure. When people read probability statistics, they are often less interested in the numbers and more anxious about whether a hazard is new or old, known or unknown, controllable or uncontrollable, voluntary or involuntary, and the degree to which it is potentially catastrophic. Fear of nuclear radiation is punctuated by each of these perceived threats.

Even though more people die in automobile accidents each year than have perished in accidents related to nuclear power, the public views nuclear accidents as being much more frightening. Much of this anxiety is due to a dreaded fear of the unknown. The risks of a nuclear accident are seen as unknown, uncontrollable, and most likely catastrophic. When people cannot be certain about the risks they are forced to confront, they at least want to have some measure of control over the activities and technologies producing them. Much to the chagrin of risk analysts, people tend to be more willing to accept higher risks in activities over which they have some individual control, such as driving, smoking, skiing, and consuming alcoholic beverages. They approve much lower risk thresholds for activities in which they have little control, such as industrial pollution, food additives, biotechnology, and commercial airlines. It is intriguing to review the differences between expert opinions and public perceptions regarding the risks of various activities and technologies. See Figure 11-11.

In order for society to judge "how safe is safe enough," there must be a widespread level of technology awareness. It is difficult to evaluate the potential for danger or disaster if one does not know what the risks are. Prior to the terrorist attacks in the United States in 2001, many Americans had perhaps never even heard of the spore-forming bacteria called *Bacillus anthracis*, or *anthrax*. In the last few months of that year, this reality quickly changed. *Newsweek*'s national poll at that time revealed only 43 percent of Americans believed the administration was providing as much reliable information on anthrax attacks as they needed. Fifty percent of those polled stated that the president's team was either not releasing information because they did not have it themselves, or were purposely holding back to avoid scaring the public. It may be said that the American people have never been very skillful in knowing what risks they should worry about, and they become both anxious and angry when they feel uninformed. We are at a time in history, however, when getting the risk estimates correct has become linked to thousands of lives, billions of dollars, and the difference between chronic anxiety (technophobia) and realistic prudence (technology accordance).

Ostensibly, the purpose of government regulation of technology is to protect the public from a wide range of technological hazards. The U.S. government has the obligation to manage risk. Legislative initiatives over the past years have been enacted in an attempt to reduce uncertainty and surprise, enhance the visibility of possible cause-effect relationships, anticipate immediate and future impacts, and illuminate the alternatives that might spur prevention and delegate responsibility to all parties involved. Somewhere along the way, the word *trade-off* was invented to describe the process of deciding whether or not society has the dexterity or ability to contain the side effects of its technological feats. In an era of sharpened security and anxiety, our government leaders seem to be acknowledging that, while our country can never be made completely safe, it can certainly be made a good deal safer.

Public Perceptions versus Expert Opinions Regarding Risks Associated with Various Activities and Technologies			
Activity or Technology	**League of Women Voters**	**College Students**	**Experts**
Nuclear power	1	1	20
Motor vehicles	2	5	1
Handguns	3	2	4
Smoking	4	3	2
Motorcycles	5	6	6
Alcoholic beverages	6	7	3
General (private) aviation	7	15	12
Police work	8	8	17
Pesticides	9	4	8
Surgery	10	11	5
Fire fighting	11	10	18
Large construction	12	14	13
Hunting	13	18	23
Spray cans	14	13	26
Mountain climbing	15	22	29
Bicycles	16	24	15
Commercial aviation	17	16	16
Electric power (nonnuclear)	18	19	9
Swimming	19	30	10
Contraceptives	20	9	11
Skiing	21	25	30
X rays	22	17	7
High school and college football	23	26	27
Railroads	24	23	19
Food preservatives	25	12	14
Food coloring	26	20	21
Power mowers	27	28	28
Prescription antibiotics	28	21	24
Home appliances	29	27	22
Vaccinations	30	29	25

Do your worries match those of experts? Paul Slovic, President of Decision Research found significant differences among three different groups surveyed. Orderings are based on the geometric main risk ratings within each group. Rank 1 represents the most risky activity or technology.

Figure 11-11.
This chart compares public perceptions regarding risks of various activities and technologies with expert opinions. (Paul Slovic, Decision Research)

Technology Accordance

In the end, it appears that science and technology are both good and bad. Samuel Florman referred to this state of reality as a "tragic view of technology." He believes people must be aware of the dangers without foolish illusions about what can be accomplished. Another view might suggest that science is neutral and is a process by which we attempt to form explanations of the natural environment. Technology then evolves as the use to which this understanding is applied: for good or bad.

Technological change is unavoidable. The method in which society chooses to manage it and respond to it is, however, quite flexible. *Technology accordance* is a label used by the authors to refer to the extent to which a social group is able to recognize and take advantage of the potential benefits of a technological breakthrough and simultaneously control its negative side effects and risks. At the root of accordance, we find one of the response modes identified earlier—accord. As a greater proportion of society selects this coping behavior to decide the fates of new technologies, civilization will be a giant step closer to long-term sustenance.

Summary

Change is perhaps the only constant. Although society has convinced itself most changes are revolutionary, it is far safer to suggest changes tend to be evolutionary. This further implies people should take the time to stop and reflect on the changes taking place right now in order to chart a course of active living for the future. This chapter examined the concept of social change from several points of view. Science and technology were defined from two perspectives: (1) Science and technology change our lives on a regular basis, and (2) We, the people, routinely demand new science and technology to meet our changing expectations. In the first view of reality, society lags behind technological change, whereas in the second, technology is perceived to lag behind social needs.

In order to cope with new and demanding scientific and technological breakthroughs, social groups rely on a number of behavioral response modes. Of the six presented in this chapter, accord is the one seeming to be most energy efficient and prudent. It is characterized by attentiveness to the risky side effects of technology, coupled with well-informed decision-making regarding the exploitation of its benefits. A technologically advanced civilization must also be technologically conscious in order to preserve its survival. Technological change is often perceived to be linear, but in actuality, it is full of cyclical patterns of regression. Sustained progress mandates the design of fallback systems to protect us against our sciences and technologies that just might fail.

Discussion Questions

1. Explain the reason the word *change* elicits emotional reactions from many people. What phrases are often used to tone down the use of the word *change* in various social settings?
2. Differentiate between the concepts labeled *system overload* and *information overload*. Provide contemporary examples to illustrate the similarities and differences between them.
3. How does the social need for instant gratification erode the busyness syndrome? How does it also help to sustain it?

4. Why do you think the American public continues to retain its belief that science and technology make our lives healthier and easier? What types of situations might persuade people to disagree with this assertion? Why do you think so many people say they are interested in science and technology, but so few feel well informed about science and technology issues?

5. Identify and describe the six behavioral response modes to illustrate the ways we cope with change. Provide your own examples symbolizing each behavior.

6. Define what the term *technology trap* means. Provide several examples to illustrate its literal meaning, as well as its abstract connotation.

7. If you were required to prepare a report responding to the question "How safe is safe enough?" with regard to an outdoor genetic engineering experiment, what type of information and data would you attempt to secure? Who would you rely on for expert advice and recommendations during the course of this exercise?

Suggested Readings for Further Study

Allman, W. F. 1985. Staying alive in the 20th century. *Science 85* 6 (8): 31–41.

Andersen, K. 1986. Pop goes the culture. *Time* (16 June): 68–74.

Barach, J. A. 1984. Applying marketing principles to social causes. *Business Horizons* 27 (4): 65–69.

Begley, S. 2001. Protecting America: The top 10 priorities. *Newsweek* (5 November): 26–40.

Bjerklie, D. 2001. Diagnosing the risks. *Time* 158 (16): 42–43.

Brauer, D. G. 1984. Survival in the "Information Age." *Parks & Recreation* 19 (5): 58–59.

Brod, C. 1984. *Technostress: The human cost of the computer revolution.* Reading, Mass.: Addison-Wesley.

Brooks, D. 2001. Facing up to our fears. *Newsweek* (22 October): 66–69.

Cawkell, A. E. 1986. The real information society: Present situation and some forecasts. *Journal of Information Science* 12 (3): 87–95.

Colburn, D. 1986. "Chicken Little" is alive and well in America: As society becomes less and less risky, people worry more. *San Jose Mercury News* C-1 (24 June): 8.

Comarow, A. 2001. Protecting yourself. *U.S. News & World Report* 131 (17): 46–47.

Creekmore, C. R. 1985. Cities won't drive you crazy. *Psychology Today* 19 (l): 46–50, 52–53.

Drucker, P. F. 1986. Management: Learning to face new challenges. *Modern Office Technology* 3 (4): 14, 18, 24.

Easterbrook, G. 1986. Ideas move nations: How conservative think tanks have helped to transform the terms of political debate. *The Atlantic* 257 (l): 66–80.

Florman, S. C. 1981. Living with technology: Trade-offs in paradise. *Technology Review* 83 (8): 24–35.

Godet, M. 1986. From the technological mirage to the social breakthrough. *Futures* 18 (3): 369–375.

Golden, F. 2001. What's next? It could be smallpox, botulism or other equally deadly biological agents. *Time* 158 (20): 44–45.

Gonzalez, A., and P. Zimbardo. 1985. Time in perspective. *Psychology Today* 129 (3): 20–26.

Harris, M. J. 1987. The theory of the busy class. *Money* 16 (4): 202–206, 298, 210, 212, 214, 216, 218, 220.

Hayes, D. 1989. *Behind the silicon curtain: The seductions of work in a lonely era.* Boston: South End.

International Technology Education Association. 2000. *Standards for technological literacy: Content for the study of technology education.* Reston, Va.: ITEA.

[Internet]. (2001). Americans asked opinions on developments after September 11th terrorist attacks. *USA Today/CNN/Gallup Poll Results.* <http://www.usatoday.com/news/attack/2001/11/28/poll-results.htm> [29 Nov 2001].

Jackson, A., Jr. 1985. High tech's influence on our lives: Adapting to change. *Vital Speeches of the Day* (1 January): 164–166.

Jacobs, J. 1986. Techno-dummies are conserving mind and energy. *San Jose Mercury News* (21 February): B-7.

Kaplan, D. E. 2001. A new state of fear. *U.S. News & World Report* 131 (17): 14–18.

Lardner, J., D. LaGesse, and J. Rae-Dupree. 2001. Overwhelmed by tech. *U.S. News & World Report* 130 (2): 30–36.

Lee, M. D. 1986. The great balancing act. *Psychology Today* 20 (3): 48–50, 52, 54–56.

Lowrance, W. W. 1985. *Modern science and human values.* New York: Oxford Univ. Press.

McBee, S. 1984. Here come the baby boomers. *U. S. News & World Report* (5 November): 68–73.

———. 1985. The state of American values. *U.S. News & World Report* (9 December): 54–58.

McDonald, Thomas F. 1983. Technostress lurks inside every manager. *Data Management* 2 (9): 10–14.

Marks, S. 1984. High tech, high stress? *Datamation* (15 April): 97–98, 100.

Meyrowitz, J. 1985. *No sense of place: The impact of electronic media on social behavior.* New York: Oxford Univ. Press.

National Science Board. 2000. *Science & Engineering Indicators–2000.* Arlington, Va.: National Science Foundation.

Ornstein, S. 1986. When our technological umbilical cord snaps. *San Jose Mercury News* P-1 (February): 5.

Pascarella, P. 1984. The corporation steps in: Where family, church, and schools have failed. *Industry Week* (25 June): 34–35.

Perrow, C. 1984. *Normal accidents. Living with high-risk technologies.* New York: Basic.

Rogers, E. and J. Larsen. 1984. *Silicon Valley fever. Growth of high-technology culture.* New York: Basic.

Saviotti, P. P. 1986. Systems theory and technological change. *Futures* 18 (6): 773–786.

Schor, J. B. 1991. Workers of the world, unwind. *Technology Review* 94 (8): 24–32.

Slovic, P., ed. 2000. *The perception of risk.* London: Earthscan.

Taylor, T. 1986. All the conveniences of yesterday. *San Jose Mercury News* (27 January): B-7.

Toffler, A. 1970. *Future shock.* New York: Bantam.

Trafford, A. 1986. Living dangerously. *U.S. News & World Report* (19 May): 19–22.

U.S. still believes "Tomorrow will be better than today." 1984. *U.S. News & World Report* (31 December): 44–45.

Walsh, E. R. 1987. Workaholism: No life for the leisurelorn? *Parks & Recreation* 22 (l): 82–84, 103.

Wenk, E., Jr. 1986. *Trade-offs: Imperatives of choice in a high-tech world.* Baltimore: John Hopkins Univ. Press.

Wilson, R. 1979. Analyzing the daily risks of life. *Technology Review* 8 (4): 40–46.

Wylie, F. W. 1984. Change... the only constant: The ability to cope with change. *Vital Speeches of the Day* (15 March): 333–335.

Chapter 12

The Stability of Traditional Social Institutions

San Jose, California

A disturbing phenomenon is evident in this city and numerous other metropolitan centers—the disappearance of the traditional family dinner. Social scientists report that the days of gathering around a huge dining room table for meals on a somewhat regular basis is certainly at risk in this country. Instead of eating evening meals with members of our families, we are eating out, taking out, ordering in to be able to eat at our desks well into the evening hours, and munching on the run while driving on the interstate.

For decades, the evening family meal represented a time when children and parents could bridge the gap of their respective days. Eating at the table in the privacy of one's own home is one of the primary ways in which growing children learn about the conventional behaviors they will need to be in touch with as adults. Mealtime once was a time when parents taught values and social graces, and children learned to communicate and express their feelings. Preschool children in particular benefit from mealtime conversations in that their vocabulary skills can be enhanced quite significantly. The phrase *family dinner* has almost become a metaphor for a commitment to family.

Many of us seem to be forgoing this opportunity to teach and learn in order to have time for fitness classes, jogging, pick-up basketball, television, and dieting. Children themselves are torn from the dinner table to fulfill commitments related to their own after-school activities. It seems spending time together is becoming less important than being busy....

Introduction

Social institutions are the structural mainstay of civilization. Throughout history, human beings around the globe have recognized the need to establish a variety of social settings in order to most efficiently conduct the business of life. Beyond their ability to order living patterns, these social institutions provide a sense of hope and security, as well as a means for survival. People have a tendency to associate specific roles and responsibilities with each of the social institutions existing in their communities. Formal institutions provide the order and organization for a culture to sustain itself. Cultural anthropologists conduct systematic and scientific studies of the typical behavior patterns of people residing in each of the world's continents. A contemporary concern among these researchers is the extent to which modern technology may be influencing the stability and shape of traditional cultural morals, values, and customs.

It is always difficult to formulate a precise definition for a generic concept. Since *social institutions* are the primary focus of this chapter's discussion, it would be very helpful to understand what they are. In all probability, most of you already have a clear sense of what this term means. You could probably provide a list of examples of these institutions evident in your hometown community quite easily. In fact, before reading on any further, take a moment to write down a few of the social institutions you believe to be currently operative in your life right now. For the purpose of our discussion, a social institution refers to a formal entity existing primarily to nurture and contain certain behavior patterns considered to be important in society. If one chooses to use either the word *traditional* or *contemporary* as a descriptive modifier, this explanation should still remain adequate.

Let's look at some of the social institutions we depend on to structure the manner in which we conduct our daily routines. If the list you just compiled includes such items as family, church, school, and government, you have identified some of the most commonly accepted social institutions throughout the world. These formal groups are readily viewed and categorized as "traditional," since they have been so widely respected for centuries. Some cultures revere their sanctity and believe in the importance of their preservation for the good of all humankind.

In recent years, a series of equally formal, but far less conventional, groups have evolved in our society. If your list identifies health clubs, corporate offices, research laboratories, golf leagues, and neighborhood taverns as social institutions, your perspective is significantly more modern. Although none of these entities has earned the level of recognition afforded to family, religion, and education, they are just as important to many career-minded individuals in today's social milieu. In the eyes of some researchers, the corporate office has achieved a higher position on the list of personal priorities than the family living room. Similarly, in many cases, individualized exercise programs have become a substitute for community involvement and civic duties. See Figure 12-1. Computer software advancements and the availability of on-line courses at both the university and high school levels continue to threaten the integrity of real live teachers. Computers seem to be influencing our religious beliefs. The corporate gymnasium is often comparable to the corporate boardroom as a setting where powerful political decisions are often heatedly debated.

Are the traditional social institutions we have relied on for years being replaced with a host of alternatives? Furthermore, is modern technology having an adverse effect on their stability? Some researchers respond affirmatively to both inquiries. They seem convinced technology is the driving force behind the introduction of these new colored threads being woven into the tapestry we call our society. This chapter provides a synthesis of some of their opinions and theories as they relate to both traditional and contemporary social settings. Although it is not possible to provide a thorough analysis of all the institutions mentioned so far, the following areas will be discussed in some detail: family life, religious attitudes, educational programs, and recreational activities. These four topics are selected to depict a cross section of "contemporary" social activities and the persistence of "traditional" social institutions in a world of technological change. The following questions will be addressed in this chapter:

- ❏ Is the traditional nuclear family becoming an endangered species in present-day society? What sorts of social groups do you think will come together to act as surrogate families?

- ❏ How has technology affected the way women structure their lives? Further, what sorts of influence has modern technology had on the socialization of the children of these women?

- ❏ What impact have science and technology had on the importance of religion in America? Are computers changing our image of God?

Figure 12-1.
Group fitness classes coordinated to popular music have become more common across the nation since the mid-1980s. Individuals who pursue this type of exercise program recognize the benefits of a complete body workout and cardiovascular health. Some health clubs offer these classes throughout the day, but the evening classes enjoy the greatest attendance among the membership.

❑ How have advancements in distance education technologies begun to affect the ways educators and school administrators structure teaching and learning communities?

❑ In what ways has technological change affected the manner in which we spend our leisure time? What sorts of high-tech developments have changed the profile of the recreation industry?

Key Concepts

As you study the ideas presented in this chapter, try to develop a persuasive definition for each of the following concepts as it relates to a cultural analysis of technology in society.

> creationism
> distance education
> expert system
> nuclear family
> parental leave
> social institution
> zapping
> zipping

The Family: Technology and Personal Relationships

In today's fast-paced world of technological and social change, it is unlikely that any other social institution elicits as much contentious debate as the family. One side argues that the movement away from marriage and traditional gender roles has seriously degraded family life across the board. On the other side of the field, the more optimistic group views family life as dynamic, resilient, and generally adaptive to new social circumstances.

The term *family* carries cultural meaning and has individualized significance for most people. For statistical purposes, the Bureau of the Census defines a *family* as two or more people living together who are related by blood, marriage, or adoption. At the onset of the twenty-first century, families maintain most households in the United States, but the percentage has declined since 1960. The U.S. Census Bureau further defines a *household* to be one or more people who occupy a house, apartment, or other residential unit that is not "group quarters" (i.e., dormitory). In 1960, families, as previously defined, occupied 85 percent of U.S. households. By 2000, that proportion had decreased to 69 percent. In the same time frame, the proportion of households run by just one person doubled from 13 percent to 26 percent.

For those satisfied with being single, the task of establishing friendships and alternative relationships (sometimes called surrogate families) is evident in contemporary social settings. Varying levels of intimacy are evident within these personal involvements. Men and women have expressed various forms of reverence for the institutions of marriage and the family throughout most of history. These age-old social institutions have been applauded and often referred to as the lifeblood of society. With specific regard for the family, we believe this institution will continue to adjust to dramatic social demands.

As we discussed in Chapter 11, some people are resistant to change, but remain willing to propose modifications to it to adapt to changing social circumstances. Economists commonly refer to the family as the fundamental unit of society. Some economists also claim to be able to assess the well being of the whole nation on the basis of what they perceive as the family's state of health. The traditionally accepted functions served by the family unit in civilization include the following:

- ❏ It is a provider of love and emotional support for its members.
- ❏ It is the principal agent of socialization for children.
- ❏ It is a primary source of economic support for children, often throughout their post-secondary school years and into early adulthood.

We have most certainly broadened our definition of family in this nation amidst great shifts in demographics and accepted lifestyles. The profile of the contemporary family unit is quite a bit different than it was in the 1960s, or even the 1970s. The long-admired traditional image of the *nuclear family,* consisting of a working husband and his stay-at-home wife with two or more dependent children, now accounts for less than 25 percent of all U.S. households according to Census 2000 data. Incredibly, however, this antiquated nuclear ideal persists in many well-publicized advertisements. For example, in a recent national campaign, Travelodge featured a little girl clutching a stuffed animal with her mother and father on either side. Chances are, suggestions to include grandparents or other adult figures in this ad were discounted in favor of tradition.

The look and feel of household arrangements continue to evolve with changes in religious, social, and legal directives. A significant change in the late twentieth century was the increase in the number of unmarried men and women living together. Also on the rise are the numbers of gay and lesbian households and multigenerational arrangements. Cohabitation and marriage have a lot in common, but they differ in critical ways. Marriage is a social institution resting upon common values and shared cultural expectations regarding behavior. It adheres to legal, moral, and social rules. Cohabitation often serves as a precursor for marriage, which we believe is a social institution that will prevail in the coming

decades. Cohabitation does not boast a widely accepted social blueprint to guide appropriate behavior. If it ultimately helps couples strengthen their commitments to one another before marriage, however, cohabitation may gain higher approval ratings in the years ahead.

Individuals' reasons for marriage in the late 1990s were quite similar to those mentioned in the past: love, commitment to growth as a couple, security, social approval, and a desire for greater intimacy. Companionship represents a slightly revised variation on the older themes; both men and women have found they can be self-sufficient, but still have a desire for the closeness of a committed relationship. Other contemporary reasons for marriage generally expressed by both sexes, are a growing disillusionment with the "singles' rat race" (short-term and often promiscuous dating with more than one person) and concern for physical health. Sexually or physically transmitted diseases, especially acquired immunodeficiency syndrome (AIDS), have had an unequivocal impact on the interpersonal relationships between men and women. "Casual sex" is becoming less common in the early stages of a couple's interaction. Also, the frenetic pace of work in a technologically progressive society does not leave most people with a whole lot of energy or ingenuity to go out and search for companionship at the end of the day. Because of this, the comfort and quietude of a marriage often seem quite appealing.

A previous perception of marriage commonly viewed the family as a business enterprise of sorts; the husband earned money and provided for the woman. In return, the wife agreed to do the housework and bear children. People who decide to get married these days are bringing a whole handful of new expectations into the marital contract. Prenuptial agreements have become much more common in recent years. Today's partners seem to be more aware of the need for their marriages to involve other dimensions in order to be both fulfilling and successful. Women expect to be included in financial matters. Men expect to have direct involvement in the nurturing of their infant children. Not all people expect their marriages to be permanent, but most enter into the union intent on making it work.

Despite the changes taking place in family living arrangements, shared activities are still considered to be very important. Even though the family dinner may not be scheduled every night of the week, this activity is generally regarded as a significant, central time for all members (see *Dateline*). There is a move toward emphasizing the quality of time spent with family to compensate for the reduced quantity. In other words, the family dinner table may not be on its way out, but actually experiencing a rebirth in some households. Families who consistently plan to have dinner together at least once or twice a week may find it to be a genuinely rewarding endeavor.

Without question, there has been a social revolution in family lifestyles over the past four decades. The extent to which technological changes have been a driving force behind this revolution is still somewhat vague. There are those who suggest that industrialization and technology have torn the family unit apart. Others predict that the advent of telecommuting as a work mode will draw families closer together. Some social observers insist this new method of going to work is straining traditional family roles and causing household conflicts. Continued advancements in household technology will stimulate further evaluation of men's and women's roles on the home front. For some people, life in today's technological world is too complicated, often scary, and unimaginable. This may ultimately cause contemporary families to pull closer

together to make it through the hard times. Undeniably, the essence of family will survive as an important source of stability well into the future.

Modern Childhood

Career-focused adults spend a significant percentage of their days attending to work-related issues; therefore, it is not uncommon for them to lament about how difficult it is to be a parent. Instead of worrying about polio vaccines, pneumonia, and ear infections, today's parents find themselves worrying about drugs, school yard violence, teenage pregnancy, and threats of worldwide terrorism. If the truth were told, their children are probably having just as much difficulty trying to be kids! In some ways, there appears to be an overall merging of childhood and adulthood happening in our culture. Children growing up now are undergoing a host of psychological and social pressures. Too many children are being forced to grow up too quickly.

During the first half of the twentieth century, our culture revered childhood. The days of youth lasted into the teen years and were generally considered to be a time of innocence, amusement, and isolation. It was a time for young people to be sheltered from the demands and alarming realities of adult life. Kids were dressed differently than adults, and each group had its own language. Certain topics like birth, death, sex, and money problems were treated as off-limits for children's ears. Schools were structured within a strict age-grading system designating what a child of a given age should know and be able to do.

As the shape of family units has undergone change, the last five decades have also witnessed a remarkable change in the image and characteristics of children. Childhood is no longer viewed as a protected and sheltered period of life, and it seems to end before the chronological period called adolescence (often by age ten). Today's children seem less childlike in a number of ways. They speak, dress, and behave more like adults than they did a couple of decades ago. The so-called "magic years of youth" do not always seem to be so magical anymore. Children are trying to cope with a world in which both parents are working, sex and drug abuse have permeated the elementary school yard, and graphic violence is commonplace both on the television screen and in video games. See Figure 12-2. Mimicking adult behaviors, more children than ever

Figure 12-2.
Numerous video game issues have been in the public eye over the years with respect to their addictive nature. Proponents of this form of youth activity suggest video games can stimulate attention, exercise memory and reaction time, and develop fine motor skills, especially hand-eye coordination. Less enthusiastic are individuals who believe video game addiction among youths causes them to skip school and waste a significant amount of money on the machines. Video games seem to mesmerize youth (and some adults as well). The extent to which any components of this techno-toyland are applicable to learning has yet to become an accepted fact among contemporary educators. (Jack Klasey)

before experiment with alcohol and other drugs at an earlier age, sometimes under ten years old. Eating disorders and depression are also growing among preadolescents and early adolescents.

Their overly status-conscious parents are inadvertently inflicting some of the stresses contemporary kids are confronting. Many of the pressures on children are directly related to America's new and novel lifestyles. The movement of women into the workplace created a need for a new assortment of childcare arrangements. In the mid-1960s, less than 10 percent of preschool children of working mothers were in a childcare center. By the mid-1990s, 29 percent of children under the age of five were registered in organized childcare centers across our nation. Child psychologists seem to be in agreement that sending children who are under three years old to day care is more stressful for the kids than staying at home with a parent or a familiar babysitter. The child must comply with the environment, and the daily program does not always respond to every child's individual needs or expectations.

There seems to be a growing concern that overworked parents will most likely beget harried children. Today's children have a higher degree of concern about time and meeting deadlines than those raised in an earlier, less hectic period. Unlike previous generations, mom and dad's workload commonly dictate many children's daily routines. Their schedules for meals, bedtime, television privileges, and baths are no longer predictable. Even the traditional summer vacation period has become incredibly structured, including many types of organized programs taking kids away from the home. It is unlikely the intermittent parcels of substitute "quality time" parents devote to their young ones can adequately compensate for a lack of "quantity time."

Modern childhood is further plagued with a much greater amount of fear. Kids now face a dangerous world. The milk carton on the breakfast table is plastered with pictures of missing children. The evening news is filled with stories of child abductions. It is no longer clear what topics should or should not be discussed with children. In any event, children now seem to know about previously forbidden topics well before they are spoken about in their homes or schools. Birth control, abortion, alcoholism, and suicide have all become children's issues in recent years.

Young children today also seem to be expressing their fears of being at risk of nuclear or other terrorist attacks. Discussions centering on this topic commonly evoke feelings of confusion and helplessness. Adolescents seem to know there is something vastly dangerous and threatening in their world; however, very few have any real understanding of the technological information about that threat. One wonders how many of the problems today's youths are facing are actually tied to their underlying fears about weapons of mass global destruction. Consciously or unconsciously, children know their future lives can be obliterated at any moment. Perhaps one reason today's kids are having difficulty working toward long-term goals is due to their lack of confidence in their own futures. It may be true the age of innocence is shorter than ever before, but all hope is not lost.

Parents and educators alike can do a lot to minimize stress on kids and reduce their fears about technology and violence. Children need to be raised with a strong realization that there are many problems in the neighborhood, the nation, and the world. They are growing up with the capabilities to help solve those problems, however. Instead of dealing with fear, adolescent students need to begin experiencing a sense of their own power as citizens with a responsibility to help eradicate serious social dilemmas.

Parents need to communicate with their children and learn to respect each child's individuality. The recent emphasis on fierce competition and getting ahead has typically gotten out of hand. Preadolescents should not be pushed so much that they lose sight of the importance of cooperation, helpfulness, and good old-fashioned love in their daily adventures. We may eventually come full circle as a result of the problems our young people are encountering. Parents may become more inclined to spend less time (when possible) on work and more time with their children, and the traditional family unit will find itself back in style.

Women's Roles: Change versus Constancy

Any discussion attempting to evaluate the stability of the family unit in our society would be considered deficient if it neglected to address the changing roles and responsibilities among women. Since the women's movements gained national momentum during the early to mid-twentieth century, numerous essayists, historians, and social scientists have published on a wide assortment of topics. Some universities also offer baccalaureate degree programs bearing the title "Women's Studies." Several prominent areas of discourse include women's rights; equality in the workplace; entrance into the labor force; recognition in the nontraditional fields of science, engineering, and technology; the development of household technologies; and use of reproduction technologies. If one makes a diligent effort to sift through all of the material, one very general conclusion can be derived. While women's roles have changed dramatically during the past several decades, many roles have also remained constant.

Household Technologies

In some circles, there is a commonly held belief that technology developed for the home can free women to participate in other non-housework tasks. This might be true if the technology could actually change the way in which household maintenance tasks are divided among the family members who reside there. Regardless of whether or not they hold a wage-earning position, the most common practice in North American homes today is still for women to do the largest proportion of these tasks.

Utilities such as running water, electricity, gas, telephone, garbage service, and wastewater with sewage treatment plants represent the technological infrastructure in the home. These technologies probably changed the nature of household work more than other technical improvements, since they eliminated a series of burdensome tasks. On another level though, utilities such as gas and electricity may have fostered other technological changes by facilitating the introduction of many in-house appliances of all shapes and sizes.

Amazing as it might sound, some studies discovered that some of these so-called "laborsaving" devices actually make tasks more difficult instead of easier! Cleaning some electrical kitchen appliances often takes more time than completing the task without the device. Two examples are the food processor and the electric can opener. See Figure 12-3. Appliance repair in one's home has become more difficult and mysterious. The service charge to fix some appliances often costs as much as replacing them with new models. As household machinery becomes more specialized, the difficulty of understanding either the mechanical or the electrical operation has increased. Further, it often appears this computerized and electrified equipment is designed with planned obsolescence in order to sustain demand for new sales.

Figure 12-3.
Food processors were introduced into the American kitchen over three decades ago as an exciting laborsaving technology. The above photo depicts a typical model consisting of a base, housing the motor; a work bowl sitting on top of the base; and a series of discs and blades designed to chop, slice, grate, puree, shred, grind, and knead a wide assortment of foods. Since all of the individual components must be cleaned separately, its laborsaving profile is best observed when large quantities of ingredients must be prepared. (Michael Dafferner)

While technological innovations may have lessened the physical effort of housework, they have not significantly reduced the amount of time involved in keeping up with it. The presence of technological gadgets and devices in the home has not lowered the psychic burdens on women, nor has the presence of technology changed the allocation of labor by gender. In fact, the greatest impacts on the manner in which contemporary housework is conducted emanate from nontechnological changes. Specifically, the pressures of inflation, along with the desire for additional goods and services, have drawn women into wage-earning positions.

Households are smaller, and women are becoming less anxious about having their homes spotless. According to one report, during the thirty-year period from 1965 to 1995, the total number of hours per week women spent doing housework tasks decreased from 30 to 17.5. Even still, women were doing twice as much of it as their male partners.[1] More recently, research completed at the University of Michigan concluded that, on average, men spend 16 hours each week on housework (up from 12 per week in 1965), and women spend 27 hours doing housework tasks (down from 40 per week in 1965). Regardless of which study one references, family activities seem to be winning over the need to get all of the housework done.

Reproduction Technologies

Birth control technology has played a critical role in stimulating and fueling women's movements throughout the twentieth century. It will most likely continue to do so into the twenty-first century. The early 1900s changed the primary role of women as mothers, with the help of barrier methods of birth control, and culminated with the passage of suffrage for women (1920). The later introduction and dissemination of the pill and a variety of intrauterine devices (IUDs) further increased the ability of women to control procreation. These methods were more effective and much more convenient to use than previous solutions. More so than technological changes in the home, birth control developments led to the greatly increased rate of labor force participation among women.

Improved, safer, and more reliable birth control technologies will continue to increase the amount of control women have over their own bodies. Women can now better control both the total number and timing of the children they bear. Birth control has

been a catalyst for a number of social changes, making the quest for long-term, intimate relationships for both males and females somewhat easier. On the other hand, many of the changes associated with their searches have made the maintenance of long-term relationships much more difficult. Even though birth control technology has altered and expanded the number and types of relationships available to men and women, most people still seem to desire a monogamous, long-term commitment at some point.

Reproduction technologies have increased the control women have over childbearing. For those couples wishing to have children, modern birth control usually makes it possible to achieve the desired number and timing of children. It enables the woman to simultaneously pursue other career objectives. The ability to delay childbearing via birth control devices, however, does not always result in the desired ideal family. Some couples who postpone childbirth decisions find themselves experiencing fertility problems. In yet another response to a social trend, new technologies to correct infertility are being tested and are becoming increasingly available.

The more important dilemma of who will rear the children remains. There is no technological fix for this responsibility. Not all duties associated with childcare can be purchased, regardless of how affluent the couple is. Kids are not robots, and they require a lot of parental attention and interaction for their mental and emotional stabilities. In the twenty-year period between 1977 and 1997, the average workweek among salaried Americans working full-time increased from forty-three to forty-seven hours. On average, in the late 1990s, employed Americans worked an astonishing eight weeks more per year than their Western European counterparts. In Norway and Sweden, it is common for

---

[1]Bianchi, S. M. and L. M. Casper. 2000. American families. *Population Bulletin* 55 (4): 33.

workers to receive four to six weeks of vacation and up to a full year of paid *parental leave.* In other European countries, thirty-five hour maximum workweeks are enforced, partially to reduce unemployment. If male and female Americans alike continue to sustain this love affair with work, questions about childrearing will remain on the table (most likely not the family dinner table!).

The Church: Technology and Religion

For centuries, people from all ethnic and educational backgrounds, as well as all socioeconomic levels, have looked (and continue to look) to their religious institutions for comfort and guidance during times of sorrow or confusion. Aspects of modern technology, however, have caused some people to change their traditional views of church, their definitions of God, and other beliefs. As modern technology and science continue to provide cures for previously incurable illnesses, some may feel they no longer need a Supreme Being who is all knowing and everlasting.

This reason and various others have led some sociologists to suggest that modern technology functions as a type of religion for some who have abandoned belief in God. At the same time, there are religious leaders who believe modern technology is causing a resurgence of humanity's quest for religion. There are no simple explanations for why the proliferation of scientific fact and technological know-how would lead one person to question the existence of God, while it would lead another to even greater belief in a deity. In a national study commissioned by the National Science and Technology Medals Foundation in the mid-1990s, two-thirds of the adults polled in the

United States said that new developments in science and technology did not have much to do with their personal religious beliefs one way or the other.[2]

Many of the prevailing religious and ethical issues of the day are rooted in the world of scientific experimentation and technological innovation. The dilemmas of euthanasia, the wonderment associated with genetic engineering, the exciting new frontiers in stem cell research, the uncharted territory related to cloning, and the medical technologies of procreation are all legacies of a contemporary techno-scientific culture.

Vatican officials delivered one noteworthy response to several of these and other contributions offered by technologists in the late 1980s. In a document titled "Instruction on Respect for Human Life in Its Origin and on the Dignity of Procreation—Replies to Certain Questions of the Day," the Vatican instructs governments around the world to ban artificial procreation, including sperm banks, embryo banks, postmortem insemination, and surrogate motherhood. It also recommends laws against embryo research and abortion. These strictures represent one of its most expansive public statements since the encyclical on birth control released in the late 1960s. See Figure 12-4.

These concerns about the impact of technology on procreation stem from the Catholic Church's view that human life is sacred from the instant of conception. A belief in that idea commands further opposition to in vitro procedures producing spare embryos to be destroyed or used later for research. Similarly, prenatal testing, sometimes viewed as genetic counseling, giving parents the chance to abort an abnormal fetus must be considered forbidden. The Vatican insists any technology separating the act of conception from the conjugal act is morally wrong. It is difficult to fully describe widespread public reaction to this ecumenical treatise. As with

[2]Ferree, G. D. 1996. American views on science and technology. *The Public Perspective* 7 (6): 60–63.

Procreation Policy Established by Vatican Congregation for the Doctrine of the Faith

Highlights of the Vatican Document

The Vatican document on artificial procreation expresses moral opposition to the following practices:

- Prenatal diagnosis, including the use of amniocentesis and ultrasonic techniques, for the purpose of eliminating fetuses that are affected by malformations or that are carriers of hereditary illnesses.
- Surrogate motherhood.
- Any artificial fertilization involving persons who are not married.
- Test-tube fertilization and embryo transfer involving a married couple.
- The artificial insemination of an unmarried woman or of a widow even if the sperm was donated by her husband.
- The collection of sperm through masturbation.
- Therapeutic interventions and methods of observation of an embryo that involve disproportionate risk to its health.
- All experimentation on living embryos not directly therapeutic.
- The keeping alive of human embryos for experimental or commercial purposes.
- Utilization of dead fetuses for commercial purposes and unnecessary mutilation of dead human embryos and fetuses.
- The voluntary destruction of human embryos obtained in vitro either for purposes of research or of procreation.
- Biological or genetic manipulation of embryos through practices like fertilization between human and animal gametes and gestation in the uterus of animals or an artificial womb.
- Efforts to obtain a human being without any connection with sexuality through methods like "twin fission," cloning, or parthenogenesis.
- The freezing of an embryo even when done to preserve its life.
- Efforts to influence chromosomic or genetic inheritance that are not therapeutic but which aim at producing human beings selected according to sex or other predetermined qualities.

The document expresses moral approval of the following practices:

- Prenatal diagnosis, including the use of ultrasonic techniques and amniocentesis, that aims at safeguarding or healing the embryo.
- Prenatal therapeutic efforts designed to heal an embryo of maladies including chromosomal defects.
- Prenatal research limited to the simple observation of an embryo.
- The use of experimental therapy on an embryo as a final effort to save its life.
- Artificial insemination within marriage when the technical means is not a substitute for the conjugal act but serves to help it achieve procreation.
- Fertility drugs.
- Medical interventions that facilitate the performance of the conjugal act.
- Medical intervention to remedy the causes of infertility.

Excerpted from the Vatican document entitled "Instruction on Respect for Human Life in its Origin and on the Dignity of Procreation—Replies to Certain Questions of the Day" (San Jose Mercury News)

Figure 12-4.
This chart summarizes highlights of the Vatican statement regarding procreation and technology. It was released in the late 1980s.

previous discussions regarding science, technology, and religion, the initial discourse may seem rhetorical. In other words, various groups will make elaborate attempts to describe a complex social issue eluding specific characterization.

In its condemnation of artificial conception, the Vatican document is heavily reliant upon its earlier restrictions against contraception. In the same way that contraceptives separate sexual union from procreation, the church argues in vitro fertilization techniques separate procreation from sexual union. Many Catholics do not agree with their church's opposition to birth control. It follows that many have been inclined to question this directive. In fact, some theological observers believe the Vatican's response to the technology of fertility may cause its present congregations to disregard some of its other teachings. On the other hand, society may ultimately view this document as a timely warning against a growing scientific indifference to the origin and vulnerability of human life.

The Debate between Science and Religion

It seems that whenever the word *Christianity* surfaces in the scientific press, its readers tend to prepare themselves for yet another round of a **creationism** debate. Stated differently, spectators will expect the fundamentalist Christians to insist upon the validity of Genesis and refute any scientific diatribes concerning the evolution of the human species. Indeed, there are a number of variations on the creationist theme, and each group is trying to reconcile Genesis with science in some particular manner. There are also many Christian scientists who regard the creationism debate as irrelevant to the essence of contemporary science policy. Their basic tenet is that faith needs science

and science needs a firm grounding in western religious heritage. Both parties seem to have a great deal to gain from one another if the primary objective is a deeper humanization of science, technology, and society.

Today's scientists should be reluctant to live out their scientific lives in a research laboratory devoid of any spiritual foundation. Likewise, religious leaders who aspire to have a voice in determining the future uses of technical information and technology in general should have at least a rudimentary understanding of the scientific method. The core of the debate between science and religion rests on the following precepts:

❑ Scientists believe religion skews objective reasoning, fuels repressive movements, and stifles freedom of thought.

❑ Religious leaders are convinced science pursued from an agnostic or atheistic base will continue to feed the ever-growing materialism in society and contribute to the violence of the modern age.

Fortunately, there seems to be an emergent group of humanists who support the goals and convictions of both camps. According to a recent poll sponsored by People for the American Way Foundation, 83 percent of Americans generally support teaching evolution in public schools, and about 79 percent believe creationism also has a place in the curriculum. Essentially, respondents say evolution should be taught as scientific theory, and creationism should be discussed as a religious theory. Although people may differentiate science from religion as two distinctly separate areas of human life, they also identify ways in which the two share common points of contact in our society.

One example of a commonality is related to mankind's relentless quest for understanding and intelligibility. Many believe the essence of science is experimentation, which produces data observable to the senses. Empirical methods of research usually yield empirical data, but the true objective of scientific inquiry is the attainment of intelligibility, not the generation of data. Scientists search for patterns isolating relationships between one bit of sensory data and another. These patterns constitute an understanding of reality, rather than an unrefined list of facts and figures.

Understanding science as the search for intelligibility suggests science and religion share the same ultimate goal. Namely, they both envision the capacity to provide tangible meaning to the world and human existence. Theologians do not use empirical methods with a great amount of scientific rigor, but they do seek to make our unique human experiences significantly more lucid. Journalists once asked Albert Einstein what he thought about the existence of God. He replied he thought it obvious there was an intelligent force in nature that put the universe in organized motion. He had no problem with science and God residing in the same space.

Both religion and science provide ways of obtaining intelligibility in contemporary society. Neither institution can speak with absolute precision. Both are constantly trying to use language to demystify realities often beyond the capacity of language to explain. Many of us look to religion for personal guidance, support, a sense of belonging, and hope for the future. At the same time, we look to science and technology for an enhanced quality of life, a sense of progress, and a hope for the future. Religion does not try to explain nature the way science does. Science, with all of its limitations, cannot give intelligibility to all of what we experience as human beings.

Computers and God's Images

The traditional view of God deeply integrated into our culture is inherited from Judeo-Christianity. Specifically, God is out there, enduring beyond time and space in a state of permanency, and not susceptible to change. God is the creator of all that exists and is the unifying force allowing the continuation of life as we know it. In addition, God determines and represents the nucleus of life after death.

Social scientists speculate as to whether computer technology will have a far-reaching impact on these traditional images we have of God as our creator who is all knowing and all-powerful. For instance, computers have become integrally involved with the medical field. They appear to be changing and improving the manner in which we are able to procreate human life. Is the view of God, as the one who creates, nurtures, and sustains human life, rendered less precise, when humans and their machines are allowed to interfere with the process?

On another front, through the aid of computers, the general population is able to tap into vast resources of knowledge from their own homes and offices. The average citizen living in the industrialized world of the early twenty-first century has access to an immense bank of information. Precomputer scholars were known to spend their entire academic careers in the pursuit of wisdom now available to the common man. Perhaps an omniscient God does not seem as awesome to the individual who can find out a great deal through a directed, computerized search of the literature. Instant communication on a global scale via computer technology might also change our perceptions of omnipresence. Corporate executives in Chicago who are engaged in a real time videoconference with engineers in Bangkok may feel they are, in some ways, approaching being everywhere at once.

Further, the field of artificial intelligence (AI) has developed such refined ways of emulating human reasoning that it may not always be possible to distinguish between humans and machines. We are bombarded with interactive programs to diagnose diseases and make suggestions regarding ways we can improve our sense of self-esteem. Will people who have difficulty differentiating between minds and machines have just as much trouble seeing the difference between humans and God? Avowedly, this last inquiry is rhetorical, but intriguing just the same.

Computers may truly be changing our traditional views of God. To suggest computers have the power to either replace God or diminish a person's religious beliefs seems fatuous. In fact, astute cognitive scientists would probably be the group least likely to make this sort of testimonial. They are quite in touch with the reality that a modern computer capable of processing the amount of information stored in the human brain would need to be thousands of times larger in volume than the human brain. From this vantage point, computer technology provides a new mode of comparison without affecting our traditional images of God. Computers are our newest and most sophisticated tools to make further attempts to comprehend the true meaning of life and human existence.

The School: Technology and Education

Computers and education have become inextricably linked. Parents worry about whether their children are computer literate, educators fret about tracking the latest trends in technology for the classroom, and students are anxious to gain experience with computers in order to launch a successful career.

The days of the little red schoolhouse constructed on the prairies of this great nation seem like ancient history as we enter the first decade of the twenty-first century. Nevertheless, the stalwart school building has retained its role as one of the most influential social institutions we encounter throughout childhood, adolescence, and early adulthood. The profile of the curriculum and the salient features of the building and its classrooms are different, but the primary objective is alive and well. However comforting that might sound, there is something a bit scary and slightly unsettling about this statement. Think for a moment about all you have learned about change as you have read through the other chapters of this textbook. Undoubtedly, a physician trained in the 1850s would be terrified if asked to perform a surgical procedure in a modern hospital operating room. The average American farmer of the 1920s would hardly recognize anything besides the cows in a twenty-first century technology-dependent dairy farm. Astronomers who studied the skies in the sixteenth century would be clueless during a tour of a contemporary space observatory, but a seventeenth century school teacher would most likely feel right at home in one of your current university classrooms.

Schools exist to educate and socialize the residents of a community, state, nation, or continent. One of the most significant goals of education is to challenge learners to think critically about issues, generate new solutions to both old and new problems, and develop top-notch written and oral communication skills. Schools house the industry we call *education*. Their productivity and quality are measured via the success of the students who traverse their hallways and learning centers. Technological advancement is certainly challenging the face of this traditional social institution, but is its stability also being threatened? How are the fruits of our technology impacting educational delivery systems in an ever-changing world?

Educational commissions, task forces, and ad hoc committees regularly issue reports warning us about the demise of our nation's educational system. These editorials indicate that the quality, value, and relevance of a major segment of our educational services have been declining over the past several years. Projections of this nature have made it clear this country's future leaders will be at risk unless more attention is devoted to our schools. Education has never been more essential to the survival of humankind than it is today.

A major responsibility of schools is to prepare students to enter a rapidly changing and constantly evolving job market and political environment. Schools are expected to mold students who are adaptable and able to respond quickly to the changing requirements of new technologies. Gone forever are the days when a high school education could get a person through life pretty well. In our society, it is expected that workers' job descriptions and task definitions will change significantly every four to eight years. This means our schools will train both youths and adults. Adult workers will require additional education and retraining each time business and industrial operations are updated. Schools will continue to be commissioned to provide a greater number of services to the local community, to area businesses, and to very young students. Educational institutions may become the primary mechanism to mitigate skills and knowledge obsolescence in a society where lifelong learning is the accepted norm.

Technology in the Classroom

Education is one of just a handful of major American industries yet to be thoroughly computerized. Its turn is on the visible horizon as computer hardware manufacturers and software developers have already identified education as a huge gold mine. An intriguing assortment of high-tech wares already exists for classroom teachers. Products in the form of computers, software packages, databases, interactive electronic media, and far-reaching telecommunications linkages all cost money. Obtaining classroom equipment is certainly an integral part of curriculum development and modification, but it should not be the first step. The two most essential questions prospective buyers should ask are: Will the new technology improve productivity and quality in teaching and learning? and Is the technology relevant to specific instructional goals?

As we enter the twenty-first century, nearly all full-time regular classroom public school teachers in this nation have access to computers or the Internet (usually both) in their schools. Although a majority of them use technology for classroom instruction, a few years ago many said they were not as well prepared for the task as they would have liked. The National Center for Education Statistics (NCES) reported recently, and not surprisingly, that teachers who have more than thirty-two hours of professional development related to instructional technology are twice as likely to use computers effectively than their colleagues who have not been trained. In 2000, the U.S. Department of Education reported most classrooms still do not have telephones installed in them, but 63 percent are connected to the Internet. This is more than twenty times the number wired in the mid-1990s.

Education Week, in a joint project with Public Agenda, found that more students today report using computers at school on a regular basis, and they are using them for serious learning.[3] In contrast to earlier findings by the NCES, most teachers interviewed here said they receive the technological support and equipment they need, and they felt comfortable using computers in the classroom.

---

[3]Public agenda reality check 2001. 2001. *Education Week* 20 (23): S1–S8.

Not only are students' and instructors' use of technology increasing during the school day, but the ways computers are being used are also more sophisticated. In this same study, 63 percent of the kindergarten through high school students questioned stated they used computers in school to gain assistance in learning new things, as opposed to using them just for typing or playing games; this represented an increase of 21 percent in just a bit over two years. Many students use the Internet to do schoolwork or homework several times a week. Computer-aided instruction (CAI) has forever changed the way teaching and learning activities unfold both inside and outside the walls of today's school buildings.

Reduced costs, improved performance, and access to incredible amounts of information are among the most compelling reasons for increasing the use of computers in instruction. In CAI, computers are used as supplements to conventional teacher-led instruction and can be used over a wide range of subjects. The price of computers has fallen dramatically. Along the way, a tremendous surge in the development of educational software has occurred. Complementing the earlier drill-and-practice programs designed to supplement the teacher's presentation are a wide variety of software packages. See Figure 12-5. These instructional aids consist of tool software, problem software, and simulation software designed for use in courses

Figure 12-5.
The presence of networked computers in elementary and secondary classrooms is commonplace today. These students are working together.

such as programming, logic, music, science, social studies, language arts, graphics, and mathematics. Computer programming specialists have demonstrated their adeptness at devising expert systems for use in our schools. This concept finds its origin in the field of AI. An ***expert system*** for educational use is a computer program condensing a huge amount of accumulated knowledge on a given subject into a form of question-and-answer routine to lead the learner through a logical progression to the correct response.

While software improvements are being made regularly, some professionals believe the resultant learning modules will soon replace some kinds of textbooks. Software can be tailored to meet individual student needs and updated more efficiently and inexpensively than textbooks. Computers can also be linked with videos to provide sight, sound, and movement. These features promise to provide a new dimension to lessons in subjects such as history, art, and political science. The one drawback to this form of delivery system is that younger students may have a tendency to view these lessons as they would a television program. In the absence of direction and instructional cues, less studious learners may not absorb a lot of the information presented.

As computers and educational software become increasingly available in contemporary classrooms, teachers are finding it necessary to figure out how to best manage these resources. The first step toward effective management is to view the computer as a tool. The acronym *CAI* itself implies instruction is *aided* rather than *crafted* by the computer. Educators who approach CAI with specific problems in mind and then ask how the computer can be used as a tool to solve them are more likely to synthesize sensible and successful applications for their computers. For instance, it may not be necessary to

house a large number of computers in the classroom. Strategic planning may reveal a ratio of one computer station per five or ten students is adequate. In selecting software for the class, the teacher must be careful to identify those units that will ultimately promote the effective use of the computer as a tool to meet his or her curricular objectives. It is especially critical to develop a management strategy that can facilitate the smooth rotation and recycling of students through the CAI package.

In the rush to join the computer revolution, school district administrators should not lose sight of the importance of careful evaluation and planning. CAI is just one of many instructional approaches. The cost-effectiveness of CAI should be assessed by the extent to which it can provide instruction more efficiently than other alternatives.

Distance Education and Distance Learning

If we closed this segment of the chapter without a brief discussion of various distance education and distance learning options, we would be remiss in our analysis of technology's impact on today's schools and universities. Across the nation, a vast assortment of classes is being taught in virtual classrooms. A commonly agreed upon definition of *distance learning* is a formal educational process occurring with the teacher and learner separated by either time or distance (often both). In the realm of instructional technology, ***distance education*** is a subset of what might be viewed as a larger world of technology-enabled teaching and learning. Despite all the hype and media attention devoted to this industry, distance learning is really not a new or novel phenomenon. Every one of you who is reading our textbook has at some time engaged in distance learning.

When you ponder the definition on the previous page, the most common, and perhaps oldest, form of distance learning is your daily homework. When you complete a homework assignment, you accomplish a learning activity that was constructed by your instructor. You achieve this without the presence of the instructor (at a distance). Correspondence courses represent another form of not so new distance learning options. A person's cumulative, acceptable completion of an entire set of assignments constitutes a course for which credit can be earned. If one strings many of these together, a degree might be the ultimate result. The unassuming independent study course is yet a third example of distance learning where students rarely meet face-to-face with their professor.

Nevertheless, when administrators talk dollars about distance in contemporary educational institutions, they are not generally envisioning these three examples. In most instances, distance education courses are identified as those being delivered to remote (off-campus) locations via audio, video, and computer technologies. Advancements in fiber-optic, microwave, compressed video, and satellite technologies allow many of our traditional face-to-face real-time classrooms to be simulated for distance learners who are separated from their teachers.

The two major types of interactive television delivery systems are satellite and non-satellite. Most satellite delivery systems allow for one-way video and two-way audio transmission, while those that are non-satellite dependent allow for two-way audio and video transmission. Synchronous two-way interactive video systems caught the attention of and excited many faculty members in the late 1980s and early 1990s. Some states invested in widely dispersed interactive video communication systems with the expectation that one faculty member could be counted on to teach many students at multiple sites across the state. Unfortunately, the interactive nature of the systems was not as promising as they had hoped. A recent study revealed that professors who were asked to deliver courses via interactive television must be informed ahead of time about the following differences from traditional teaching:

- They will have less control over the delivery of their courses.
- It may be impossible to replicate the responsiveness found in their more familiar face-to-face classrooms.
- They must be extremely keen listeners.
- The lack of students' nonverbal cues may adversely affect their teaching and the quality of their teacher-student interpersonal relationships.

The use of the personal computer and associated networked communication technologies represents the latest, and perhaps most convenient, form of distance education. At the close of the twentieth century, nearly every two- and four-year college and university in the United States had developed (or was developing) courses and complete degree programs to be delivered and completed using the Internet as the primary vehicle for communication. Distance learning courses are designed to replicate face-to-face classroom and laboratories through the use of World Wide Web–based electronic mail, bulletin boards, listservs, and chat rooms. When these technologies are combined with other website-based information sites, videotaped materials, audio recordings, online testing applications, and the old-fashioned telephone, many instructional objectives can be accomplished through distance education courses.

In a recent *New York Times* report, it was estimated more than a million people were enrolled in on-line courses for credit during just one academic year. This did not include numerous adults who enrolled in and completed noncredit courses. Avowedly, on-line distance education is still in its infancy, and the debates about how to best blend technology and pedagogy have only just begun. In 2001, just a few examples of cyberspace instruction included the University of Illinois's master's program in library and information science, a master's in nursing from the University of Phoenix, Stanford University's on-line engineering program, and an environmental science collaborative project between the University of Virginia and several universities in southern Africa. See Figure 12-6.

Will distance education and distance learning totally supplant school and university campus-based teaching and learning? We believe it is safe to say this will not occur in the immediate future. In each of the on-line programs just mentioned, some amount of on campus instruction or another form of face-to-face dialogue between students and the instructor was apparent. We must continually reexamine educational priorities and directions and not lose sight of the fact that the learning goals of education are diverse and complex. Likewise, the array of instructional strategies and assessment

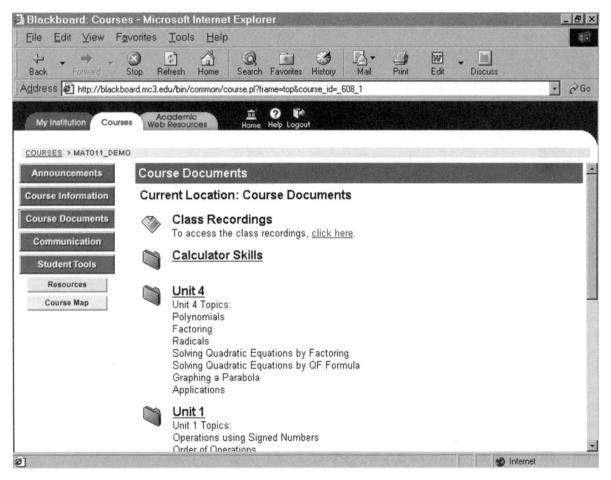

Figure 12-6.
This Internet image depicts one example of a distance education course.

approaches we use to accomplish those goals must be equally diverse and complex. Technology can either enhance our educational programs or undermine their long evident stability. People who profess to be educational leaders have an obligation to keep their curricular goals just as current as the technology-laden classrooms being designed to achieve them. In other words, the stability of educational institutions in our society will be more directly influenced by curricular relevance than by technical modernity. Further, educators must not lose sight of the fact that teaching is not enough. Contemporary educational professionals must provide numerous opportunities for authentic learning to occur.

The Playground: Technology and Recreation

Few people would have difficulty accepting the fundamental fact that play is an essential part of the human condition. The concept of a playground, however, is usually more recognizable among the residents of the world's industrialized nations. Play is central in children's lives. Children make their world into a play world at every possible opportunity. The urban movement, born of the industrial revolution, has progressively and drastically reduced the opportunities for play. Busy streets pose serious hazards, vacant lots have disappeared, and high-rise living arrangements eliminate backyards. It is not surprising that the need to set aside special spaces where children could find at least a minimal opportunity for growth and development through play first became evident in the technology-oriented sector of the world. See Figure 12-7.

The playground movement in the United States was initiated during the late

Figure 12-7.
From sand gardens to high-tech facilities, we have witnessed dramatic changes in the design and construction of playground equipment during the past century. Playground manufacturers report the number one trend shaping the design of equipment is a concern for safety. There is also a growing demand for durability and an interest in the modular concept. While there are still many examples of traditional playground equipment, such as swings, slides, and climbers in use, there is a trend away from purchasing one-use-only pieces. Recreational planners have the option of buying several components that can stand alone or fit together to form a variety of shapes serving to enhance and stimulate more creative play activity. (Michael Dafferner)

nineteenth century with the sand gardens in Boston. Its proponents asserted city life was ill suited for the normal growth of children. Crowded cities stimulated a basic social concern for the provision of adequate play space for children. This became a catalyst to spearhead the formation of a new national organization. Around the turn of the century, the Playground Association of America was established to collect and distribute information concerning playgrounds and to promote interest in their development across the country. The provision of adequate open space for recreation has become one of several public policies that originated with and grew out of the early playground movement. Its primary advocate today is the National Recreation and Park Association (NRPA).

While not everyone envisions recreation or playgrounds as social institutions, few would deny the importance of leisure in contemporary society. Questions concerning how individuals spend their leisure time have led to new fields of study, formal degree programs, and a whole new technological discipline devoted to playthings. What does leisure look like in a technological era? It is unlikely that the early playground pioneers, who initiated an organized response to a critical social concern, could ever have imagined the magnitude and diversity of organized recreation in today's society.

Throughout the twentieth century, Americans witnessed significant changes in their leisure patterns. In the early 1900s, workers put in, on the average, a fifty-six hour week. By the late 1970s, that number had decreased to forty-three hours per week; however, as was noted earlier in this chapter, the trend back upward was apparent at the dawn of the twenty-first century. Also noted in Chapter 11, many of us seem to be busy with work-related tasks, even away from our places of employment. As we all work harder, the need for leisure is even more compelling. As you will read in the following section, this translates to opportunities for leisure time for employees during their workdays.

Technology has influenced the types of leisure activities we enjoy. The term *sports* is used to describe some of the universal leisure pastimes. Developments in technology have shaped the manner in which everyone, from the serious athlete to the hobbyist, is able to participate in sports.

Free time is sometimes used to describe leisure. Nevertheless, the word *leisure* somehow conveys the idea time is being used constructively. In some instances, it even implies the individual is engaged in upper class, cultural activities, such as attending the ballet, seeing the opera, going to the theater, or reading the classics. When a person uses his or her free time to watch football on television while "slugging down" a few beverages with friends, the word *idleness* is sometimes substituted for *leisure.* This degrading expression indicates to some that the person is not doing anything worthwhile, but is simply wasting the free time available. Of course, there are arguments for both definitions, and neither leisure pattern is necessarily desirable for everyone. If anything definite can be said about our leisure time, it is individualistic. What some people consider leisure, such as washing the car, others consider work. What other people consider leisure, like watching televised football games, their friends may consider a boring waste of time. See Figure 12-8.

Besides giving us more time to enjoy leisure activities, our high-tech society has also provided a number of new ways to use technology, including electronic games, VCRs and DVD players, affordable air travel, large-screen televisions housed in local neighborhood taverns, demolition derbies, recreational vehicles, and lots of high-tech athletic equipment. Of these, the most seductive leisure format is offered by the electronic media, which absorbs staggering amounts of free time in contemporary American households. Some writers have actually compared the television screen to the more traditional fireplace hearth. Analogous as they may seem, the latter example promotes conversation and romance, while the former seems to stifle both.

Survey statistics indicate that watching television is perhaps the most common form

Figure 12-8.
People devote time to leisure activities for many reasons. Some people find it worthwhile to spend time with their families, even while doing necessary tasks, such as washing the car. Others use modern athletic equipment to stay in top physical shape. Other activities, such as reading a good novel and horseback riding, offer therapeutic value and relaxation.

of leisure today. The proliferation of VCRs during the 1980s and DVD players in the late 1990s provided the television-viewing pastime with even more room to multiply. Television has been credited as the primary source of accessible newsworthy information and as an educational tool to aid in the socialization of children. More often, television tends to be chastised as a technological intruder that has destroyed the personal interaction among family members and stimulated the growth of violence in our society.

Regardless of the label ascribed to it, television demands very little from our minds and bodies. Its mesmerizing presence allows us to absorb as much or as little as we desire. The introduction of remote control devices has made it possible to change channels, adjust volume, advance the tape in the VCR, or cue a spot on the DVD without even getting out of the chair. As with other techno-gadgets, the remote control feature has created new terms associated with its use. Two such terms were coined in the mid-1980s, and while the use of these words may no longer be apparent, the behaviors remain prevalent. *Zapping* refers to the practice of switching back and forth among channels, generally to avoid being exposed to product advertisements and commercial interruptions. *Zipping* describes the use of the remote control buttons to fast-forward a prerecorded tape to watch a one-hour show in about fifty-three minutes. This zipping strategy yields the same results as zapping—zero commercials. As you might well imagine, large corporations are becoming more reticent to spend huge sums of money on television ads people zip over or zap out as they are being broadcast. Some companies have taken to imbedding advertisements directly in network programs to counteract these consumer tendencies; think about this the next time you see a favorite actor drinking from a can of Pepsi® soft drink.

As an increasing number of Americans retreat to their electronic cocoons for their leisure, some appear to becoming psychologically dependent on the television set. In all likelihood, you probably know people whose free time is almost completely engrossed by video in its various forms. Meanwhile, another electronic device has become quite common in American homes. Even though the personal computer may be considered to be more intellectually demanding than the television, it still fits the electronic cocoon scenario of leisure patterns. As a type of leisure device, personal computers provide all sorts of electronic games. They also allow users to converse with other people via bulletin boards, chat rooms, and instant messaging. Household computers, combined with televised home shopping networks, have already made it easier for us to do our shopping, banking, and even our work, without ever leaving the house. Do you think this trend will lead to the resurgence of the long lost delivery boy of the early twentieth century?

In a previous section of this chapter, it was stated there appears to be an overall merging of childhood and adulthood in our society. As children try to act more like adults, adults often behave a lot like kids. This behavioral tendency is most obvious in the recreational interests among a growing population of middle-aged baby boomers who refuse to slow down as they approach retirement age. In fact, the Sporting Goods Manufacturing Association reported a few years ago that its largest growth market is coming from fifty- to seventy-four-year olds. An interesting social trend that not only supports this adult need for play, but also recognizes the importance of organized leisure activities, might easily be found in corporate wellness and fitness centers. Sometimes listed under the discipline heading "Industrial Recreation," these programs have surfaced around the country, and many have been in existence for more than thirty years.

Employee Recreation Programs

It is difficult to accurately pinpoint a time period when industrial recreation really got started. Perhaps the roots of this discipline can be found at the first company picnic given for the employees of Johnson Wax in Cincinnati, Ohio in the late 1800s, or maybe the catalyst was formed during an employee baseball game between the

Equitable Life Assurance Company and the Metropolitan Fire Insurance Company in New York City in the mid-1800s. Some recreation professionals make reference to the World War II era when the nation called for increased productivity. The need arose to relieve the tensions of war and enhance employee unity in the midst of family separation and loss. In any case, employee recreation, services, and fitness (ERSF) programs have experienced tremendous growth in the United States over the past several decades.

With each passing fiscal year, the number of companies offering some type of recreation program as a component of their employee benefits package expands significantly. Many corporations have installed exercise facilities in their office complexes. Others have turned to private health spas and fitness centers as places for their employees to exercise. In other words, while some firms have actually built an on-site facility, others have found it more convenient to purchase corporate memberships from nearby health clubs. Some of the predominant features offered by both types of programs are basketball, racquetball, tennis, weight lifting equipment, stationary bicycles, fitness classes, and volleyball. Also available, on a less regular basis, are amenities like golf course access, company swimming pools, softball fields, massage parlors, and suntanning salons. The awareness of the importance of recreational facilities convenient for employees to use during the workday has spread beyond corporate executives. Concern for a holistic business environment seems to prevail among forward-thinking architects and developers who are designing contemporary office and industrial parks. In a bona fide attempt to attract corporate tenants, builders have included jogging trails, lakes, bicycle paths, fitness centers, and diversified landscaping in their plans.

If you study the construction documents for one of these industrial parks, you might wonder when anyone would find time to get any work done. On the contrary, since that first company picnic years ago, the philosophy has remained the same. Many executive officers firmly believe recreation services can cement employee loyalty to the company, reduce tensions, lower absenteeism, and promote employee camaraderie and morale. As you probably already guessed, numerous studies have been engineered to determine the validity of these sorts of convictions.

Most of the research results on this topic have been quite positive so far. It seems safe to conclude many important organizational advantages can be gained through employee recreation programs. For example, the National Aeronautics and Space Administration corroborated the findings of the Heart Disease and Stroke Control Program. They both recently concluded that a program of regular exercise reduced absences, improved stamina and work performance, and enhanced powers of concentration and decision-making. Along the way, they also found the average office worker's efficiency decreases 50 percent during the final two hours of the workday. By comparison, those who exercised on a daily basis were found to work at full efficiency all day long.

The most fascinating thing about recreational activities is their dichotomous ability to either arouse and stimulate participants or relax them and calm their nerves. Essentially, most recreational activities act to optimize and equalize a person's stress level. A certain amount of stress can actually increase productivity and creativity. In sum, ERSF programs have a lot to do with reduced absenteeism and turnover, increased job satisfaction, increased performance, improved general health, and reduced cost of care. Quite simply, in this day and age of complex technology and change, employee recreation programs can really increase profits.

High-Tech Recreation Equipment

The profile of athletics, in an era when technology can ostensibly fix anything, seems to go through one transition after the next. Today's athletes come better equipped than the chariot riders in ancient Greece.

More interesting, however, is the extent to which contemporary athletes, professionals and amateurs alike, are dependent upon a whole array of paraphernalia. We have almost reached a point when it is not enough to say you lack the necessary coordination to be a good skier or halfway decent tennis player. The chances are strong your ski instructor or tennis pro will recommend that you go out and purchase a new "whatever" to improve your performance.

Sports equipment has always been a crucial link between the athlete and the game. Now, more than ever, a host of technical refinements are introducing a whole new line of "improved" basketballs, tennis rackets, aerobic shoes, golf balls, golf clubs, baseball gloves, bicycle gears, and safety helmets. The improvements in sports equipment are being made so fast that, in some cases, they are actually revolutionizing the sport itself. On the teaching side of the games, technology is changing the look of the lesson. Two excellent examples are found in Florida.

The Grand Cypress Academy of Golf in Orlando and the IMG Bollettieri Tennis Academy in Bradenton both use computers, high-speed video cameras, and heart monitors to diagnose the problems and improve the performance of duffers, hackers, and tournament players alike.[4] Instructors at the former institution aim to develop the perfect golf swing for every registrant. Within various program formats, the students

receive daily forty-five minute lessons in front of video cameras. While the tapes are played back, a computer generated stick figure is superimposed over the image of the learner to illuminate deviations from the ideal movements. See Figure 12-9. At the tennis school, the heart monitor is one of the chief teaching aids designed to train players to perform competitively on a consistent basis. These professionals believe players function at their best skill level within a narrow range of heart rate.

The manufacture of leisure equipment is a major industry. One of the most common sales promotion gimmicks used by sporting goods firms is to sign well-known athletes on to use certain products exclusively. Sporting goods companies are subsequently under tremendous pressure to constantly improve product lines so their sponsored athletes maintain an edge over the competition. Also adding fuel to the fire are governments and national athletic organizations, coaches, and athletes who are usually clamoring for better equipment. Consumer tastes and a slight increase in the amount of discretionary income people have to spend on their leisure have also altered the shape of some traditional products.

Manufacturers use the latest in space-age materials and computer-aided techniques to improve the life, power, and controllability of tennis rackets. Shoe designers boast their particular aerobic model represents the latest technology in attire for the feet. The adult segment of the bicycle market has grown considerably. This has motivated manufacturers to develop an elite line of bicycles aimed at the quality conscious adult buyer. For the backpacking devotee, the formulation of advanced fibers and fabrics is starting to eliminate some of the discomforts caused by inadequate clothing design and construction. DuPont's fiber called Thermax® clothing combines the

---

[4]At the time of this writing, these agencies were found at the following Internet addresses, respectively: www.grandcypress.com/academy/index.htm and www.sportsladders.com/bollettieri.asp.

Figure 12-9.
Dr. Ralph Mann, former Olympic medallist and now President of CompuSport, uses Model Golf® Teaching System software to view a computer generated model golf swing. He researched this model swing from the performance of over fifty top Professional Golfers' Association golfers on the professional tour. Dr. Mann tailors the model to a student's body dimensions and overlays the model on the student's actual golf swing. (CompuSport)

softness of cotton with a wicking ability that pulls moisture away from the body for evaporation. The wearer stays drier and is much more comfortable. The list of examples to portray the use of technological genius to build new toys or improve the ones we already possess goes on much further. Regardless of your favorite sport or recreational hobby, in one way or another, technology will surely modify its profile in the near future.

Summary

Social institutions exist in all cultures around the world. Some are highly respected due to their longevity and coincidental ability to preserve tradition and nurture local customs. Others are recognized as more contemporary social groups serving as a type of substitute for conventional institutions. In either case, people tend to compartmentalize their lifelong objectives and aspirations into specific activity areas. A few of these compartments were discussed in this chapter, including family, religion, education, and recreation.

Technological developments have certainly affected the general profile of these social institutions. Each one continues to be operative in our society at varying degrees of stability. One group of sociologists reports the family unit is existing only on shaky ground, technology is replacing religion, real live teachers are becoming obsolete, and serious adults believe buying more toys may be the answer to their loneliness. On the other side of the fence, a more optimistic group of social scientists contends that marriage and family life are both quite healthy, the strength of religious affiliations is on the incline, and technology in the classroom can assuredly enhance curricular goals. A lot of adults believe recreational activities should be a productive, rather than a passive, financial investment. Regardless of which group you can most readily agree with, social institutions remain the structural mainstay of our civilization.

Discussion Questions

1. Identify and describe several characteristics of a nuclear family, an extended family, and a surrogate family. Explain how they are similar to each other in contemporary society.
2. What are some of the personal expectations you might list for your prospective spouse to peruse at the onset of your marital agreement? How are these expected outcomes different than the

ones your parents may have had when they got married?

3. Which, if any, contemporary medical procedures do you believe to be acceptable for use by couples who find they are unable to conceive a child?

4. This chapter reviewed western religious beliefs and customs and their alignment with science and technology. Select another religion and investigate its views about modern technology.

5. Explain how the responsibilities of teachers in our society are being modified as a result of the rampant pace of technological change.

6. Identify several ways the toys manufactured for the adult population in this country seem to perpetuate their visions of childhood. Give specific examples. Conversely, what sorts of toys are marketed for children to enhance the process of socialization?

Suggested Readings for Further Study

Beercheck, R. C. 1986. Recreational equipment: Engineering the winning edge. *Machine Design* (12 June): 26–27, 30, 32, 34.

Bianchi, S. and L. M. Casper. 2000. American families. *Population Bulletin* 55 (4): 2–43.

Bleier, R., ed. 1986. *Feminist approaches to science.* Elmsford, N.Y.: Pergamon.

Brody, R. 1985. Well-equipped: Choosing the right tool can improve your athletic performance. *Esquire* 103 (l): 34–35.

Brophy, B. 1986. Children under stress. *U.S. News & World Report* (27 October): 58–64.

Bureau of the Census. 1980. *Current population reports, marital status and living arrangements.* Washington, D.C.: U.S. Department of Commerce.

Cetron, M. 1985. Schools of the future: Education approaches the twenty-first century. *The Futurist* 19 (4): 18–23.

Coffey, S. 1984. Lockheed's employee recreation program. *Parks & Recreation* 19 (8): 47–49.

Cohn, A. W. 2000. Federal probation. *Juvenile Focus* 64 (2): 77.

Corn, J. J., ed. 1986. *Imagining tomorrow: History, technology and the American future.* Cambridge, Mass.: MIT Press.

Cornish, E. 1986. Future free time: How will people use it? *Parks & Recreation* 2 (5): 57–60.

Davey, K. B. 1999. Distance learning demystified. *Phi Kappa Phi Journal National Forum* 79 (1): 44–46.

Doherty, K. M. and G. F. Orlofsky. 2001. Student survey says: Schools are probably not using educational technology as wisely or effectively as they could. *Education Week* 20 (35): 45–48.

D'Onofrio-Flores, P. M. and S. Pfafflin, eds. 1982. *Scientific-technological change and the role of women in development.* Boulder: Westview.

Feemster, R. 1999. Going the distance. *American Demographics* 21 (9): 59–64.

Ferree, G. D., Jr. 1996. American views on science and technology. *The Public Perspective* 7 (6): 60–63.

Finney, C. 1984. Corporate benefits of employee recreation programs. *Parks & Recreation* 19 (8): 44–46, 71.

Fleisher, P. 1985. Teaching children about nuclear war. *Phi Delta Kappan* 67 (3): 215–217.

Florman, S. C. 1984. Will women engineers make a difference? *Technology Review* 87 (8): 51–52.

Gardyn, R. 2001a. A league of their own. *American Demographics* 23 (3): 12–13.

———. 2001b. The new family vacation. *American Demographics* 23 (8): 43–48.

Geelan, T. J. 2000. When creationism goes to school: A teacher's perspective. *Free Inquiry* 20 (2): 15–17.

Giancola, J. 1985. Introducing the child to the information age. *Design for Arts in Education* 86 (6): 15–17.

Goldbort, R. C. 1995. On classrooms, playgrounds, and our children's health and safety. *Phi Kappa Phi Journal National Forum* 75 (4): 6–7.

Golden, D. 1992. Fall from grace: How the Christian Science Church gambled on electronic salvation—and lost. *The Boston Globe Magazine* (17 May): 16–17, 30, 31, 34–43.

Gushee, D. P. 2001. The biotech revolution: A matter of life and death. *Christianity Today* 45 (12): 34–40.

Harmon, A. 2001. Cyberclasses in session. *New York Times Education Life* 4A (11 November): 30–32.

Hartsoe, C. E. 1985. From playgrounds to public policy. *Parks & Recreation* 20 (8): 46–49, 68.

Hatiangadi, A. U. (2000). Bringing up baby: A comparison of U.S. and European family leave policies. *Fact & Fallacy* [Internet]. <http://www.epf.org> [18 Nov 2001].

Henderson, C. P., Jr. 1986. *God and science: The death and rebirth of theism*. Atlanta: John Knox.

Hornig, L. S. 1984. Women in science and engineering: Why so few? *Technology Review* 87 (8): 30–36, 38–41.

Hymowitz, K. S. 1999. Cheated out of childhood. *Parents* 74 (10): 174–179.

Hynes, H. P. 1984. Working women: A field report. *Technology Review* 87 (8): 37–38, 47.

Jackson, G. A. 1986. Technology and pedagogy: Making the right match is vital. *Change* 18 (3): 52–57.

Johnson, D. 2002. Until death do us part. *Newsweek* (25 March): 41.

Johnson, M., D. R. Hoge, W. Dinges, and J. L. Gonzales. 1999. Young adult Catholics: Conservative? Alienated? Suspicious? *America* 180 (10): 9–13.

Kaplan, D. 1985. The world according to television. *Instructor* 94 (9): 52–54, 56.

Keller, E. F. 1984. Women and basic research: Respecting the unexpected. *Technology Review* 87 (8): 44–47.

King, W. H. 1986. Science and religion: Getting the conversation going. *The Christian Century* (2–9 July): 611–614.

Langer, J. 1982. *Consumers in transition: In-depth investigations of changing lifestyles*. New York: American Management Associations Membership Publications Division.

Lardner, J. 1999. World-class workaholics. *U.S. News & World Reports* 127 (24): 42–53.

Larick, K. T. and J. Fischer. 1986. Classrooms of the future: Introducing technology to schools. *The Futurist* 20 (3): 21–22.

Laver, R. 1985 Taking stock of the family. *Macleans* (7 January): 20–22.

Lepkowski, W. 1984. Scientists discuss ways to integrate science with Christianity. *Chemical & Engineering News* (27 August): 36–38.

Levin, H. M. and G. Meister. 1986. Is CAI cost-effective? *Phi Delta Kappan* 67 (10): 745–749.

Lord, L. J. 1987. The brain battle. *U.S. News & World Report* (19 January): 58–64.

Lystad, M. 1984. *At home in America: As seen through its books for children*. Cambridge, Mass.: Schenkman.

McDonald, M. 1999. Wrinkles in the spandex. *U.S. News & World Report* 127 (4): 55.

———. 2001a. Call it 'kid-fluence'. *U.S. News & World Report* 131 (4): 32–34.

———. 2001b. Forever young. *U.S. News & World Report* 130 (13): 36–38.

Menso, M. 1986. Leisure in the age of technology. *Parks & Recreation* 2 (12): 42–43.

Meyer, L. 2001. New challenges: As computer access spreads, schools look for ways to use technology to raise achievement. *Education Week* 20 (35): 49–63.

Meyrowitz, J. 1984. The adultlike child and the childlike adult: Socialization in an electronic age. *Daedalus* 113 (3): 19–48.

Mottet, T. P. 2000. Interactive television instructors' perceptions of students' nonverbal responsiveness and their influence on distance teaching. *Communication Education* 49 (2): 146–164.

Murphy, J. W., A. Mickunas, and J. Pilotta., eds. 1986. *The underside of high-tech: Technology and the deformation of the human sensibilities*. Westport, Conn.: Greenwood.

Murray, D. 2000. Unnatural selection of origin themes. *The Christian Science Monitor* 12 (130): 11.

Myers, H. 1986. The future of religion in America. *The Futurist* 20 (4): 20–23.

Neal, E. 1999. Distance education: Prospects and problems. *Phi Kappa Phi Journal National Forum* 79 (1): 40–43.

Niemiec, R. P. 1986. CAI can be doubly effective. *Phi Delta Kappan* 67 (10): 750–751.

Noble, D. F. 1992. A world without women. *Technology Review* 95 (4): 52–60.

Nudel, M. 1984. Employee recreation around the country. *Parks & Recreation* 19 (8): 40–43.

Peaceful coexistence for Darwin and religion? 2000. *American School Board Journal* 187 (5): 20.

Perelman, L. J. 1986. Learning our lesson: Why school is out. *The Futurist* 20 (2): 13–16.

Profile of tomorrow's family, A. 1985. *Children Today* 14 (l): 3–4.

Public agenda reality check 2001. 2001. *Education Week* 20 (23): S1–S8.

Robinson, H. 1986. More "religion," less impact. *Christianity Today* (17 January): 14–15.

Rogers, E. M. and J. K. Larsen. 1984. *Silicon Valley fever: Growth of high-technology culture*. New York: Basic.

Rosenberg, R. 1991. Debunking computer literacy. *Technology Review* 94 (l): 58–65.

Rossman, P. 1985. The network family: Support systems for times of crisis and contentment. *The Futurist* 19 (6): 19–21.

Rubin, V. 1986. Women's work: For women in science, a fair shake is still elusive. *Science* 7 (6): 58–61, 64–65.

Snitow, A. 1986. The paradox of birth technology: Exploring the good, the bad, and the scary. *MS.* 15 (6): 42–46, 76–77.

Spence, L. D. 2001. The case against teaching. *Change* 33 (6): 11–19.

Tobie, J. D. 1986. Are computers changing our image of God? *The Futurist* 20 (4): 19.

Treas, J. and E. D. Widmer. 2000. Married women's employment over the life course: Attitudes in cross-national perspective. *Social Forces* 78 (4): 1409–1436.

Turkle, S. 1984. Women and computer programming: A different approach. *Technology Review* 87 (8): 48–50.

Vahanian, G. 1977. God and Utopia: *The church in a technological civilization*. New York: Seabury.

Vockell, E. and D. Luncsford. 1986. Managing computer-assisted instruction in the classroom. *The Clearinghouse* 59 (6): 263–268.

Wakefield, R. A. 1986. Home computers and families: The empowerment revolution. *The Futurist* 20 (5): 18–22.

Wallace, C. M. 1999. The new family dinner. *Parents* 74 (5): 106–110.

Walton, A. 1986. Attitudes toward women scientists. *Chemtech* 16 (7): 396–401.

Weiss, M. J. 2001. The new summer break. *American Demographics* 23 (8): 49–55.

Wellemeyer, M. 1987. Sports schools go high-tech. *Fortune* (16 February): 119–120.

Westland, C. 1985. A worldwide perspective on playgrounds. *Parks & Recreation* 20 (8): 55–59, 69.

Whicker, M. and J. Kronenfeld. 1986. *Sex role changes: Technology, politics, and policy*. New York: Praeger.

Winner. L. F. 2001. MeetingGod@beliefnet.com, *Christianity Today* 45 (14): 70–74.

Woodward, K. L. 1987. Rules for making love and babies. *Newsweek* (23 March): 42–43.

Workman, B. 1986. Dear professor: This is what I want to know. *Phi Delta Kappan* 67 (9): 668–671.

Yanish, D. L. 1986. Despite gains, women engineers seek parity. *San Jose Business Journal* (27 October): 1, 8–9.

Zimmerman, J., ed. 1983. *The technological woman: Interfacing with tomorrow*. New York: Praeger.

Chapter 13

The Technologist's Responsibility for the Future of Society

DATELINE

Atlanta, Georgia

The Influenza Branch of the Centers for Disease Control and Prevention (CDC) conducts surveillances for influenza in the United States each year from October through May. This work is designed to determine when influenza viruses are circulating, identify strains, and detect changes. Health officials are also attempting to monitor influenza-related illness in the United States and measure the impact of influenza on deaths in the United States. Approximately seventy-five World Health Organization collaborating virology laboratories and some fifty laboratories for the National Respiratory and Enteric Virus Surveillance System located throughout the United States report the total number of respiratory specimens tested and the number that are positive for influenza by type and subtype each week. A subset of the influenza viruses isolated is sent to CDC to be tested for antigenic characteristics.

For most of the year, these immunization experts work like sleuths stalking a shifty opponent. By the time spring arrives, they should be more or less in agreement on their forecasts. Full-scale vaccine production must get underway by early April if they expect the serum to be available by the end of autumn and the start of the flu season. Their predictive forecasts involve a certain measure of guesswork, since the process of isolating a flu virus is very much like hitting a moving target.

Even still, these labs form the world's cornerstone of the surveillance effort. The influenza virus is a formidable foe. There are two major types, known as A and B, and numerous subtypes or strains. Flu viruses travel quickly and change radically in a span of a few short weeks or months. The present surveillance system is adequate to guard the world against a disaster like the 1918–1919 epidemic that killed 20 million people, but it is far from perfect....

Introduction

Speculation about the future has been a favorite social pastime throughout all of human history. Greek and Roman myths are full of tales of supernatural beings who could foretell the future. The early biblical prophets were believed to possess divine inspiration and the "God-given" ability to predict the future. Tribal cultures around the world have been ever trusting of their local "witch doctors" when there was a need to forecast future events. Most everybody, at one time or another, has taken a stab at making a prediction concerning his or her own destiny. Think about this for a moment. Have you ever had your fortune told at a local carnival or county fair? Have you ever noticed roadside advertisements for palm readers in small tourist towns? Have you ever scanned the Yellow Pages for a listing that might put you in touch with a spiritual consultant? See Figure 13-1. Have you ever read your daily or annual horoscope? Have you ever asked your own physician how long it might take for you to heal after a surgery? Have you or someone you know ever consulted a stockbroker to determine which financial investments would be most lucrative? Each of these simple examples presents a case of curiosity about one's life at some point in the future,

Figure 13-1.
The Yellow Pages may use the subheading "Spiritual Consultants" to advertise for a wide assortment of businesses professing to predict or forecast future events in your life.

and each of these inquiries requires some method of forecasting.

Prior to the mid-twentieth century, forecasting activities were seemingly left to prophets and philosophers who were thought to enjoy special powers of insight. Today, however, when business organizations and government agencies need an image of tomorrow or a projection about next year, they submit their request to a whole new breed of intellectuals. Although several titles have been used to describe these individuals, they are highly respected as professional futurists. Unlike the old-fashioned soothsayers, these consultants are respected for their intuitive abilities to analyze recent events and study the laws governing social, political, and economic trends in order to provide future forecasts.

Futures studies and research did not really have an organized existence much before the mid-1940s. The birth of professional futurology coincided with the realization that history itself did not follow any ironclad laws and, therefore, the future must belong to humankind. It also seemed to occur at about the same time that our destructive capacity created a new source of anxiety about the future—nuclear weapons and the later evolution of the Cold War. The Air Force and Douglas Aircraft established one of our nation's first major "think tanks," the RAND Corporation in Santa Monica, California, in the mid-1940s (*RAND* is a pronounceable contraction of the term *research and development*). Its initial mission was to do a more creative job of dreaming up military technology than the Soviets. Today, RAND's work is more diversified. They now assist all branches of the U.S. military community, and they apply their expertise to a wide range of social and international issues (for example, education, environment, energy, population, and regional studies).

Since the birth of RAND as a nonprofit entity, hundreds of private research institutions and laboratories have been launched into operation. Think tanks have essentially become an American phenomenon. It is unlikely that any other country awards such a great amount of significance to private institutions that are, in many cases, in business to stimulate creative thinking and influence public decision-making. These future-oriented firms are responsive to the needs of industrial corporations, as well as government agencies, and regularly pull in multi-million dollar contracts.

Regardless of the future time frame (six months, two years, five years, or twenty years down the road), the land of the future has become the home of futurology. A large and still growing group of future inspectors has been keeping its watchful eyes on the nation and the world over the past several decades. These inspectors routinely observe and report on the significant trends they have detected in society, industry, technology, and international relations. In the process, they create a variety of observational forecasting techniques, including futures wheels, content analyses, environmental scanning, trend impact analyses, and Delphi studies. Their ongoing investigations have transformed the study of the future into a profession with its own specialist institutions and several widely circulated journals.

More has been written, discussed, debated, and revealed about coming events over the past few decades than has been even mentioned in the entire brief history of futures thinking. The future has been a central topic of a great many books and articles, television programs, feature films, journals, societies, congressional committees, institutes, and government commissions. By the end of the 1960s, the various permutations had caught the interest of a wide range of readers, while a network of communicators spread the latest predictions about tomorrow's world. One major source of information in the United States was, and continues to be the World Future Society and its journal titled *The Futurist.* Chartered in the mid-1960s, this community of future-minded citizens came from a diversified lot of professions. Headquartered in Washington, D.C., membership in this still active organization nearly forty years later exceeds thirty thousand people residing in more than eighty countries.

Shortly after the publication of Alvin Toffler's notorious *Future Shock,* the U.S. Congress passed the Technology Assessment Act of 1972. This legislative decision established the Office of Technology Assessment (OTA) as a congressional support agency. Over its twenty-three-year history (1972–1995), its staff analysts provided congressional committees with nearly 750 objective analyses of the emerging and often highly technical issues of the late twentieth century. Since OTA's activities were based on the need for the social consequences of technological applications to be anticipated before public policies were enacted, technology assessment became a branch of futurology. It, along with several other forecasting strategies, is discussed in this chapter with reference to future prospects for various technologies.

Studies and investigations are constantly being set up to project or predict the future state of a particular technology so its impact on society can be assessed. Data accumulated by these researchers is also being used to determine ways in which the United States can maintain and continually strengthen its edge in technology development. Futures studies are invaluable to contemporary corporations requiring reliable estimates of where their industries will fit in a changing global economy. Strategic technology management seems to be an essential feature in the personality profiles of scientists and technologists who aspire to assume responsibility for the future of humanity. The following questions will be addressed in this chapter:

- ❏ As other nations continue to build up their scientific and technical skills, will American dominance in science and technology be significantly threatened?
- ❏ How can government agencies and private citizens most effectively utilize the disciplines of technology assessment and risk assessment?
- ❏ Should those who manage technology inside the plant also be responsible to oversee its successful implementation in the external social environment?
- ❏ What types of forecasting techniques should "responsible" scientists and technologists rely on for critical moves and decisive actions regarding specific, and possibly controversial, technologies?

Key Concepts

As you complete your reading of this final textbook chapter, you should already be familiar with most of the concepts being presented here. You will come across the terms and phrases included on this list. If they are new to you, make an effort to understand their meanings.

content analysis
data mining
Delphi study
epidemiology
futures wheel
heuristics
microphotonics
natural language processing (NLP)
optoelectronics
risk assessment
technology assessment

Technology Research: On the Leading Edge

Few economists would be willing to deny that technical innovativeness and scientific curiosity have stimulated a large portion of the growth in the U.S. gross national product (GNP) during the final third of the twentieth century. These are the main ingredients in the formula responsible for the amazing surge in the U.S. standard of living since the end of World War II. Technological innovation, combined with capital and an abundant supply of natural resources, made a significant contribution to this nation's overall level of productivity throughout the last half of the twentieth century. The power of American ingenuity is legendary, but the rest of the world seems to be ready and determined to catch up. Our international trading partners are always challenging American dominance in science and technology. Our government officials must remain cognizant of the fact that a significant weakening of our lead position could threaten the future prosperity and well-being of all U.S. citizens.

Technological change is exciting, but it worries a lot of companies who fear they will not be alert to react quickly enough to stay in the race. These days, they seem to be scrambling even more diligently in anticipation of the accelerating twists and turns of novel technologies. Large multinational corporations are streamlining their internal research projects. Smaller companies, wary of the huge financial commitment technology development requires, are joining research consortiums or hiring outside consultants to keep them on track. Firms not even involved with the manufacture of high-tech products have started to set up internal technology units to keep abreast of computer refinements and software enhancements that might affect their own industrial vitality.

Technology advances and changes so quickly that companies find themselves aiming at a series of elusive moving targets. They no longer have the luxury of planning years in advance and assuming the rules of the technology contest will remain unaltered. As product life cycles become shorter, businesses continue to become obsolete. A few corporations have indicated their belief in decentralization as a key to long-range technology planning. For example, back in the early 1980s, 3M Company decided to establish separate product divisions. One division handles short-term research. Another layer of laboratories oversees research that might produce benefits five to ten years hence. A third research team watches for developments even further in the future. Flexible planning for technology change is crucial to staying ahead in the global high-tech race. 3M's growth through the 1990s and into the twenty-first century evolved through a desire to participate in many markets in which they can make a significant contribution from core technologies. They have resisted the urge to be dominant in just a few markets. Their current slogan reads, "We are forever new."

Whatever form of crystal ball a company, or even a nation, uses to plan for the future, it cannot anticipate every technological twist of fate. Aggressive technology development programs can be quite risky. Corporate executives are often fearful of making hasty decisions leading to heavy financial losses. On the other hand, the price of advice to help guide their long-term speculations is no cheap investment either. One example of a technology-focused research and consulting firm, to which more than eleven thousand organizations worldwide have subscribed for assistance, is Gartner, Inc. Founded in the late 1970s and headquartered in Stamford, Connecticut, Gartner works with its clients to help them understand and capitalize on regional market opportunities within a global business context, with an eye for future entrepreneurial success. Gartner consultants serve as independent advisors on technology assessment and procurement, project development, technology management, and spending recommendations. Boasting revenues of $952 million in fiscal year 2001, their services are cost intensive.

TECHNOLOGY ASSESSMENT

The process of making projections about the future state of specific technologies, so we can assess and plan for their impacts on society.

Technological prowess is a direct result of a long chain of efforts ending with new and useful products able to compete successfully on international markets. During the early years of the twenty-first century, the United States will either remain close to the leading edge of technology development, or it will fall behind its competitors to become a technologically stagnant nation. Its residents will either continue to be the beneficiaries of economic prosperity or become the victims of dependency on foreign nations for advanced products and services. In order to prevent the latter situation, tomorrow's technologists and industrial leaders should already be thinking about and learning how to sharpen the country's competitive edge.

National Science and Technology Policy

The quality and diversity of U.S. research in science and technology is quite impressive. The major contributors to the U.S. research and development (R & D) efforts are industrial firms, the federal government, colleges and universities, federally funded R & D centers (FFRDCs), and other nonprofit institutions. The U.S. industrial sector funds and performs a significant portion of the R & D projects in science and technology (more than 70 percent). The federal government accounts for the next largest share. With regard solely to the performance of basic research in the late 1990s, universities and colleges accounted for the largest portion of this type of work; however, a major portion of their funding was procured from the federal government. See Figure 13-2.

Trends in our nation's support for technological and scientific research are traditionally analyzed by examining the amount (or percentage) of funds devoted to three categories of activity, which are more or less chronological in nature: basic research, applied research, and development. The table and graphs shown in Figure 13-3 present a profile of our R & D expenditures in the late 1990s. Research projects are often classified as being basic or applied, and development activities follow.

Basic research is fundamental and often stems from a desire to fill gaps in the existing knowledge base, rather than a desire for a practical application. In industrial settings, basic research is viewed as research that advances scientific knowledge, but does not have specific or immediate commercial objectives. On the other hand, it may very well exist in a field or area of study where the company already has a current or potential commercial investment.

Applied research is directed to gain knowledge or understanding to meet a specific, well recognized need. Applied research in industry is generally aimed at the discovery of new scientific knowledge that can be used to achieve the corporation's commercial objectives—both immediate and long-range. In contemporary venues, the line of distinction between basic and applied research is actually quite porous. Academic scientists whose promotions in rank are often contingent upon new discoveries tend to describe their research endeavors as *basic*. Similar, and occasionally identical, projects may be labeled *applied* by a government agency or

National R & D expenditures, by source of funds, performing sector, and character of work: 1998

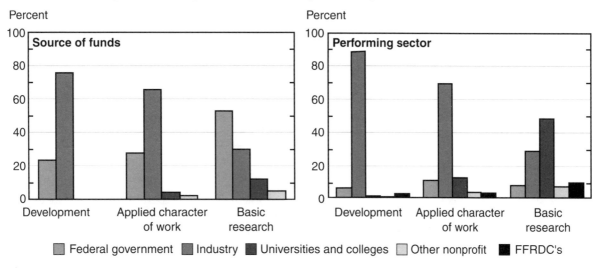

Figure 13-2.
These are graphs reflecting national research and development (R & D) expenditures by source of funds, performing sector, and character of work as a percent of total R & D expenditures. (Adapted from *Science and Technology Indicators*)

U.S. Research and Development (R & D) Expenditures, by Character of Work Shown as Percentage of Total R & D Expenditures and Amount of Funding						
	1960	1976	1980	1985	1989	1998
	(Percent)					
Total	100.0	100.0	100.0	100.0	100.0	100.0
Basic Research	8.9	12.8	13.9	12.5	17.3	16.7
Applied Research	22.3	23.2	21.7	21.5	16.6	22.5
Development	68.8	64.1	64.3	64.6	66.1	60.8

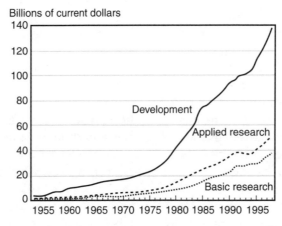

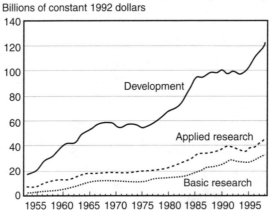

Figure 13-3.
The table at the top reflects national research and development (R & D) expenditures by character of work as a percent of total R & D expenditures. These graphs illustrate national R & D funding by character of work. (Adapted from *Science and Engineering Indicators*)

industrial firm in order to justify their credibility to a funding source or their board members.

Development is the next step beyond applied research, in which the results of earlier studies and investigations are molded to fit a particular application. Development represents the systematic use of the new knowledge gained from previous research. Development projects are usually directed toward the full-scale production of useful materials, devices, systems, or methods. The initial design and development of prototypes and processing techniques also fit beneath this heading.

National investment in R & D has been increasing incrementally in constant dollars over the past three decades, despite a brief period of stagnation in the early 1990s (following the Persian Gulf War). Federal R & D budgets have also fared reasonably well throughout the past few administrations. Through the 1980s, the defense component was, by far, the largest beneficiary of increases in federal R & D spending. In 1990, defense accounted for 63 percent of government R & D spending in the United States; less than ten years later, however, this proportion had fallen to 54 percent of the government's $74 billion investment in national R & D initiatives. By comparison, at the close of the twentieth century, government allocations to defense-related projects were considerably lower in the United Kingdom (38 percent), France (28 percent), Japan (5 percent), Canada (4 percent), Italy (4 percent), and Germany (2 percent).

The United States also places a much more pronounced emphasis on health-related research than any of these other nations. In the late 1990s, 19 percent of the U.S. government's R & D investment was devoted to health, especially to the support of life sciences in academic and similar nonprofit institutions, making this discipline second

only to defense. The National Institutes of Health (NIH) is presently our country's largest source of federal funds for university research, followed by the National Science Foundation (NSF). Nevertheless, it is complicated to make direct comparisons between health and defense R & D because most of health-related projects are research, and nearly 90 percent of defense projects today are development. (This was not always the case; during the Manhattan Project, when atomic weapons were still new, academic scientists held a near monopoly on knowledge of nuclear physics.)

Nevertheless, we seem to have reached a point where it is critical to examine the nation's science and technology research priorities. In some ways, the terrorist attacks on our nation in 2001 may have actually shocked the nation into rethinking its assumptions about our scientific superiority and technological invincibility. The challenges of wartime have certainly given birth to numerous scientific advancements in the past. It is apparent that defense, health, and energy will continue to occupy a prominent position when it comes to distributing federal research dollars. The authors are optimistic that these emphasis areas will undoubtedly be balanced with other well organized research projects in areas such as genetics, civil engineering, fiber optics, materials science, magnetic transportation, and intelligence. Since the traditional boundary lines between scientific disciplines have become much less distinct, attention and energy must be focused on multidisciplinary pursuits for knowledge and understanding. For example, breakthroughs in computer science continue to inform the biotechnology discipline. Similarly, R & D in manufacturing to improve robotic assemblies has the potential to improve our work onboard the International Space Station.

The complexion of research methodology is also undergoing significant change.

Disciplines that have long been dominated by sole investigators in self-contained laboratories insulated from the outside world now often require larger, more elaborate, and more expensive instrumentation. The cost factor alone makes it necessary for these pieces of equipment to be shared by several researchers working in a team on a variety of (perhaps even competitive) projects.

Research efforts around the world are producing new knowledge and information at an accelerating pace. The time lag between basic research and applied technology is substantially shorter now than it was even a decade ago. Subsequently, cooperation between universities and industries is expanding to perform work in areas like biotechnology, biometrics, microelectronics, microphotonics, computer hardware and software, natural language processing (NLP), materials science, and data mining. Cooperative arrangements like these are highly beneficial to both parties, but they are not often executed without a certain amount of strain. Essentially, private industry's concern over proprietary rights or government's concern over the confidentiality of its projects must be balanced against the importance of unobstructed communication within the academic sector.

We must continue to nurture working relationships among universities, industries, and governments to enhance the flow of personnel and research findings. This assertion prompted the NSF to implement several creative programs permeating traditional institutional and disciplinary boundaries. A number of U.S. universities have been able to establish engineering and technology research centers as a result of federal funding through the NSF and monies from groups of company sponsors.

Two other federal technology partnership programs were initiated in the early 1990s. Their purpose was to accelerate the development and deployment of innovative, often risky, technologies for broad national benefit through cost-sharing, collaborative partnerships between government agencies and industrial firms. The Department of Defense (DOD)'s Technology Reinvestment Project (TRP) lasted only a few years, but the Department of Commerce's Advanced Technology Program (ATP) is still functioning. Although the ATP was set up to assist U.S. businesses specifically, its legislation also authorizes it to aid industry-led U.S. joint R & D ventures, which may include universities and independent research organizations. Since its inception in 1990, a little less than 10 percent of the $1.65 billion in ATP funding has been awarded to more than 150 different universities across the country. Leading the list of top ten universities that have seen the most action in ATP projects are Stanford, University of Michigan–Ann Arbor, and the Massachusetts Institute of Technology (MIT). Through the past decade, these nonprofit organizations have recognized that ATP can provide critical funding for state-of-the-art laboratory equipment and the risky development of innovative technologies for commercial gain. The ATP also exposes university faculty and students to the ways industry carries out its cutting edge R & D. Collaboration via an ATP award further extends and deepens the relationships nonprofit organizations have with competitive companies, which can lead to future sponsored projects.

Despite these efforts in the right direction, a gap still remains between the universities and industry in this nation's R & D arena. The parties are learning how to work together, but neither has yet been able to completely master the exercise. Universities need industry because the nation's economic future depends on industry. Academic scientists and technologists need to be informed about industrial problems in order to more succinctly design their own research projects and assignments for their students.

On the flip side, industry needs both the graduates the universities produce and access to the talent and creative ideas flourishing in their free and open climates. It is certainly not enough for corporations to recruit universities' graduates each year and tempt away their professors with larger salaries. Industrial leaders must assume an active and ongoing advisory role in shaping both educational curricula and applied research programs.

Furthermore, the extent to which scientists and technologists are able to articulate their expertise to inform national policy is questionable. It is alarming to note that, for the past three decades, academic scientists have mostly remained on the sidelines, while the White House and Congress have ignored them. In some cases, our legislators have seemingly made decisions running counter to scientific advice. For example, many scientists argued that the International Space Station will never produce enough valuable research to justify its exorbitant cost; they have argued since the mid-1980s that this huge project had more to do with foreign relations and national pride than with scientific discovery. As you may recall from Chapter 4, this most recent space exploration project does have its share of dissenters.

As was mentioned earlier, Congress abolished its own OTA in the mid-1990s. Since then, members of the Congress have directed numerous appropriations to specific universities in their home districts for research projects. Academic scientists have reasoned to no avail that since such funds are not subjected to merit-based scrutiny via scholarly competitions, the dollars might very well be supporting science projects of dubious quality. As a result, many scientists and even some lawmakers believe Congress needs to re-create a regular source of

bipartisan advice about all policy issues involving advanced science and technology.

Similar to the compelling national R & D objectives driven by World War II, the resultant Cold War, *Sputnik,* and the resultant Apollo Program, our nation's contemporary aspiration to eradicate global terrorism may become a primary unifying issue that will guide future science and technology policy. Prevailing social problems, including water shortages, malnutrition, tropical deforestation, global warming, public health, and the side effects of information technologies, must also receive significant and justified attention. Given these realities, it may be even more difficult in the coming years to reach a national consensus with reference to our future research priorities. Future strength, success, and trust in science, technology, and engineering endeavors will be accomplished to the extent that our nation's research community can create and sustain partnerships. Along the way, let's hope their responsiveness to public concerns will be fortified, and the integrity and excellence of standards in their research will be heightened.

Leading Edge Technologies

Attempting to predict which technologies will be front-runners in the years ahead versus those that will simply plod along in the middle of the pack is not an easy task. In fact, if you were to travel around to interview scientists at work in a variety of research labs across the country, each of them would easily contend that their efforts are on the leading edge. Who should say the people working on isolating a cure for various forms of cancer are more important than those trying to devise a stream of new space age materials? Who should argue that research related to desalinization processes is more critical than designing more energy efficient motor vehicles? No one should, of course. Labeling one technology as being any more useful or significant than any other is second-guessing reality. Such second-guessing may lead to the curtailment of lifesaving breakthroughs. In any case, there are those scientific experts among us who insist on identifying which groups of sciences and technologies are paving the way into the twenty-first century and are the ones to be viewed as most critical or leading edge.

Advances being made in the fields of energy systems, medicine, information technology, materials science, and genetics are totally re-charting the course of human history and transforming the way human beings live on this planet. As you have read throughout the previous twelve chapters of this textbook, some of the consequences associated with these technological changes are not always pleasant. Regardless, scientific breakthroughs will continue to take place in the abstract world of theory, as well as in the concrete achievements of sophisticated machines. Several emerging technologies having the potential to transform the world as we know it today are briefly described in the following list:

❑ Neurobiologists are pioneering the use of neural implants to study the human brain. These hybrid brain-machine interfaces may allow human brains to control artificial devices designed to restore lost sensory and motor functions. Paraplegics who are confined to wheelchairs may actually regain partial control over their own limbs.

❑ Flexible transistors based on organic molecules (carbon-based) or polymers will have advantages over silicon-based electronics. New hybrid materials that can be dissolved and printed onto paper, plastic, or fabric may pave the way for digitized newspapers, product labels, and clothing.

❑ The emerging field of **data mining,** also labeled *knowledge discovery in databases (KDD),* is based on elaborate pattern recognition algorithms. Exciting applications are projected in many fields, including automobile maintenance, astronomy, geology, medicine, quality assurance, and military intelligence.

❑ The world of intellectual property protection will never be the same as it was in pre-Internet years. The nascent discipline calling itself *digital rights management (DRM)* looks like a prime catalyst for electronic content protection. Encryption schemes combined with methods to collect payment for use may be the solution corporate officers are seeking, through which they can protect their property.

❑ The field of biometrics has already emerged such that it is obvious to us as laypeople, and it involves technologies allowing us to be identified by specific biological traits. Under development and in early stages of use are capacities for face recognition, iris recognition, fingerprint recognition, voice recognition, and dynamic signature recognition (see *Dateline* in Chapter 1).

❑ Advancements in robot design may lead to more affordable robots that can be efficiently deployed in hospitals, households, and grocery stores. Leading robotic researchers are devising ways to fully automate the design and manufacture of robots by programming computers to conceive, test, and even build configurations of robotic systems. In actuality, robots will be building robots.

❑ Those of you who have seen the classic 1968 film titled *2001: A Space Odyssey* will surely remember HAL 9000—the friendly (and later crazy) computer that conversed quite easily with human astronauts. The world of *natural language processing (NLP)* may be taking us several steps closer to HAL and making it possible for us to verbally interact with our computers. (On a more trivial note, the label *HAL* consists of the three letters before *IBM* in the alphabet!)

❑ The exciting field of *microphotonics* reveals an array of technologies for directing light on a microscopic scale that will significantly impact the future of telecommunications. Researchers are studying the utility of photonic crystals that may lead to the creation of the world's smallest laser and electromagnetic chamber—both of which are essential components for integrated optical circuits.

❑ Actor Robert DeNiro, in the 2001 action film titled *The Score,* used the statement "It's just simple physics" to describe his strategy for breaking into a heavily guarded, indestructible safe (the authors do not want to spoil the film!). The forces of physics move ocean tides, geological formations, and entire galaxies. Applied physicists are investigating ways to manipulate tiny volumes of fluids thousands of times smaller than a raindrop. Microfluidics is the name of this promising new branch of biotechnology that may enable us to perform instant diagnostic tests and build implantable drug delivery devices.

As scientists and technologists throughout the world continue to decipher the few remaining mysteries of the atom and the gene, considerable advances will be made in the fields of physics and molecular biology. Ongoing and often interdisciplinary research in the nine fields identified on the short list above will also make lasting and far-reaching contributions to the well-being of humanity.

The authors describe three of these emerging technologies in the following paragraphs.

Data Mining

Most of you, at this point in your academic careers, are quite familiar with the term *information overload.* In today's world, few of us can say we suffer from a lack of information. We are deluged with facts, opinions, news releases, formulas, statistics, and lengthy reports on a daily basis. The problem is that information does not equal knowledge. In order to understand and fully grasp more complex issues in contemporary society, especially those related to science and technology, we need more than bare facts and figures or brief news bites. With access to so much information, we yearn for a mechanism that can sort, integrate, and analyze information in a way that makes sense and allows us to extract useful knowledge and insight.

The emerging discipline of data mining may ultimately be the answer to our prayers. Data mining is the process of finding patterns and correlations in large amounts of data. This technology combines pattern recognition with statistical and mathematical techniques to root through millions of data values in search of discernable patterns. Take a moment to think about the last time you launched an Internet search in order to complete a research paper—perhaps even for this class. Using conventional software, let's say you typed in the following key words: *laboratory, animal, research.* Once you entered the "search" command, millions of documents were reviewed in a few seconds. The "results" line states that 275,172 "hits" were located during the search. Where do you start? The problem here is that without more specific pattern recognition algorithms, conventional databases are limited.

Several promising applications for emerging data mining technologies are found in genetics, medicine, quality assurance, and customer service. Biological research has become very database driven, due to large-scale functional genomics experiments measuring gene expression. New challenges in the field called *bioinformatics* call for integrated databases connecting disparate information and performing complex studies to collectively analyze many different data sets. In practice, data mining normally takes the form of statistical surveys, almost like demographic censuses. When studying different populations of genes and proteins, the goal is to identify certain outstanding features possessed by a given population. Some of the exciting breakthroughs discussed in Chapter 2 (Biotechnology) and Chapter 5 (Medicine) are largely dependent on integrative data mining.

A peripheral area of development is called *text data mining*, which entails the extraction of unexpected relationships from huge collections of free-form text documents. Textual formats provide a very flexible way to store large amounts and various types of information. Analyzing huge amounts of textual data, however, takes a tremendous amount of time and energy, since we must read and organize the content. Text data mining investigators are experimenting with NLP, statistical word counts, concept extraction, and creation of semantic dictionaries. These researchers hope to develop technologies that can analyze more detailed information in the content of individual documents and extract information that can only be provided by multiple documents viewed as a whole set (for example, a stack of customer service call documents can be analyzed to reveal trends and isolate areas of concern and complaints). Several organizations presently engaged in work related to data mining are identified in Figure 13-4.

Natural Language Processing (NLP)

Information scientists have been experimenting with retrieval programs simulating natural human language queries for several decades. Work in natural language years ago took notice of the fact that human language is not an abstract code. It is a very complex and subtle form of communication. Getting computers to understand ordinary language when it is either written or spoken has been a long-time goal of computer scientists and is at the core of natural language processing (NLP) research.

Language cannot be understood without a perception of context and an extensive knowledge of beliefs and expectations. In the 1980s, artificial intelligence (AI) research utilized **heuristics** to engineer programs that could understand simple sentences with limited vocabulary. The term *understand* was used loosely, since the machines could not really understand ordinary language as humans do. A "smart" machine can electronically translate the sound of a human voice into symbolic patterns. It can next identify individual words and phrases, analyze syntax, and contextualize them in its memory. Finally, it can make thousands of logical inferences and respond in the user's native language.

In the two decades since these earlier natural language processor applications were introduced, a new generation of interfaces has begun to emerge from research laboratories. This breakthrough presents a complex integration of speech recognition, natural language understanding, discourse analysis, world knowledge, reasoning ability, and speech generation—the sum total of which will enable those of us who are so inclined to engage our computers in extended conversations. Applications for coherent, interactive human-to-machine communication are endless. A few areas where these technologies have value include machine translation (for

Leaders in the Leading Edge Technologies of Data Mining, Natural Language Processing, and Microphotonics		
Field	**Organization**	**Project**
Data Mining	Usama Fayyad (CEO, DigiMine)	Corporate Data Mining Services
	Marti Hearst (UC Berkeley)	Automated discovery of new information from large text collections
	Nokia Research Center (Helsinki, Finland)	Finding recurrent episodes in event sequence data
	Raghu Ramakrishna (University of Wisconsin)	Visual exploration of data on the Web
	Jessica Fridrich (SUNY Binghamton)	Steganalysis—searching for hidden messages in video images
Natural Language Processing	Karen Jensen (Microsoft)	MindNet, a system for extracting hyper linked concepts
	Larry R. Harris (CEO, EasyAsk)	Internet content searching; on-line shopping applications
	Victor Zue (MIT Lab for Computer Science)	Conversational interfaces
	Alexander I. Rudnicky (Carnegie Mellon)	Verbal interaction with small computers
	Ronald A. Cole (University of Colorado)	Domain-specific conversational system
Microphotonics	John Joannopoulos (MIT)	Photonic crystals integrated into optical chips for telecommunications and ultrafast optical computers
	Eli Yablonovitch (UCLA)	Photonic crystals for optical and radio frequencies
	Axel Scherer (Caltech)	Optical switches, waveguides, and lasers
	Clarendon Photonics (Boston)	Filters for wavelength division multiplexing (WDM)
	Nanovation Technologies (Miami)	Integrated devices for telecommunications

Figure 13-4.
This table lists forerunners in the leading edge technologies of data mining, natural language processing, and microphotonics.

example, English-Spanish), automated medical coding, cultural tourism, and monitoring stock transactions and other financial indices.

The NLP group at Microsoft Research is currently at work on a unique system called MindNet® software. This applied technology work seeks to develop a system that will automatically extract a massively hyper linked web of concepts from something like a standard dictionary. If, for example, the dictionary defines *cyclist* as a person who rides or races a bicycle, MindNet software is designed to use its parsing technology to decipher the underlying logical structure for that definition. It would next identify *cyclist* as a type of person, and then *rides* and *races* would be seen as verbs, thus taking *cyclist* as a subject and *bicycle* as an object. The eventual result is a conceptual network linking human understanding together with words. The subsequent act of placing this type of conceptual network into a computer takes the machine many steps closer toward understanding natural human language. Figure 13-4 identifies other organizations conducting research in the field of NLP.

Microphotonics

Photonic crystals made of a few ultrathin layers of nonconducting material represent yet another new discovery in materials science. On the cutting edge of microphotonics, photonic crystals are able to reflect various wavelengths of light almost perfectly. Years ago, a technology labeled **optoelectronics** emerged to become a marriage between electronics and optics. Researchers learned in the 1980s that photons could be employed as more compact carriers than electrons to achieve greater efficiency in data processing and transmission. Photons are discrete units of electromagnetic energy having zero mass and no electric charge, and they are essential to optical communication networks.

In our contemporary world where it is now a challenge to keep pace with the Internet's increasing need for more bandwidth, microphotonics show potential for being able to break up the logjam created by the somewhat difficult marriage between fiber optics and electronics (optoelectronics). At present, photons go whizzing through the network's optical core, only to run smack into bottlenecks, at which time they must be converted into a much slower stream of electrons handled by electronic switches and routers. If these electronic switches could be replaced with faster, miniature optical devices (photonic crystals), an incredibly fast all-optical Internet could be the huge payoff. Several large telecommunications equipment manufacturers, including Lucent Technologies, Corning, Agilent Technologies, and Nortel Networks, are already at work developing new optical switches and other miniature devices. Regardless, some of the most provocative and promising work so far in this field is taking place at MIT under the direction of physics professor John Joannopoulos. In his laboratory, photonic crystals are providing the mechanism to create optical circuits and other small, inexpensive, low power devices that can carry, route, and process large quantities of data at the speed of light. Most assuredly, this leading edge research merits careful scrutiny over the next several years as we become more enamored with audio and video messages on demand via the omnipresent Internet. Others whose work shows promise in the field of microphotonics are identified in the chart in Figure 13-4.

Life on the Edge

The above three still emerging disciplines of data mining, NLP, and microphotonics were highlighted to illustrate the competitive nature of scientific research and technological advancement. Although being at the cutting edge of a discovery sounds rather exciting, scientists' lives are often full of stress and anxiety, and sometimes even guilt, over their projects. They may actually describe life at the edge as long periods of demanding, exhausting, and commonly frustrating labor, interspersed with occasional moments of euphoria.

Once an individual earns elementary status as a scientist by completing the Ph.D. with a creative dissertation, he or she is likely to enter some type of research laboratory or academic institution. The world of research is characterized by a constant search for funding, in which scientists submit a number of proposals each year. The funding opportunities may come from agencies having specific guidelines and expectations. Rarely does a research grant come with no strings attached. Research specialists must then make choices as to whether to pursue unknown areas out of curiosity (basic research) or to construct their research designs toward the money sources (more applied research). These types of choices are especially objectionable when researchers, who are personally committed to research for human progress, discover funding for their specialty is available primarily from the DOD. Unfortunately, in order to be a successful scientist these days, one must possess a certain amount of entrepreneurial drive.

Other sources of anxiety at the front line come from the realization that your colleagues might get there first. The relentless pace of the entire research enterprise makes it almost impossible to even keep up with the new literature. Since there are often no prizes for second place, scientists usually live with the pressure to succeed quickly—in order to gain the rewards of precedence. Some people feel this frenetic race to win first prize might diminish the quality of the research being conducted.

Life at the edge is filled with criticism and peer evaluations. New discoveries must stand up to professional confirmation first and public scrutiny later. There is always the chance your new found theory is either incorrect, incomplete, or invalid. For whatever excuse, society seems to take a certain amount of pleasure in witnessing scientists "scrape the egg off their faces." Perhaps this is one justification for the existence of an insulated scientific community, in which scientists and technologists can occasionally lose touch with scientists in other disciplines and also get away from the general public.

Scientists determined to stay with it until they witness a breakthrough may find life on the edge to be quite stimulating. For others, the work schedule may lead to insomnia and deep periods of depression. The struggles of the solitary scientist go largely unnoticed by the rest of the world. Even though a large portion of the science and technology projects are being monitored by research teams, the individual scientist still plays a central role in the process of inquiry leading to benchmark findings. Although there is no magic formula for creating or becoming an outstanding scientist, one thing seems clear. To survive on the edge, these individuals must be competitive, curious, and brilliant. Furthermore, they must take extreme satisfaction in being able to solve one of nature's riddles before anyone else does via high caliber research, whatever the discipline.

Technology Management

Embarrassment is an experience most of us prefer to avoid. Many people fear they will be humiliated if they do not know the answer to a question or, worse yet, give an incorrect response to a question. Well, when it comes to our contemporary world of science and technology, even the most highly trained, well-educated experts among us must accept the fact that what they do not know is many times greater than what they do know. Ironically, a high state of ignorance seems to be a universal condition in today's society. While this state of affairs may sound rather bleak on the surface, the balance of knowledge and understanding seems to average out favorably among the individual technical disciplines in our midst. To a great extent also, interdisciplinary breakthroughs have led to lower levels of ignorance among the experts, who are often called upon to work outside their immediate areas of specialization.

Somewhat related to the dialogue surrounding technological ignorance versus scientific specialization is a discipline known loosely as technology management. For many years, American business students have been taught a good manager can manage anything and it is not necessary to have a technological understanding of the process one is attempting to manage. In some settings, we seem to have reached a point where American managers have lost contact with the technologies they are supposed to be directing and monitoring for productivity. By comparison with the rest of the world, a higher percentage of American managers come up through nonproduction jobs in business, law, accounting, and marketing. The number of managers who have an applied technological background is much lower than found among management personnel in both Europe and Japan.

Nontechnically-oriented managers may very well *understand* the technologies being used and developed by their own firms, but they generally lack the *intuitive* capacity to determine which of the possible technologies out on the horizon should be further developed and which should be ignored or discarded. They are challenged to plan for progress in an intelligent and strategic fashion. This level of intuition and type of insight only comes from firsthand experience. Engineers, technologists, and scientists are often reprimanded for believing too strongly in untried technologies and hastily plunging ahead where caution is warranted. Sometimes this charge is true, but it seems that managers who possess a technological

background might end up with a more successful betting record than those who lack an understanding of technology.

Does the field of technology management say every master of business administration graduate should also possess an undergraduate degree in science, technology, or engineering? Not exactly, but it does suggest effective managers should be able to do the following:

- ❑ Understand scientific arguments.
- ❑ Interpret computer-generated documents.
- ❑ Synthesize statistical information.
- ❑ Understand the dynamics of technological change in the global economy.
- ❑ Articulate an understanding of the chronology of events from scientific discovery to technology applications to engineering development.

The art and science of technology management encompasses a long list of responsibilities, including risk assessment, environmental scanning, measurement and control, competitive analysis, strategy formulation, technological forecasting, organizational climate, and corporate structure and culture. Corporate managers who are well versed in technological issues represent an essential profile characteristic in successful institutions and the necessary human element long overdue in this country's business environment.

Technology Assessment

Managerial shortsightedness, with reference to technological decision-making, is generally a result of working within a very narrow perspective of one's own business or industry. One form of the problem surfaces when industrial leaders fail to broaden their visions to analyze how technological progress in their own or related industries will affect their future status. Another type of crisis situation might occur when a company or industry actually has advanced technology available to them, but does not recognize how to apply it in a strategic manner. The first example illustrates the failure or even reluctance to stay informed about emerging technology. The second example amounts to failing to utilize technology in a timely manner with regard to market expectations and economic conditions. In either case, the managers in question are somewhat shortsighted in their approaches to technology assessment.

The former U.S. OTA defines this concept as it relates to a projection of the possible future state of a technology and an assessment of its potential impact on society. Over its twenty-three-year history, the bulk of OTA's work focused on comprehensive assessments taking one to two years to complete. They dealt with all sorts of technology issues earmarked by congressional members as areas of concern or controversy. The OTA staff consisted of skilled professionals who had advanced training in physical science, life science, engineering, and social science. Throughout each project, OTA made use of advisory panels of experts on the particular topic being investigated.

Its library of some 750 reports ran the gamut of urban ozone reduction, safety of medical devices, information technology, reliability of polygraph tests, and ethics of engineering.[1] Essentially, these panels helped to formulate OTA studies by defining them initially, critiquing them while in progress, and reviewing the reports before they were released. OTA assessments were subsequently published by the Government Printing Office and were frequently reprinted by commercial publishers. Interestingly, even though the OTA's official function was to do analysis for congressional committees, its reports were widely disseminated and ultimately used by interest groups, academics, and the general public.

Less than ten years after it decided to eliminate its OTA, Congress finds itself grappling with highly controversial issues like stem cell research, energy policy, anthrax, Internet monitoring, human cloning, air traffic security, and greenhouse gas emissions. Several legislators have been vocal with regard to the need for a new organization to fill the huge void left behind when the OTA's doors were closed. In fact, Representative Rush Hold (D-N.J.) recently introduced legislation (H.R. 2148) to actually revive the OTA; despite its bipartisan support, the bill has not shown promise for success. A major criticism of the former OTA was that it took too long to complete its studies. Therefore, a newly constituted organization must consider making improvements in its design and operation. Lessons can be learned from the success of several European technology analysis units, many of which were inspired by the U.S. OTA. M. Granger Morgan, head of the School of Engineering and Public Policy at Carnegie Mellon University, even suggests a new name for this U.S. agency—the Office of Science and Technology Analysis. Further, this new enterprise should be set up to provide advice to individual members or small groups of congressional members within a time frame of three to six months for most studies. At the time of this writing, the idea is still in the early stages.

Technology assessment reports may be respected as documents guiding public policy making decisions. In an industrial setting, however, technology assessment is an important component of managing current as well as emergent technology developments. Strategic decisions to establish a national telemarketing center, require 40 percent of your employees to become telecommuters, switch over to a more efficient bottling line,

---

[1]The OTA was closed in 1995 due to the passage of a Congressional reform plan. As of 1996, all twenty-three years' worth of OTA publications are available from the Federal government on CD-ROM. As of 2002, the OTA's electronic archive (1986–1995) is available at http://www.ota.nap.edu/pubs.html.

or totally change the container specifications for your company's product should not be made without a comprehensive assessment of the potential consequences of these actions. For instance, failure to recognize that the price of glass bottles may soon become prohibitive compared to plastic containers would make the investment in a brand new glass bottling line somewhat faulty and very costly for the firm. The word *strategic* implies foresight, logic, statistical analysis, and an awareness of trends in the making. When it is used as a modifier for the activity known as *decision-making,* technology assessment becomes part of the equation.

Technological forces have begun to play a much greater role in corporate strategic planning. Within a company's annual strategic formula, emerging technologies can present seemingly endless opportunities for a whole range of new products. On the downside, new technology is also a gamble that eats up precious research time and money without any real assurance of success. Fierce global competition in many industries has led to a new zeal to optimize the return on investment in new or modified technology.

In an effort to reduce the percentage of bad risks, an increasing number of corporate executive officers are recognizing the value of technology assessment. Business managers view technology assessment as a management discipline allowing for the systematic determination of which technologies are most appropriate for their companies' long-term objectives. The process further identifies the avenues best suited for exploiting the technologies ultimately selected. Each potential technology shift receives a rigorous appraisal, and its probable degree of success is estimated by the projected margin of profit over a specific period of time.

One large corporation that has made very effective use of technology assessment procedures over its two hundred year history is the Delaware-based DuPont Company. Two hundred years ago, DuPont was primarily an explosives company. One hundred years ago, its focus turned to global chemicals, materials science, and energy. Today, as it enters its third century of existence, DuPont continues to deliver science and technology-based solutions affecting businesses like agriculture and nutrition, nylon, performance coatings and polymers, pigments and chemicals, polyester, specialty fibers, specialty polymers and electronics, bio-based materials, and safety resources.

Over the years, DuPont made wise investments in its R & D, and it was quite selective as it moved into new product directions. It has successfully adapted to changes along the way and was able to establish a long-term leadership position by choosing new product venues in which there was still a considerable amount of technology to be perfected. They have become, via clear and careful technology assessment strategies, one of the world's most innovative companies. Throughout its long history, DuPont's technology game plan has adhered to several core values, including commitment to safety, health, and the environment; integrity and high ethical standards; rapidity of response; sensitivity to customers; service to the marketplace; quality; and treating people with fairness and respect.

The range of new technologies is so expansive that no single company currently in existence can hope to master them all. Technology transfer and the rapid (often immediate) exchange of information makes it ever more difficult for a company to protect its trade secrets and other intellectual property. Technological development is expensive, and the gains it produces can often prove to be quite short-lived. Technology assessment strategies are critical across the entire range of product design and development. Today's companies do not have the luxury to say, "We have invented a wonderful new composite material. Now where is an optimal market for it?" R & D teams must be savvy in their attention to market demands and expectations. They must also realize that without strong patent protection, and even with it, the most sophisticated technology can be outmoded quickly, almost overnight. The technology race has no finish line—its contenders just need to keep on running if they want to stay in the black, ahead of the pack. DuPont now operates in seventy countries worldwide, utilizing 135 manufacturing and processing facilities. They maintain forty R & D and customer service labs in the United States and operate more than thirty-five labs in eleven other nations. Undoubtedly, you will find a DuPont imprint somewhere in your household. (Some examples of DuPont's products are Teflon® coating, SilverStone® cutlery, Lycra® spandex fiber, Stainmaster® carpet, Kevlar® synthetic fiber, and Tyvek® fabric.) See Figure 13-5.

Technology assessment entails making projections about the future state of specific technologies in order that their impacts can be projected and planned. Technology assessment studies, if they are done well, take time and money, but they can result in much more efficient technology management. Even if the decision is to scrap the innovation, the company may have saved hundreds of thousands of dollars in developing a product that did not sell. When corporate managers who are in charge of technology start to think about entering a new research area,

Figure 13-5.
The versatility of the DuPont Company's science and technology solutions is ever present in our lives. A building construction site shown here features Tyvek® fabric, a widely used weatherization system made by DuPont. (Edward Antonelli)

they truly must make an honest evaluation of how their company stacks up against the competition. Further, even if a company has targeted specific areas of technology to focus on, it does not hurt to know what is happening in adjoining areas. Technology management entails technology assessment via an external environmental scanning system to permit timely decisions toward effective implementation of the new technology.

Ethical Dilemmas in Technology

You may recall our discussion of ethics as it relates to quality of life in Chapter 5 (Medicine). We made it a point to give examples in which certain medical interventions might actually sustain life, but the quality of life may not be deemed acceptable to the patient who is suffering. In other words, just because we *can* prolong life does not necessarily mean we *should* do so. Earlier in this chapter we provided a listing of DuPont's core values, and you may have noticed its commitment to "integrity and high ethical standards." It is difficult to ignore the importance of ethics and responsible research in today's science, technology, and engineering community. Unfortunately, many of the discussions in the literature written about science, technology, and engineering ethics tend to focus on negligence, wrongdoing, mistakes, and oversights—and then the dialogue turns to how these events could have been avoided or prevented. In discussions related to ethics, much less coverage is devoted to the good work and responsible contributions made by scientists, technologists, and engineers. Perhaps this is because we generally take for granted the good and reliable work of these professionals (for example, we assume the bridge will bear the weight of our personal automobile plus several others if we are stopped in traffic over a waterway, we trust the brakes and various other stopping devices to work as the pilot lands the Boeing 737 aircraft and we arrive as passengers in Chicago, and we believe the elevator car will lift us safely to the seventy-fifth floor of our office building). All too often, it seems like we take special note only when these feats of science, technology, and engineering fail to do what we expect them to do. These types of failures commonly receive more media coverage as well, making it difficult for them to be ignored. Perhaps the phrase "no news is good news" has relevance here.

Even though many of us have limited understanding of the highly complex work going into the design and construction of today's technology and engineering artifacts and even less knowledge about specialized efforts that may have prevented failures or improved reliability, we still trust the experts. On the other hand, we also find comfort in the belief that these professionals take their responsibilities seriously to ultimately protect public safety, health, and welfare. We depend heavily on the expertise of scientists, technologists, and engineers, but a significant portion of their collective work is anonymous. Few of us ever actually meet the people who designed, developed, and built the tools and devices we routinely use each day; however, somewhere in our minds we discover a sense of faith and trust that these people are both professionally competent, and further, they possess an intrinsic belief in a set of strong professional ethics.

Many universities offering degrees in engineering include courses in engineering ethics in their curricula. The Accreditation Board of Engineering and Technology (ABET) requires that its accredited engineering programs provide opportunities for students to gain an understanding of the engineering profession and practice. The ABET 2000 standards now require programs to ensure students have a major design experience including ethical, as well as economic, environmental, social, and political, factors. Similar expectations exist within the professional associations aligned with science education and technology studies.

Throughout this textbook, the authors have strived to give both the good news and the not-so-good news about a wide range of established and emerging technologies, including genetics, medicine, manufacturing, and the military. Science and technology are dominant in our lives. Decisions being made across the country in research labs, university facilities, government centers, and industrial firms are replete with ethical dilemmas. Each day, scientists, technologists, and engineers are making choices about which way to go next in their research investigations. These investigations are often sponsored by large corporate entities or nonprofit foundations with whom the people doing the actual research may never have met face-to-face. You might be in a position like this one day in the future. For those of us who will be depending on you to make the right choice and to be ethically and morally responsible, never forget the importance of character and virtue as you proceed in your career. A recent study among practicing engineers and engineering managers revealed their opinions about responsible engineering and exemplary engineering practice. The results of this study clearly apply to science and technology professions. See Figure 13-6.[2]

Technology Risk Assessment

As mentioned earlier, another responsibility falling under the jurisdiction of the technology manager is related to a procedure labeled *risk assessment.* At least one, if not several, facet(s) of a technology assessment study should examine the potential risks to humans, if the technology is allowed to enter full-scale development and production.

> **RISK ASSESSMENT**
> A method of studying the risks of a new technology, material, or device so they may be better avoided, reduced, or otherwise managed.

Dispositions	
Engineering Ethics	**Engineering Practice**
• Integrity	• Competence
• Honesty	• Ability to communicate clearly and informatively
• Candor	• Cooperativeness, good team player
• Civic-mindedness	• Habit of documenting work clearly and thoroughly
• Courage (to speak up)	• Commitment to objectivity
• Courage (to stick to one's guns)	• Openness to correction and admit mistakes
• Willingness to make self-sacrifice	• Commitment to quality
• Willingness to assume some personal risk	• Being imaginative
	• Seeing the big picture as well as the details of the smaller domains

Figure 13-6.
In a survey of engineers and managers, researchers at Western Michigan University assembled a list of commonly mentioned dispositions; these dispositions reflect ethics and behaviors associated with successful engineers. (Pritchard, M. 2001. Responsible engineering: The importance of character and imagination. *Science & Engineering Ethics.*)

---

[2]Pritchard, M. S. 2001. Responsible engineering: The importance of character and imagination. *Science and Engineering Ethics* 7 (3): 394–395.

Risk implies uncertainty. It follows then that risk assessment is largely concerned with uncertainty. The related concept of probability is often hard to grasp. The effective illumination and subsequent minimization of risks requires knowing *what* they are and how *big* they are. Knowledge of this sort is commonly gained through experience over a sustained period of time. The essence of risk assessment features the application of this knowledge of both previous mistakes and deliberate actions in an effort to prevent new mistakes in new situations.

The results of risk assessments are most commonly presented in the form of an estimate of probabilities for various events that are usually injurious or dangerous. (For example, for women over thirty-five years old, there is less than one chance in two hundred that an amniocentesis procedure will result in any pregnancy complication). Although the goal of conducting a risk assessment is to obtain this type of estimate, the major value in performing a risk assessment may actually be the exercise itself. Ideally, the people who are executing the study will explore all aspects of the action or technology under scrutiny. Refined knowledge about all angles and ramifications of the technologies will serve to make their estimates considerably less subjective.

The way we perceive various risks is strongly influenced by the manner in which they are calculated. When risk estimates are based on historical data, they are fairly easy to understand and tend to be perceived as reliable reports. Historical data provides the estimators with the frequency of events over a certain period of time. This facilitates the derivation of trends.

In order to make further predictions about the future probability of risks, researchers must employ some type of statistical model. The historical approach to estimating risks is only useful when the particular hazard has been around for a while, the risk is apparent, and it is large enough to be empirically measured (for example, cigarette smoking). If there is no historical database for an individual hazard in question, risk assessors may choose to consider it in separate parts. In doing so, they calculate the risks from each segment and add them together to estimate the risk for the whole system. A well-known instance in which this risk assessment technique has been employed is in the calculation of the probability of a severe accident at a nuclear power plant. Some carcinogenic risks may be estimated from historical data. This process is complicated, however, by the time delay between initial exposure and the final incidence of a cancer. This is the difficult field of *epidemiology,* in which causality is very hard to prove when the risk is small or seemingly insignificant.

Cause and Effect Relationships

Risk assessment with regard to new technologies is a relatively youthful discipline. While many agencies, consulting firms, and technology managers have been performing these assessments over the past decade and a half, there still doesn't seem to be a consensus on which basic rules and procedures to follow in order to solve problems involving risk.

The process of evaluating an environmental hazard to predict how many people may be at risk is indeed a challenge. As time goes on and we continue to develop more sophisticated detection technologies, scientists are able to measure the effect of toxic substances in smaller quantities. These assessments produce results that may actually contradict, rather than reinforce, the results of earlier studies. Risk analysts must take the time to follow the hazard through whatever twists, turns, and shifts it makes in the real world (refer to the *Dateline* segment of this chapter; flu experts are constantly attempting to do this in order to formulate an effective vaccine). Risk assessors must be able to determine how potential threats are released into and move through the ever-changing environment. Next, they have to figure out how much of the substance people might eat, breathe, or swallow and then estimate how much of it they could ultimately absorb. Knowing all this, they finally make an effort to determine the severity of the hazard associated with the absorbed level of the substance.

Risk analysts rely heavily on computer modeling to get a handle on typical risk estimates, but it is extremely difficult to incorporate all the important information into these models to make their findings reliable for the population at large. Consider the previous exercise. Even if the analysts knew exactly how much of a substance is in the environment (phase one), they can not necessarily predict how much of it people will absorb (phase two). People breathe at different rates, depending on their levels of health. Individuals possess widely disparate breathing rates and dietary habits. Each of these factors will affect the dose of specific substances taken in from food and air. People also absorb substances in varying amounts, depending on the thickness of their skin and the properties of their nasal mucus. These are just a few of the discrepancies making it difficult to complete phase two of the risk assessment.

Phase three entails a determination of how much of a hazard the absorbed substance poses (whether it will cause symptoms such as acute nausea or chronic loss of hearing). This is another complicated problem. Epidemiological studies are frequently insensitive in detecting health hazards from relatively low levels of exposure. In other words, the rates of specific illnesses from a given substance must often be several times above average before analysts can conclude they are not simply random fluctuations. By definition, risk assessments are officiated to reduce the amount of risk that might be *caused* by a particular technology, but if causation is so tricky to infer, one must wonder about the extent to which the whole activity is an exercise in futility.

Analogous Relationships— Animal Studies

In cases where the epidemiological evidence is either tentative or ambiguous, risk assessors may choose to use animal studies to make their projections. Studies done by analogy suffer from their own uncertainties. Laboratory animals are commonly exposed to high concentrations of chemicals to ensure that if there are any toxic effects, they will be evident at statistically significant levels. The use of a mathematical model called a *dose-response curve* must be fitted to the resulting data to assess the probability that humans will get cancer at much lower levels than they might realistically encounter. Still, serious limitations plague the process of interpreting animal studies and extrapolating them to humans.

The initial use of high doses of toxic substances may actually interrupt the recipient's internal metabolic reactions toward the initiation of carcinogenesis. Molecular biologists have discovered the existence of certain enzymes in cells that convert chemicals to more toxic metabolites—starting the march toward cancer. When the toxic chemical agent is introduced in large quantities, the enzymes are fully occupied and cannot generate the toxic by-products at a greater rate. When risk assessors extrapolate downward to realistic levels of human exposure, they generally do not take this saturation effect into account. Therefore, they often underestimate the risk to human beings.

Another limitation is that animals and humans metabolize substances differently. For this reason, the level of the test chemical reaching various organs in the animal and the human body can differ a great deal. Subsequently, they may suffer from different health afflictions or not at all. For instance, a laboratory chemical called *bis-chlormethyl ether* has been found to produce nasal tumors in rats, but lung tumors in people. In other cases, dogs and humans have developed bladder cancer from some aromatic amines, such as benzidine (used to make various dyes), while rodents displayed cancer of the liver. Correcting for these differences is not easy because scientists still lack critical information regarding human and animal systems across the board. A significant portion of risk assessment studies rely on data accumulated through tests conducted among various animal species.

Risk Communication

Technology risks are not only found in large industrial plants and hazardous waste facilities. They are also present in the air in your homes, the chemicals used by your neighborhood dry cleaner, and the wood-stoves found in many kitchens. Since controlling these risks has the potential to seriously disrupt individual lives and livelihoods, people insist on being more fully informed about them. They also expect to have a voice in the methods of control. Risk communication is therefore becoming more necessary, but not any easier, than it has ever been in the past.

Risk analysts should not attempt to overstate or understate threats, but rather to give their best estimate of possible occurrence and the specific range of uncertainty. It also seems important for the various federal regulating agencies we rely on for our information to coordinate and standardize their risk assessment procedures so they will arrive at similar estimates for given technological activities.

Progress in the discipline called *risk assessment* has been commendable. If technology managers are to be expected to make enlightened decisions about whether or not to go with a new or modified technology, they must be able to operate with a reliable set of data. This expectation is predicated on improving our contemporary forecasting techniques.

Forecasting for a Responsible Future

When does the future begin? Who is or should be responsible for the way it turns out? Both of these questions are central to the ongoing debates among scholars who study the future. As noted previously in the introduction segment of this chapter, the field of futures studies, or futurology, has grown considerably over the past several decades. The pictures these professionals paint of tomorrow's world illustrate a wide selection of scenarios. Some views are exciting, others are hopeful, and still others are rather grim. These images are not so much predictions of the future as they are reflections of society's contemporary concerns about the quality of life on Earth.

Advertising agencies play an interesting role in future forecasting these days. They are not directly involved in the futurology profession per se. Their futuristic slogans are really created to convince people to support their clients' products or services; however, they also seem to be making a business out of defining the future. The manner in which the essence of time is used tells consumers about the product's image with reference to the past, present, and future.

Consider some of the following expressions found in a variety of contemporary advertisements:

"The future is now."

"The future is right around the corner."

"We've been working on the future for two hundred years."

"We represent tradition with a future."

"Tomorrow's technology, available at today's prices."

"The future is already here."

"Our products let you experience the future before its time."

"This is the wave of the future."

"Don't let the future pass you by!"

"New pathways to the future, where creativity knows no bounds."

Each of these slogans implies the future is something to look forward to with a smile on your face and a positive attitude. See Figure 13-7. The statements define the future via a clear linkage with the past. They are

Figure 13-7.
Founded in the early 1900s, this youth organization proudly displays its belief in the future course of its social agenda. Still very active in community projects, the Girl Scouts of Santa Clara County used a slogan about the future, denoting an alliance with the past. Even though this photo was taken a while ago, the Girl Scouts' current website states "Girl Scouting has held on to its traditional values while maintaining a contemporary outlook—a dual focus expected to continue into the twenty-first century." (Michael Dafferner)

meant to be comforting, from the perspective that the future is not only going to be better, but it will also be familiar.

Present-day advertisers may be described as marketing soothsayers, but their primary objective is to sell goods. The business of technology forecasting is much more comprehensive with regard to its justification for existence.

Forecasting Techniques

The rise of the predicting profession of futurology has simultaneously popularized several forecasting tricks of the trade. Three techniques that have earned respect among the futures studies community include the futures wheel, content analysis, and the Delphi study.

The *futures wheel* projection technique is a small-group brainstorming session in which the participants examine the consequences of a particular social trend or a technological event. This type of activity may precede important policy making decisions. It may also be used to format the early discussions within a technology assessment exercise. In developing a futures wheel, the group members physically prepare a visual diagram of their opinions and hypotheses about the event or trend being evaluated. Refer to Figure 13-8.

In the first phase of this procedure, the group leader is required to identify and describe the technology action, event, or social trend the group members are supposed to discuss. This event becomes the hub of the wheel in the diagram. Participants are asked to describe the immediate probable consequences of the trend, with the proviso that any member of the group may veto any consequence perceived to be unlikely. The search is for a lineup of first-order consequences the group can agree on unanimously. Each of these first-order consequences next becomes the

center of its own wheel, as the group continues to make projections about second-order consequences, while the veto rule remains in effect. The exercise proceeds until the wheel has expanded out to include a list of third- and fourth-order consequences. This type of discussion activity enables the experts to review a vast selection of cause and effect relationships in a relatively short period of time. The entire list of consequences of an action, event, or trend can be further analyzed to make an educated forecast about future possibilities.

Content analysis is a forecasting method in which reviewers keep track of events as they occur, in order to make an accurate projection of major trends one to two years down the road. Although it had its initial use by military strategists during World War II, content analysis gained popularity during the 1980s and into the 1990s by John Naisbitt's group. This group published a tri-annual newsletter called *The Trend Report*, which compiled much of its information through an in-depth analysis of daily newspapers printed in some two hundred U.S. cities. The premise behind content analysis is this: All you need to obtain a preview of the future is an awareness of the changes taking place right now. Companies endorsing this technology forecasting strategy are able to *reconceptualize* the nature of their own businesses in anticipation of changes taking place in the external business environment. In other words, they are not forced into quick fixes, bailouts, or crisis reactions to changes for which they were not prepared.

Another analyst who takes what might be called a synoptic approach to future studies is Joseph Coates, who founded his Washington, D.C.–based consulting firm of Coates & Jarratt in the late 1970s. His approach, similar to content analysis, is to study the current trends and developments

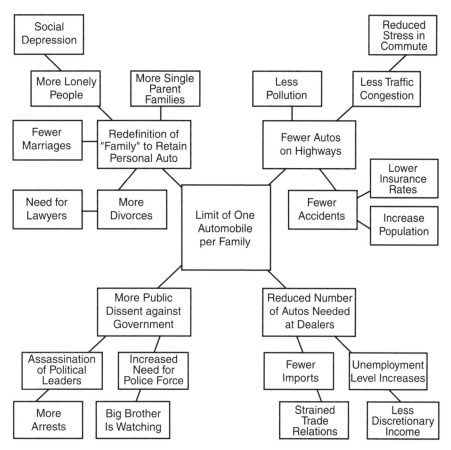

Figure 13-8.
Here is an example of a futures wheel. This is a brainstorming technique used to project probable consequences of a particular social trend, legal action, technological development, or other such topic. (Prepared by students in Technology and Civilization, offered through the Department of Aviation and Technology at San Jose State University)

in science and technology and project them into future scenarios. The scenarios his firm forecasts include economic trends, political agendas, and social issues. Coates began his futures work professionally at the NSF, and he later moved to the OTA. Today, scenarios development is a major component of the futures research he takes on at Coates & Jarratt, and the firm's studies generally look ahead five to fifty years. His firm works closely with clients, as they are shown new opportunities and choices for dealing with changes in science, technology, and society.

The *Delphi study* process involves an anonymous exchange of opinions through an intermediary who controls the feedback of opinion in sequential rounds of interchange. Delphi studies have proven to be useful for long-range forecasting of technological and societal developments. Corporations and government agencies have conducted future-oriented Delphi studies dealing with topics like political affiliations, technological advancements, technology transfer, environmental degradation, medical developments, military strategy, and economic indices.

When reasonable consensus had been reached, the potential events described provided a basis for subsequent planning and action. The Delphi study is essentially designed to overcome some of the difficulties encountered when experts try to arrive at a consensus during face-to-face interaction in a committee meeting.

In a typical Delphi study, the participants are initially selected on the basis of their expertise or interest in a particular subject area (such as nuclear physics, meteorology, religion, education, or futurology). They are sent a series of questionnaires through the mail. In the first round, for example, they might be asked to provide their *judgment* as to the probable dates of occurrence of a group of events, technological developments, or scientific discoveries (2005, 2010, 2020, after 2050, or never). The intermediary collates the responses to reveal the spread of opinions. This data is sent back to the participants in a second questionnaire. In round two, the respondents are given the opportunity to *revise* their estimates in view of the overall group response. Respondents whose answers are either above or below the responses given by the majority are asked to *provide reasons* for their opinions. These narratives, along with the new figures of the group as a whole are collated and presented to the participants in a third questionnaire. In the third round, they are again asked to reassess their earlier estimates in light of the new group response profile and the reasons given by the deviants. As the study progresses to a possible fourth and fifth round, there is a convergent tendency toward a total group consensus.

Each of these forecasting techniques is reliant on the judgment and opinions of the people involved with the exercise. The credibility of the results achieved is therefore contingent on both the imagination and technical competency of the forecasters employed. You may one day be asked to participate in a forecasting activity as part of a technological assessment, environmental impact analysis, or even a Defense Department planning session. Your input into such a forecast, based on careful judgment, might serve to provide a coherent structure for testing alternative actions being contemplated, warning that certain actions are too risky and should be avoided, or defining the scope of reasonable expectations in the technological world of tomorrow. If this sounds like a tremendous responsibility, your perception is right on target.

The Technologist's Responsibility

As time passes, science and technology are becoming less a matter of hardware and more affixed to the structure of our social network. Scientists and technologists are therefore enjoined to be responsible to the social groups in which they operate. Where are we headed with our science and technology? Where should we be headed? Where do we go from here? Where is here? We are in the midst of a host of technological evolutions that generate much to hope for and much to fear. Tomorrow's world will present more advancements and just as many setbacks. Ten years from now, many of our current speculations will have become clear to us, but they will be replaced with a whole new set of unknowns as we move further ahead into the twenty-first century.

A significant portion of contemporary society has become mesmerized by technology. In search of answers regarding what to do about the future course of events on their planet, social groups often look to the experts for guidance. In other instances, they place their trust in computerized expert systems for immediate solutions to long-term social and political problems. In many ways, the technologists and scientists have come to the rescue. Their contributions to humanity, however, present many paradoxes:

- ❏ We have more knowledge and information, but not a lot more understanding.
- ❏ We have more comforts and conveniences, and we also have more risks.
- ❏ We have more arms and military prowess, but no greater sense of security; we have more communications technology, but less community cohesion.
- ❏ We have more education, but less appreciation of the efforts needed to secure the degrees affiliated with it.
- ❏ We have hundreds more choices about how to spend our time, but less time to choose among them.

These realities of the day seem to make us want to ask two questions:

- ❏ To whom do we owe our appreciation for these miraculous achievements?
- ❏ Who should we hold responsible for these ominous threats to our general well-being and survival?

The obvious first response to both questions is to say the people in charge of science, technology, and engineering should be both thanked and blamed for their numerous innovations. If one delves a bit deeper, however, it should become apparent that every person in society operates in more than one social circle and is no less responsible for the profile of the future than any other person. The social management of technological change is enmeshed in a complex network of objectives, including international cooperation, political stability, economic growth, renewed community vitality, scientific vision, technological responsibility, and a social conscience regarding the future of civilization on this planet. Regardless of educational training or technical background, every person has the unique capacity to make a contribution toward at least one of these objectives.

You have certainly had the experience at one time or another of being called on to make an important decision affecting the lives of other people around you. Perhaps you found it necessary to step away from the situation to collect your thoughts on more objective footing before taking any action. This mental relaxation strategy illustrates the human awareness of the need to use caution when other people's lives are involved. Nearly twenty years ago, Joseph Weizenbaum, an esteemed pioneer in the field of AI and Emeritus Professor of Computer Science at MIT, was quoted with the following sentiment:

> What's needed more than anything else, I believe, is an energetic program of technological detoxification. We must admit that we are intoxicated with our science and technology and that we are deeply committed to a Faustian bargain that is rapidly killing us spiritually and may eventually kill us all physically.[3]

In response to the question "where do we go from here?" even though this recommendation was made some time ago, it seems to make a whole lot of sense in today's world of uncertainties. Human beings have the responsibility to each other to act as leaders in society's technological R & D pursuits. Each of us must make a contribution toward the improvement of humanity's existence to bring it into accordance with both the natural environment and the technological environment.

As each of the chapters in this book has shown, responsible citizens must keep abreast of new and emerging technologies in order to intelligently determine whether or not the benefits outweigh the risks of its development. In other words, decisions made under the influence of an overdose of

---

[3]Long, M. 1986. The West Interview with Joseph Weizenbaum. *West Magazine* (19 January): 4–5.

faith in technology may not always be the ones most responsive to the needs and expectations of the populace.

End Line

If this seems a rather ambiguous ending to an otherwise concrete presentation, you are once again quite correct in your perception. The fact is, we can *always* pick up from wherever we left off, as long as we can still *find* that end point.... And here we are! Each of the preceding chapters, and this one as well, presents a list of suggested readings for further study; in fact, the list of suggested readings for this final chapter is especially lengthy, given the diverse nature of the topics discussed. These readings provide an excellent starting point if you are interested in increasing your understanding of technology. Perhaps they will also pique your curiosity enough that you will become personally committed to the construction of a responsible technological future. Stay with it!

Discussion Questions

1. Explain the organizational concept known as *decentralization*. How can it be used to help a corporation keep abreast of technological change?

2. Where do you believe the most significant advances in leading edge technologies will take place in the next decade? Provide your rationale.

3. Identify and describe the roles and functions served by the former Office of Technology Assessment. Generate a list of five to ten contemporary topics or issues in technology you believe warrant investigation by a replacement agency, should our government establish one. Explain why you included these topics on your list.

4. Work with two or three members of your class and study several current newspapers to locate an incident related to science, technology, or engineering ethics. As a group, do an analysis of the case to determine how ethical and moral values did or did not play a role in the event.

5. Explain what is meant by the term *risk assessment*. Why do people attribute a greater amount of credibility to historical risk assessments than to those based on computer models? Provide examples to highlight your answer.

6. Develop a futures wheel illustrating a variety of first-, second-, and third-order consequences related to an issue or event in contemporary science and technology.

Suggested Readings for Further Study

Adams, L. 2001. Mining the world of quality data. *Quality* 40 (8): 36–38.

Aldridge, S. 1998. Ethical dilemmas. *New Scientist* 160 (2156): supplement 1–4.

Asimov, I. 1985. The policy of confrontation: A path to certain destruction. *State Government* 59 (4): 139–140.

Bak, D. 2001. Optics on-board. *Design News* 56 (14): 60–62.

Bearman, T. C. 1987. The information society of the 1990s: Blue skies and green pastures? *Online* 1 (1): 82–86.

Benditt, J. 2001. Bankrolling the future. *Technology Review* 104 (8): 78–83.

Bertone, P. and M. Gerstein. 2001. Integrative data mining: The new direction in bioinformatics. *IEEE Engineering in Medicine & Biology* 20 (4): 33–40.

Blankenship, J. and J. Kenner. 1986. Images of tomorrow: How advertisers sell the future. *The Futurist* 20 (3): 19–20.

Bloch, E. 1985. The challenge of science and technology in the states. *State Government* 59 (4): 144–145.

———. 1986. Managing for challenging times: A national research strategy. *Issues in Science and Technology* 2 (2): 20–29.

Brainard, J. 2001. The waning of influence of scientists on national policy. *The Chronicle of Higher Education* (23 November): A19–A22.

Brody, H. 1991. Great expectations: Why technology predictions go awry. *Technology Review* 94 (5): 38–44.

Brooks, A. A. 1995. Test animals and risk. *Science* 270 (17 November): 1101–1102.

Brown, A. 2001. Sometimes the Luddites are right. *The Futurist* 35 (5): 38–41.

Bylinsky, G., et. al. 1986. The high-tech race: Who's ahead? A *Fortune* Special Report. *Fortune* (13 October): 26–37, 39–40, 42–44.

Cetron, M. 1985. Technology will shape the way we live. *State Government* 59 (4): 148–150.

Cetron, M. J. and O. Davies. 2001. Trends now changing the world: Economics and society, values and concerns, energy and environment. *The Futurist* 35 (1): 30–43.

Clarke, I. F. 1986a. American anticipations: The first of the futurists. *Futures* 18 (4): 584–592.

———. 1986b. American anticipations: Is the future what it used to be? America views, 1945–1985. *Futures* 18 (6): 808–820.

Coates, V. T. 2001. The need for a new Office of Technology Assessment. *The Futurist* 35 (5): 42–43.

Cope, R. 2001. Manufacturing with 2020 intelligence. *Manufacturing Engineering* 127 (2): 120.

Cornish, E. 2001. How we can anticipate future events. *The Futurist* 35 (4): 26–33.

Crawford, A. 2000. The ethical dimension. *Chemistry and Industry* 18 (18 September): 590.

Crisis in science funding, The: An interview with Robert M. White. 1991. *Technology Review* 94 (4): 40–47.

Dalton-Taggart, R. 2001. The Internet is dead: Long live the Internet. *Machine Design* 73 (11): 196.

Davies, O. 1985. Scenes of future perfect. *Omni* 7 (8): 76–78.

De Gortari, A. and M. Waissbluth. 1990. A methodology for science and technology planning based upon economic scenarios and Delphi techniques: The case of Mexican agroindustry. *Technological Forecasting & Social Change* (July): 383–398.

Eddison, E. B. 1987. Choreography for technology and humans. *Online* 1 (1): 49–50.

Elsner, R., C. A. Martin, and C. M. Delahunty. 2001. Ethically responsible research. *Food Technology* 55 (3): 36–42.

Fairley, P. 2000. The microphotonics revolution. *Technology Review* 103 (4): 38–44.

Fresco, J. and R. Meadows. 2002. Engineering a new vision of tomorrow. *The Futurist* 36 (1): 33–36.

Garcia, K. 1987. Physicist is returning to Livermore lab. *San Jose Mercury News* (20 January): B-1, B-5.

Gibson, W. D. 1986. A maze for management: Choosing the right technology. *Chemical Week* (7 May): 74–78.

Hamilton, D. 1986. Technology and institutions are neither. *Journal of Economic Issues* 20 (2): 525–532.

Hattis, D. and D. Kennedy. 1986. Assessing risks for health hazards: An imperfect science. *Technology Review* 89 (4): 60–71.

Hecht, J. 1999. Wavelength division multiplexing. *Technology Review* 102 (2): 72–77.

Heinze, D. T., M. L. Morsch, and R. E. Scheffer. 2001. LIFECODE: A deployed application for automated medical coding. *AI Magazine* 22 (2): 76–88.

Hill, R. 1984. Today's thorniest management problem: New technology. *Inter-national Management* 39 (12): 58–59, 62, 65.

Hong, J., et al. 1990. Tomorrow's technologies at your doorstep. *Design News* (12 March): 103–105.

Howard, J. A. 1985. Taking the blinders off: A hopeful view of America in the years ahead. *Vital Speeches of the Day* (1 March): 315–317.

Huber, P. 1984. Discarding the double standard in risk regulation. *Technology Review* 87 (l): 10–11, 14.

[Internet].(1999). *Non-profit and university participation in ATP.* <http://www.atp.nist.gov/alliance/npa_part.htm> [8 Jan 2002].

Kiplinger forecasts: The new American boom. Exciting changes in American life and business between now and the year 2000. 1986. Washington, D.C.: Kiplinger Washington Editors.

Latour, B. 1998. From the world of science to the world of research? *Science* 280 (5361): 208–209.

Lave, L. B. 1987. Health and safety risk analyses: Information for better decisions. *Science* 236 (17 April): 291–295.

Leonard-Barton, D. and W. A. Kraus. 1985. Implementing new technology. *Harvard Business Review* 63 (6): 102–110.

Lepkowski, W. 1997. Projecting the future rationally. *Chemical & Engineering News* 75 (14 July): 33–35.

Long, M. 1986. The West Interview with Joseph Weizenbaum. *West Magazine* (19 January): 4–5.

Malakoff, D. 2001. Memo to Congress: Get better advice. *Science* 292 (5525): 2229–2230.

Marien, M. 1984. Hope, fear and technology: Changing prospects in the 1980s. *Vital Speeches of the Day* (15 December): 137–142.

Markusen, A. and J. Yudken. 1992. Building a new economic order. *Technology Review* 95 (3): 22–30.

May, R. M. 2001. Science and society. *Science* 292 (5519): 1021.

McDaniel, M. F., R. D. Siegel, and K. J. Leuschner. 2001. Develop a strategy for residual risk. *Chemical Engineering Progress* 97 (1): 31–36.

McKelvey, J. P. 1985. Science and technology: The driven and the driver. *Technology Review* 88 (l): 38–47.

Meadows, D. H. 1985. Charting the way the world works. *Technology Review* 88 (2): 54–63.

Menkus, B. 1986. Predicting the future is a chancy thing at best. *Journal of Systems Management* 37 (11): 5.

Morgan, M. G., A. Houghton, and J. H. Gibbons. 2001. Improving science and technology advice for Congress. *Science* 293 (5537): 1999–2000.

Naisbitt, J. 1984. *Megatrends: Ten new directions transforming our lives.* New York: Warner.

Nasukawa, T. and T. Nagano. 2001. Text analysis and knowledge mining system. *IBM Systems Journal* 40 (4): 967–984.

National Science Board. 1991. *Science & Engineering Indicators—1991.* Washington, D.C.: U.S. Government Printing Office.

Ojala, M. 2001. Easy content searching with EasyAsk. *Econtent* 24 (3): 62–63.

O'Neill, G. K. 1983. *The technology edge: Opportunities for America in world competition.* New York: Simon and Schuster.

Parkinson, S. and V. Spedding. (2001). *An ethical career in science and technology? Scientists for global responsibility.* [Internet]. <http://www.sgr.org.uk> [9 Jan 2002].

Patton, C. 1986. Abdicating responsibility for leadership can have disastrous consequences. *Electronic Design* (3 April): 223.

Pritchard, M. S. 2001. Responsible engineering: The importance of character and imagination. *Science and Engineering Ethics* 7 (3): 391–402.

Quinn, J. J. 1985. How companies keep abreast of technological change. *Long Range Planning* 18 (2): 69–76.

Ruckelshaus, W. D. 1984. Risk and society. *Consumer Research* 67 (6): 37–38.

———. 1985. Managing risk: The people's choice. *Across the Board* 22 (12): 5–7.

Russell, M. and M. Gruber. 1987. Risk assessment in environmental policy-making. *Science* 236 (17 April): 286–290.

Salerno, L. M. 1985. What happened to the computer revolution? *Harvard Business Review* 63 (6): 129–138.

Satish, S. C. 1986. Futures research: The need for anticipatory management. *Futures* 18 (3): 366–368.

Schulz, W. 2000. Report puts a magnifying glass on federal R & D effort. *Chemical & Engineering News* 78 (26): 13.

———. 2001. Advising Congress: Holt introduces legislation to reestablish Congress' Office of Technology Assessment. *Chemical & Engineering News* 79 (30): 29.

Seeman, B. T. 2001. Events of September 11th prod scientists to innovation. *The Post-Standard* (2 November): A-1, 1–14.

Shefrin, B. M. 1986. Images of the future, futuristics and American politics. *Technological Forecasting and Social Change* 30 (3): 207–219.

Sheridan, P. 1990. Risk assessment: The path of uncertainty. *Occupational Hazards* (October): 63–68.

Sivy, M. 1985. What we don't know. *Money* 14 (11): 209–210, 212.

Slaughter, R. A. 1987. Future vision in the nuclear age. *Futures* 19 (l): 54–72.

Slovic, P. 1987. Perception of risk. *Science* 236 (17 April): 280–285.

Steele, L. 1983. Managers' misconceptions about technology. *Harvard Business Review* 83 (6): 133–140.

Sterling, T. 2000. Managing the risks of going online. *American City & County* 115 (13): 12.

Stock, O. 2001. Language-based interfaces and their application for cultural tourism. *AI Magazine* 22 (1): 85–97.

Sulc, O. 1986. Integration of scientific forecasts: Methodology for integration of scientific forecasts in the process of national science policy making. *Technological Forecasting & Social Change* 30 (3): 251–260.

Szakonyi, R. 1986. Don't concentrate on traditional R & D management. *Research Management* 29 (5): 6–7.

Tangley, L. 1984. Life after 2000—The debate goes on. *Bioscience* 34 (8): 477–479.

Taylor, A. 1985. Talk, future shock and future schlock: Real education teaches problem solving. *Vital Speeches of the Day* (1 September): 680–683.

Technology Review Ten, The. 2001. 10 emerging technologies that will change the world. *Technology Review* 104 (1): 97–113.

Thomas, L. M. 1986. Risk communication: Why we must talk about risk. *Environment* 28 (2): 4–5, 40.

VanRavenswaay, E. 0. and P. T. Skelding. 1985. The political economics of risk/benefit assessment: The case of pesticides. *American Journal of Agricultural Economics* 67 (5): 971–979.

Wagschal, P. H. and L. Johnson. 1986. Children's views of the future: Innocence almost lost. *Phi Delta Kappan* 67 (9): 666–669.

Wenk, E., Jr. 1986. *Tradeoffs. Imperatives of choice in a hightech world.* Baltimore: John Hopkins Univ. Press.

Wilson, R. and E. A. C. Crouch. 1987. Risk assessment and comparisons: An introduction. *Science* 236 (17 April): 267–270.

World trends and forecasts. 1987. *The Futurist* 2 (1): 40–48.

Wozniak, G. D. 1987. Human capital, information, and the early adoption of new technology. *Journal of Human Resources* 22 (l): 101–112.

Wyman, J. 1985. SMR Forum. Technological myopia—The need to think strategically about technology. *Sloan Management Review* 26 (4): 59–64.

Glossary

acid rain: [C-9] A serious environmental problem resulting from the presence of sulfur oxides and nitrogen oxides in the atmosphere. When these compounds react with moisture, they form dilute sulfuric and nitric acids, which are then carried back to the ground.

aerobraking: [C-4] A technique used by NASA scientists to slow down a spacecraft by taking advantage of a planet's atmosphere.

agile manufacturing: [C-6] An industrial construct that focuses on being fastest to market, with the lowest total cost and greatest ability to meet varied customer requirements.

agroecology: [C-8] An approach to agriculture promoting sustainability and linking ecology, socioeconomics, and culture to the design of agricultural systems.

agroforestry: [C-9] A technology management strategy proposing creative farming techniques, combining trees (forest preservation) with other crops or livestock.

Air Quality Index (AQI): [C-9] The updated version of the Pollutant Standards Index. It includes data on several commonly occurring pollutants and enables comparisons to be made between different regions of the country.

amniocentesis: [C-5] A medical test normally performed during a woman's sixteenth week of pregnancy to determine if the unborn child has certain birth defects. Amniotic fluid is withdrawn through a thin hollow needle inserted into the abdomen. Fetal cells are then cultured in the laboratory and chromosomes are carefully studied. Accuracy of the chromosome results is greater than 99.5 percent.

appropriate technology (A.T.): [C-7] An intermediate level of technology development, as opposed to one too sophisticated or complex for the problem in question. A.T. is also an attitude regarding an overreliance on high technology. It refers to technology that fits the cultural situation for which it is proposed without causing more problems than it solves.

artificial satellite: [C-4] An artificial object, such as a telecommunications satellite, orbiting a larger astronomical object, usually a planet.

balloon angioplasty: [C-5] Formally referred to as balloon dilation therapy. It is a medical procedure that can replace coronary bypass surgery for clearing clogged arteries in a person's heart. Other angioplasty procedures have the potential to clear clogged arteries in the neck, brain, kidney, and abdomen.

bar coding: [C-6] A primary component of automatic identification systems in industry. The bar code industry uses several different code languages, or symbologies, for which uniform symbol descriptions have been published. Each code is generally made up of binary digits, in which bars and spaces in different configurations represent numbers, letters, or other symbols. The code is scanned and transmitted directly into a computer to identify the part or product.

bioastronautics: [C-4] The study of life in space. Bioastronauts study how living organisms adapt to conditions of weightlessness and then readapt to life on Earth. They also research and develop technological innovations to make this cycle occur more smoothly.

biological weapon (BW): [C-10] A weapon that uses a living organism, usually a pathogenic microorganism, for hostile purposes. Biological agents can be bacterial, fungal, viral, rickettsial, or protozoan.

biomass: [C-8] Organic material used to supply energy.

biomining: [C-2] A technique in mining that uses bacteria to eat away any unwanted iron and minerals, leaving the valuable mineral behind.

biotechnology: [C-2] The manipulation of biological organisms to make products benefiting human beings.

bioterrorism: [C-10] The use of biological organisms by terrorists to attack a country and its people.

blastocyst: [C-2] The hollow sphere of cells resulting from the division of the single cell formed by the union of a sperm cell and egg.

busyness syndrome: [C-11] A concept suggesting it is almost a status symbol in our society to be constantly busy. People who methodically check their date books or palm pilots to find they have no time to meet a friend for lunch are viewed as more successful than those who have lots of free time on their hands.

cell fusion: [C-2] A union involving the artificial joining of cells to combine the desirable characteristics of different types of cells into one cell, known as a hybridoma.

chaff: [C-10] A penetration aid included in the deployment of nuclear missiles. It consists of thin, confetti-like metallic strips released to confuse the opponent's radar system.

chemical weapon (CW): [C-10] A weapon that uses poisonous, asphyxiating, or other gases affecting humans when they are either inhaled or absorbed through the skin.

clone: [C-2] An exact genetic copy. It is an individual grown from a single somatic cell of its parent. Clones are genetically identical to their parents.

cluster bomb unit (CBU): [C-10] A common workhorse weapon that can deliver lethal explosives to personnel over expansive areas. Instead of having one large explosion, these bombs contain dozens, hundreds, or even thousands of bomb live units.

command, control, communications, and intelligence (C^3I): [C-10] Pronounced "see-cubed-eye" or "see-3-eye," this is a military acronym. In the U.S. military, C^3I provides information about communications, counterintelligence, security, information surveillance, and reconnaissance to the military forces so they can succeed in every mission.

command, control, communications, computers, intelligence, surveillance, and reconnaissance (C4ISR): [C-10] The military's research and information plan that uses improved information technologies to enhance military operations.

common spaces: [C-9] Regions of the world entirely or partially considered to be international territory. These areas—Antarctica, the open sea, and outer space—are beyond national jurisdiction.

compellance: [C-10] The deployment of military power in order to convince another country to change its behavior to make it more satisfactory to the aggressor.

computer-aided design (CAD): [C-6] The use of a computerized drafting and design program to design a new product to be manufactured.

computer-aided manufacturing (CAM): [C-6] A manufacturing strategy using computers and information technologies to control, manage, operate, and monitor manufacturing processes.

computer blacklists: [C-3] Databases containing the names and identification codes for individuals who should not be given credit, automobile insurance policies, or the like, due to some legal problems in their past. These are regularly compiled and updated by financial institutions and government agencies. This form of electronic record keeping has caused many people to question their rights to privacy.

computer integrated manufacturing (CIM): [C-6] The umbrella acronym (pronounced "sim") for a host of automation technologies in the manufacturing environment. It does not refer to one specific technology, but to the integrated use of computers in all sections of enterprise, from the planning of production, through the design and manufacture of a product, up to the assurance of good quality.

computer mediated communication (CMC): [C-3] An area of study including the use of computers and telecommunications technology to enable people to communicate with one another over real or perceptual distances. Two widely used forms of CMC include e-mail and computer conferencing.

concurrent engineering (CE): [C-6] The central emphasis is "getting it right the first time" rather than "redoing it until it is acceptable." It requires a systematic and collaborated effort to simultaneously design and develop a product and its manufacturing processes. The creation of an integrated product team should include the customer and people from engineering, management, manufacturing, and quality assurance.

content analysis: [C-13] A forecasting technique in which reviewers keep close track of events as they happen, in order to make an accurate projection of major trends one to two years down the line. Content analysis often makes use of daily newspapers as its primary source of reference material. It had its roots among military strategists during World War II.

cookies: [C-3] Files placed on users' computers when they visit a specific website. They generally store personal information about the users, such as passwords and screen names, and are stored on the hard drive of a user's computer.

Cooperative Threat Reduction (CTR) Program: [C-10] A Defense Department initiative that has the primary goal of preventing the widespread acquisition of nuclear, biological, and chemical weapons.

counter-force targeting: [C-10] A defense strategy in which weapons are aimed at the military forces of the opponent, particularly at strategic or theater nuclear forces.

counter-value targeting: [C-10] A defensive strategy in which weapons are aimed at soft targets, such as urban population centers, industrial areas, and economic enterprises.

creationism: [C-12] A theological doctrine ascribing the origin of all matter and living forms as they now exist to distinct acts of creation by God. Creationists generally refute theories of evolution, which hold that humankind evolved biologically from a less sophisticated species.

cyberterrorism: [C-10] The electronic disruption of military information or information in our homeland.

data mining: [C-13] This emerging field is also labeled *knowledge discovery in databases (KDD).* It is the process of finding patterns and correlations in large amounts of data. It lies behind the future of a fully personalized World Wide Web on the Internet and much more.

decoy: [C-10] A penetration aid included in the deployment of nuclear missiles. It is a dummy warhead that looks like the real thing, but carries no explosive power.

Deep Space Network (DSN): [C-4] An international network of antennas that keep in constant touch with spacecraft. NASA's current Deep Space Network has three antenna facilities 120 degrees apart around the world—at Goldstone, in California's Mojave Desert; near Madrid, Spain; and near Canberra, Australia.

Defense Advanced Research Projects Agency (DARPA): [C-3] An organization established by the U.S. Department of Defense in the late 1950s. It regularly funds research projects in high-tech fields, most notably those looking at the applications of information technology in the military.

Delphi study: [C-13] A forecasting strategy in which a group of preselected experts engage in an anonymous exchange of opinions through an intermediary who controls the feedback of ideas in sequential rounds of interchange. There is a tendency, over time, toward a converging consensus among the group members.

demand lag: [C-7] Related to the rate of diffusion of technologies being transferred. The time it takes for a technology to gain acceptance in foreign markets after it is first introduced into the innovating country's domestic markets.

design for manufacture and assembly (DFMA): [C-6] An engineering tool for product simplification conveying an evaluation of the product's complexity and subsequent development of design alternatives to simplify product features and improve the firm's capacity for defect free manufacturing.

deterrence: [C-10] The deployment of military power to threaten an adversary and thereby protect a country.

diagnosis-related group (DRG): [C-5] Written into federal law in the early 1980s as the funding mechanism of Medicare. Adopted as a payment scheme, DRGs work this way: After a patient has been diagnosed, the hospital receives a fixed fee reflecting an average cost of curing that condition. If the patient is worse than average, the hospital must pay any cost beyond the DRG allowance. If the patient is better than average, the hospital keeps any money left over.

distance education: [C-12] A subset of technology-enabled teaching and learning. It refers to a formal educational process occurring with the teacher and learner separated by either time or distance (often both).

DNA chips: [C-2 & C-5] Researchers at Stanford University and Santa Clara, California–based Affymetrix led science in building DNA microarrays in the late 1990s. They are more generally called *DNA chips.* A DNA chip is a wafer of silicon, glass, or plastic, onto which DNA strands are arrayed or placed. They are capable of analyzing thousands of genes at a time to identify which ones are active in a particular sample or specimen (such as saliva, blood, or water).

durational expectancy: [C-11] A concept introduced by Alvin Toffler in his *Future Shock.* It refers to a culturally-based attitude or expectation regarding how long things are usually supposed to last.

When these expectancies are disrupted, people are forced to cope with unplanned changes.

e-commerce: [C-3] Sales or purchases of goods and services using the Internet.

e-finance: [C-3] An aspect of e-commerce specializing in money management and investments.

electronic mail (e-mail): [C-3] A form of correspondence in which messages can be composed and sent anytime by way of an Internet connection.

entrepreneurial science: [C-7] A trend especially apparent in university research laboratories in which there is a convergence between basic and applied research, such that commercial opportunities emerge. Scientists who are engaged in pure research actually assess the potential of their patented findings to bring in substantial income.

environmental impact analysis: [C-9] The process of conducting a comprehensive study and review of a broad range of environmental features, such as topography, hydrology, geology, and cultural status for a specified land area. The results of this study can aid in decision-making and creation of a plan that is environmentally sound.

environmental management: [C-8] Managing environmental issues through appropriate business strategies.

environmental release: [C-9] The accidental or deliberate release of a contained substance into the natural ecosystem (air, land, or water). It represents a central point of debate in genetic engineering experiments in which genetically altered organisms are allowed to be tested outside the laboratory.

epidemiology: [C-13] The systematic study of widespread epidemics (contagious diseases) or physiological maladies that have reached epidemic proportions.

Medical researchers often use laboratory animals in this field of research, as it relates to risk assessment for the human population.

Escherichia coli (E. coli): [C-2] The single-celled bacterium commonly found in the human intestinal tract. It is used in recombinant DNA research because of the ease with which it can be engineered. *E. coli* bacteria reproduce themselves every twenty-five minutes, even after their genetic code has been altered via splicing.

ethnobotany: [C-8] An initiative that emerged from the scientific field studying how people from a particular culture use indigenous plants.

evapotranspiration: [C-9] A naturally occurring process akin to Earth's tropical forests and their role in global climate management. As tropical forests breathe, they release moisture into the air, which later accounts for certain rainfall patterns. Depletion of the forests could result in a drying out of the equatorial region and a northward shift of agricultural lands.

expendable launcher: [C-4] A throwaway booster rocket used to launch payloads into an Earth orbit. The word *expendable* means they burn after they are disengaged from their cargo.

expert system: [C-12] An intelligent computer program using knowledge and inference procedures to solve problems challenging enough to require a significant amount of human expertise for their solution. It portrays a model of the expertise of the best minds in a particular discipline (such as medicine, prospecting, or air traffic control).

extranet: [C-3] An intranet a company has made accessible to customers and other users, even if only on a limited basis.

false negative/positive: [C-5] A negative consequence associated with the purchase of self-diagnostic kits for use in the home. These may occur when the user does not perform the test correctly or misinterprets the results and finds, erroneously, that he or she does or does not have the condition for which the test is being administered.

faster, better, cheaper (FBC): [C-4] A policy adopted by NASA to reduce costs in the early 1990s. Its goal was to shorten development time, reduce costs, and increase the scientific output of NASA by having more missions in less time.

fiber optics: [C-3] A technology that converts data, sound, or images to light impulses, which are propelled by a laser across a thin strand of glass at very high speeds. Many of these glass threads are bundled together and buried in the ground or hung from existing power or telephone lines.

firewall: [C-3] A combination of hardware and software placed between a private intranet and the Internet to keep the intranet private and protected. It examines all information entering and leaving the intranet to make sure there are no unauthorized messages, thus, stopping hackers from breaking into a company's computer system.

first strike: [C-10] Generally refers to a disarming attack on the enemy. Land-based MIRVed missiles are very accurate—they are excellent first strike weapons that can be used to destroy the opponent's missiles before they are even launched. They are also quite vulnerable. Second strike refers to the retaliatory efforts of the opponent using any arms surviving the first strike. *MIRV* is an acronym for multiple independently targetable reentry vehicle. This feature allows one rocket to deliver several warheads with great accuracy to different targets.

flexible manufacturing system (FMS): [C-6] A fully automated, computer controlled production system offering substantial advantages in comparison to the conventional job shop. An FMS is a set of machines linked by a material handling system. All are under central computer control. Flexible machining centers (cells) can produce a variety of parts by a simple change of software. They also allow multiple operations to be performed on a workpiece.

foreign direct investment: [C-7] A channel for technology transfer in which a U.S. manufacturing plant or other type of industrial firm establishes an offshore or overseas subsidiary in order to reduce distribution costs, avoid import tariffs, or decrease labor costs. The parent firm claims either full or partial ownership of the foreign-based plant.

Futures Wheel: [C-13] A forecasting technique first devised by the World Future Society. It entails a small-group brainstorming session in which the participants examine interrelated chains of events that might be consequential to a particular social trend or technology.

gene therapy: [C-2] A field of study in health science, which researchers hope will enable them to cure diseases genetically transmitted from one generation to the next. The procedure entails transplanting healthy genes into the cells of patients who are suffering from disabling diseases in an attempt to increase their chances for survival or decrease the chance that the defect will be passed on to their progeny.

genetically modified (GM) food: [C-2] A new food product created through biotechnology including genes or characteristics not seen in a traditional plant.

Recently, these foods have begun to acquire the negative label of "Frankenfoods."

genome: [C-2] A description of an organism's genetic material, including all the DNA.

geostationary orbit: [C-4] A geosynchronous orbit. It is approximately twenty-two thousand miles away from Earth's surface and directly above the equator. Satellites in this orbit are always in the same position over Earth's surface.

glass cockpit: [C-4] A cockpit that has computerized or digital controls and displays.

global positioning system (GPS): [C-4] An application of satellite technology becoming popular with consumers, although it was first developed for military purposes. It can give an exact position of a person or object, by using a system of twenty-four satellites positioned so a person on Earth can receive a signal from four of them at one time.

global warming: [C-9] A widespread and long-term trend toward warmer global temperatures caused by increasing levels of carbon dioxide and other gases in Earth's upper atmosphere.

green building: [C-8] A building constructed in a way that minimizes waste and includes recycled, renewable, and reused resources to the maximum extent possible.

greenhouse effect: [C-9] The heating of Earth's surface due to increased levels of carbon dioxide in the upper atmosphere, which traps solar energy. This phenomenon is also called *global warming*. It has dramatic ramifications with reference to the stability of the planet's ecosystem.

group technology (GT): [C-6] An approach to manufacturing that segregates parts according to their design or manufacturing characteristics. When similar parts are gathered together, each group can share setups and machine tools, which reduces the cost of production. A GT database is a computerized filing system enhancing the efficiency of work performed in manufacturing cells.

gyroscope: [C-4] A scientific instrument that has a rapidly spinning wheel in a frame, which allows it to tilt freely in any direction or to rotate about any axis. Gyroscopes are used in compasses, on ships and aircraft, and in guidance systems. When used in spacecraft, a gyroscope usually monitors motion in one or more of the three spatial movements: vertical, longitudinal, and lateral.

health maintenance organization (HMO): [C-5] Essentially a merger of an insurer and a provider of health care. It is an alternative to traditional fee-for-service providers. It utilizes a fixed payment medical plan by representing a prospective payment on an annual basis. In the HMO form of managed care, consumers buy memberships; when ill, they go to HMO-run clinics to see physicians who are salaried employees. As long as members stay within the HMO sphere, they spend little or nothing beyond the annual fee.

heuristics: [C-13] Rules of thumb based on a compilation of personal experiences over time. They are used to help guide problem solving exercises. In the field of artificial intelligence, they are built into computer programs to enable machines to make educated guesses. Unlike algorithms (mathematical computation), they do not guarantee a solution.

imitation lag: [C-7] Related to the rate of diffusion of technologies being transferred. The time it takes foreign countries to obtain the production capabilities to produce a new device, after the lead nation begins producing it.

information overload: [C-11] An individual's inability to think clearly due to an excessive amount of cognitive stimulation.

integrated pest management (IPM): [C-8] A system of managing pests to keep them at levels at which they cause minimal damage to crops. It attempts to use ecologically sound, non-pesticide methods to reduce and manage pests, such as insects, diseases, and weeds.

intelligent software agent: [C-3] Software acting autonomously on behalf of the user, based on the user's interests and needs. It has the capacity to learn from experiences and respond to unexpected situations. Such a program strives to achieve a known goal and can sense the current state of its environment in order to make continual progress toward the goal.

International Space Station: [C-4] An orbiting space station that will serve as an international laboratory for research. Its construction has relied heavily on the use of the space shuttle to transport materials and personnel into space.

International Organization for Standardization (ISO) 14000: [C-8] A management tool with the focus of managing environmental issues through appropriate business strategies. It describes how a company can manage its business in an environmentally friendly manner.

Internet: [C-3] The largest information packet switching network in the world. A mammoth network of networks, its roots can be found in the academic research community with original funding from the Department of Defense, but its current constituency has extended to the commercial world and into homes and continues to grow at an exponential rate.

intranet: [C-3] A private corporate computer network.

in vitro fertilization (IVF): [C-5] A procreation option for women who are unable or unwilling to conceive with the conventional method of intercourse. The father's sperm fertilizes the eggs outside of the woman's body, in a laboratory dish. The fertilized egg is then reimplanted in the woman's uterus.

just-in-time (JIT): [C-6] A manufacturing philosophy attempting to eliminate waste throughout the system, including inventory at both ends of production and all machinery and manpower that do not add value directly to the product. JIT has its roots in the Japanese automobile industry, which sought to get rid of excess, waste, and unevenness.

lean manufacturing: [C-6] A system meeting high manufacturing or service demands with a small amount of inventory. To become leaner, manufacturers must remove obstacles preventing them from manufacturing with high velocity. Obstacles, such as complicated setups, excessive material handling, poor physical flow, and production interruptions, interfere with an organization's ability to design and build the best quality products in the shortest time possible.

licensing: [C-7] A formal channel for the transfer of technology tied to a legal agreement between parties. The purchaser of the license normally pays royalties to the party or parties who have ownership of the technology in demand.

lithotripter: [C-5] A device supporting a minimally invasive medical intervention performed in a hospital. It is a machine that uses sound waves from outside a person's body to destroy kidney stones.

local area network (LAN): [C-3] A network of cables and switching boxes providing high-speed data communication links between computers or computerized machines operating in the same geographic area.

low-intensity conflict (LIC): [C-10] The dominant military motif that includes incidents such as border skirmishes,

internal battles, police work, civil uprisings, and regional conflicts. These tend
to occur in less developed areas of the
world, and they rarely involve regular
armies on both sides.

manufacturing resource planning (MRP II):
[C-6] An application software structure
that allows for integrated planning of a
manufacturer's material, equipment,
facilities, and people. The integration
requires that the same information (for
example, the sales forecast, bill of materials, and actual orders) is used throughout the company. Some of the new
applications added to MRP II software
in recent years includes field service and
warranty tracking, marketing support,
and engineering change control.

maquiladora: [C-7] A foreign owned and
operated assembly plant, most often
along the U.S.–Mexican border, that
imports components into Mexico,
assembles them into products ranging
from televisions to auto parts, and reex-
ports finished products. Mexican taxes
are paid only on the value added by
Mexican worker assembly activities.

material requirements planning (MRP):
[C-6] A technique for planning future
purchase orders and manufacturing lots,
according to what is required to complete a master production schedule.

microphotonics: [C-13] This emerging array
of technologies represents systems combining electro-optics with fiber optics
with semiconductor micro-machining
technology. The emphasis in research
labs is to develop reliable and cost-
effective integrated optical systems.
These integrated systems include the
use of lasers, photo detectors, optical
fibers, waveguide devices, photonic
crystals, and micro-mechanical structures.

mid-intensity conflict (MIC): [C-10] The
Pentagon's label for periodic battles the

United States may be forced to wage
against well armed regional powers.
Future MICs (like Operation Desert
Storm in the Persian Gulf) are likely to
be fast-paced and high-tech, dependent
on unrestrained use of the most sophisticated weapons systems.

minimally invasive surgery: [C-5] A group
of new medical technologies employing
endoscopic devices, laser beams, or
shock waves instead of a scalpel to penetrate the damaged or affected area of
the patient's body. Robotics often assist
in these minimally invasive procedures.
For instance, arthroscopic (needlelike
devices using a fiber-optic light source)
surgery has transformed many types of
orthopedic injuries from calamities to
brief inconveniences. In some venues,
the term *noninvasive surgery* is used to
describe the processes.

modern environmentalism: [C-9] An
information-based perspective supporting the need for harmony and balance
between industrial objectives and environmental protection.

monoclonal antibody (MAb): [C-2] A protein in the bloodstream generated by an
organism's immune system in reaction
to foreign proteins and polysaccharides.
It neutralizes the foreign agents and
produces immunity against certain
microorganisms or their toxins. The
clones of a single hybridoma cell produce
MAbs. A hybridoma is a modified cell
resulting from the fusion of two different
types of cells. They are homogeneous,
and their production is predictable and
repeatable. MAbs can potentially be
used to diagnose diseases and purify
proteins.

most favored nation (MFN): [C-7] MFN
trading status is that in which one country gives another country the tariff treatment enjoyed by its least restricted

trading partner. Most recently, the term *permanent normal trade relations (PNTR)* has been used instead of *MFN*.

mutual assured destruction (MAD): [C-10] A military defense policy attempting to deter opponents from a first strike against each other. Where both sides have the ability to absorb a first strike and deliver a retaliatory second strike capable of inflicting an unacceptable level of damage, a situation of MAD is said to exist.

natural language processing (NLP): [C-13] The central goal of these investigations is to design and build computer systems that can analyze, understand, and generate language that humans use naturally. At some point, we may eventually be able to talk to our computers like we talk to each other. In this field of study, natural language analysis includes lexical, morphological, syntactic, semantic, and discourse processing. NLP is critical to research in the field of data mining.

new urbanism: [C-8] The movement of middle and upper class people living in the suburbs into town, where they typically purchase or rent space in what are called *infill developments.*

nongovernmental organization (NGO): [C-8] A nonprofit or for-profit organization established to promote a particular cause. A NGO is not affiliated with any particular country; rather, it works as a quasi-governmental agent to affect change in the world.

nuclear family: [C-12] This concept gained popularity across post–World War II U.S. villages, towns, and cities. The nuclear family unit includes a working father, a stay-at-home mother, and two or more dependent children who live in their purchased home. Although largely outdated, its image persists in a significant number of advertisements.

nuclear winter: [C-10] A variety of environmental consequences that may result from a full-scale nuclear war. Researchers have theorized that Earth's surface will be very dark and cold due to a massive temperature inversion caused by the nuclear debris scattered in the upper atmosphere. Any life-forms that do not perish as a direct result of war will die soon after in the cold snows of a long nuclear winter's night.

optical network: [C-3] A network using optical fibers to transmit data.

optoelectronics: [C-13] A field of technological research and development that has given us, among other things, fiber optics. It represents an interdisciplinary blend of electronics and optics technologies.

parental leave: [C-12] The Family and Medical Leave Act was passed in the United States in the early 1990s. It requires all employers with fifty or more paid employees to provide up to twelve weeks of job protected leave for both the employee's own health needs (including pregnancy) and for the care of newborn children. The law provides parental leave of the same duration for the care of a new foster or adopted child or the serious health condition of a child, spouse, or parents.

pharmacogenomics: [C-5] A new science in the field of medicine combining knowledge of a person's genome with what is being done in the pharmaceutical industry. The notion that one drug fits all people is being supplanted by theories suggesting we can manufacture drugs more accurately suited to a person's genetic makeup.

pharming: [C-2] The use of genetically altered livestock to produce drugs.

Pollutant Standards Index: [C-9] Developed by the Environmental Protection Agency with the Council on Environmental Quality to comply with the Clean Air Act. It is used to convert the concentration of pollutants in the air to a number between zero and five hundred. These numbers equate with a series of descriptive air quality terms (such as *good, unhealthful,* and *smog alert).*

preferred provider organization (PPO): [C-5] This form of managed medical care has an advantage over the HMO in that it can be imposed on the present network of hospitals and doctors without having to build clinics or convert doctors into employees. In a PPO, a group insurance buyer agrees to steer a company's employees to particular hospitals or doctors in return for volume discounts.

quality function deployment (QFD): [C-6] A systematic way for a manufacturing firm to identify customer requirements and convert them into design and manufacturing needs. It is a management-planning tool that can be used in any phase of production to help concerned personnel identify the critical design parameters. These parameters are then optimized through quality engineering to minimize variation during production.

rapid prototyping (RP): [C-6] An emerging technology associated with customized fabrication. RP machines are computer driven units that fabricate parts directly from design data. Stereolithography is an example of an RP process that builds a prototype by depositing layer upon layer of powdered metal.

recombinant DNA (rDNA): [C-2] Presently one of the most refined genetic engineering technologies. It involves splicing specific codes along the DNA chain of a host organism in the laboratory; the engineered organism will reproduce itself with the new DNA sequence. The term *clone* is often used with reference to this procedure.

renewable energy: [C-8] Sources of energy that are always available because, in reality, they cannot run out, and they can be replenished.

response modes: [C-11] A variety of human behaviors. They are influenced by emotions, expectations, experiences, and the nature of the stimulating incidents. Response modes may equate behavioral responses to changes usually thought to be out of one's direct personal control.

reverse engineering: [C-7] An astute capacity to carefully disassemble an imported device, examine it, and eventually copy it. This is a form of an unplanned transfer of technological know-how taking place through a formal channel (export of manufactured goods).

risk assessment: [C-13] A relatively recent field of inquiry sometimes carrying the title *risk-benefit analysis.* Professionals in engineering, science, and social science are actively conducting both publicly and privately funded studies. Their projects are rigorous and replete with systematic research designs and mathematical models to determine the extent to which society is at risk because of certain modern technologies or their side effects. Some of the most creative minds of our day are probably at work right now on the problem of "How safe is safe enough?"

risk-benefit analysis: [C-11] Attempts to and must ultimately answer the question "How safe is safe enough?" Experts use historical accident and fatality records to reveal patterns of acceptable risk-benefit ratios. Generally, an acceptable risk for a new technology equates to the level of safety associated with ongoing activities providing similar benefit(s) to society.

science: [C-1] A stream of human events involving a mathematical or systematic study of nature resulting in a body of knowledge that is practical, as well as theoretical.

Single Integrated Operational Plan (SIOP): [C-10] The U.S. plan for the conduct of a nuclear war incorporating weapons arrival and survival probabilities, weapons characteristics matched to targets' damage criteria, routing and timing of weapons, and modeling to assure the plan's viability under various conditions.

Skylab: [C-4] A preliminary version of a U.S. space station. It was a huge orbiting space laboratory launched by NASA in the early 1970s. The project was scrapped a few years later, and Skylab burned in the late 1970s after its orbit had decayed.

smart cockpit: [C-4] A cockpit of an aircraft or spacecraft that has computerized or digital controls and displays. A smart cockpit also includes computer software that reduces the workload of the pilot and assists the pilot in flying.

smart growth: [C-8] Development accommodating the needs of a community without sacrificing the environment. It aims to balance development and environmental protection by creating new developments that are centered more in the towns and cities, include alternative transit options, and have mixed use development.

smart weapon: [C-10] A missile or bomb that can be steered toward a target through a laser or video guidance system. Smart weapons used by the U.S. military today are all autonomously guided and include capabilities for functioning in adverse weather. They have been added to almost every U.S. aircraft flown today and include laser guided bombs, precision cruise missiles, and air-to-surface GPS-guided glide bombs.

social institution: [C-12] A formal entity existing in society primarily to nurture and contain certain behavior patterns, rituals, and beliefs considered important to a group of human beings. Several that have been preserved for many centuries include family, church, marriage, government, and schools.

social lag: [C-11] Related to the phenomenon of social readiness, with respect to new inventions in technology. Lag refers to the length of time required for society to adapt to or accept new technological ideas, processes, or devices, once they have been introduced. Sometimes called *cultural lag*.

space adaptation syndrome: [C-4] A malady commonly referred to as *space sickness*. It continues to be a number one medical concern among space agencies internationally. It is a generic label ascribed to a series of undesirable physiological and psychological side effects associated with weightlessness or Z-gravity in space.

space junk: [C-4] Pieces of spacecraft, nuts, bolts, gloves, and other debris from space missions forming an orbiting garbage dump around Earth. There are an estimated 4 million pounds of space junk in orbit around Earth.

spectrograph: [C-4] A scientific instrument that separates the light from a telescope into different components so, for example, the composition and temperature of a planet or star can be analyzed.

spectrometer: [C-4] An instrument that separates light into its different wavelengths to create spectra. It allows scientists to determine what elements are present in an object.

statistical process control (SPC): [C-6] An approach to quality assurance differing from the traditional policy of product control via inspection. The focus of process control is on individual operations and the roles they perform in manufacturing. SPC is prevention-oriented versus inspection-driven. It involves statistical analysis and increased personnel responsibility to ensure all equipment and processes are operating within acceptable limits. SPC implementation focuses initially on two or three critical parameters in each process that must be controlled—operating ranges are established.

Strategic Defense Initiative (SDI): [C-10] A military strategy defined to be a comprehensive research and development program to explore the possibilities of strategic defense. Claims regarding the efficacy of these exotic *Star Wars* projects have yet to materialize.

Superfund: [C-9] Formally labeled *Comprehensive Environmental Response, Compensation, and Liability Act,* this program earmarks federal funds to clean up the nation's more critical hazardous waste sites.

sustainable agriculture: [C-8] A philosophical and practical approach to farming and agriculture built upon three broad goals: farm profitability, improvement of the environment, and increased quality of life for farmers and their communities. It is future-oriented and has the principle that we should meet our food needs for the present without limiting the ability of future generations to meet their food needs.

swaggering: [C-10] The deployment of military power by a government to enhance prestige, national pride, or the personal ambitions of its rulers.

technological literacy: [C-11] A recent topic of discussion in social studies and technology education. Although numerous definitions have been proposed, technological literacy is generally thought to

involve an effective level of knowledge and understanding, with general reference to events and issues in science and technology, and an above average grasp of technical terminology in one's field of specialization.

technology: [C-1] The cumulative sum of human means developed in response to society's needs or desires to systematically solve problems.

technology accordance: [C-11] This social behavior or attitude refers to the extent to which a social group is able to recognize and take advantage of the potential benefits of a technological breakthrough and simultaneously control any of its negative side effects or risks.

technology assessment: [C-13] The process of making projections about the future state of specific technologies, so we can assess and plan for their impacts on society. An outgrowth of federal legislation that created the congressional Office of Technology Assessment in the early 1970s.

technology equity: [C-8] An issue implying that all technologies are available to and used by all socioeconomic segments in a society.

technology transfer: [C-7] The development of a technology product or process in one setting, which is then transferred for use in another setting. The transfer of intellectual property (such as a patent or trademark) from Columbia University to Monsanto Corporation is an example of this type of exchange.

technology transfer channel: [C-7] Within the international network of technology exchange and transfer, the channel is the formal or informal route through which technological artifacts travel from the supplier to the end user.

technology trap: [C-11] A contemporary social phenomenon describing our dependence on a technological system or device that we either do not understand or cannot control.

technophobia: [C-11] Refers to a general tendency to shy away from or refrain from using the tools of technology due to (1) a lack of faith and trust in common technological systems, (2) a low level of confidence in the use or operation of technology intensive products, and (3) the perception that one is not technologically literate. In this definition, trust is a confident belief that technology will work the way it is supposed to, when it is supposed to, and faith is an intrinsic belief that does not necessarily rest on logical proof. It has more to do with honesty and sincerity among the scientists, technologists, and engineers who design, develop, and introduce us to new technologies.

technostress: [C-11] A modern affliction referring to a person's inability to adapt to or cope with new computer technologies in a healthy manner. It manifests itself in two distinct, but related ways: in the struggle to accept computer technology and in the more specialized form of overidentification with computer technology. People who suffer from the latter adaptation problem are sometimes called *hackers.*

telecommuting: [C-3] The concept of using telecommunications technology to transport work to employees who therefore avoid the daily commute to a distant job site. It fits well within the "electronic cottage" phenomenon first coined by Alvin Toffler in *The Third Wave*—a projection that more people will spend time working in their homes instead of going out to an office or factory. Also related to this term is a construct labeled *virtual transportation.*

teleconferencing: [C-3] The newest video-conferencing option found in desktop systems connected to a network utilizing digital video transmission.

telematics: [C-3] The use of wireless voice and data technologies between a car and some other place.

terrorism: [C-10] The ability to use readily available technologies as tools of mass destruction.

testbed: [C-3] Limited access test networks that can be used to try new technologies and applications before they are released to the entire Internet.

transgenic: [C-2] A characteristic describing genetically modified (GM) foods or animals containing genes outside their species.

Trojan Horse: [C-3] A computer file masquerading as something else. Trojan Horses contain computer programs that, when started, can cause the loss or theft of data. Trojan Horses will not start automatically; you must start them by, for example, opening an e-mail attachment.

turnkey plant: [C-7] A formal channel for the transfer of technology from one nation to another. A contractual sales agreement is established, by which the technology supplier constructs a fully operational production plant for the recipient nation. Figuratively, when the facility is finished, the recipient need only "turn a key" to initiate production.

universalism: [C-11] A characteristic of modern technology making it very difficult, if not impossible, for people to live in cultures not at all interested in advanced technology. It describes a high level of interdependence among nations and the fact that technological changes, advancements, and accidents do not occur in a vacuum. The world is an open system heavily reliant on an international network of technology transfer.

videoconferencing: [C-3] The ability to communicate visually using computer technology, television monitors, cameras, microphones, and special modems. An interactive group communication takes place through an electronic medium. A substitute for face-to-face meetings, videoconferences are synchronous to real time—all participants are present at the same time.

virtual private network (VPN): [C-3] A new type of network that is a low cost alternative to a private WAN. It is a secure connection between two points on the Internet and is available through an Internet Provider.

virtual reality: [C-3] A realistic, three-dimensional, computer-based simulation of reality that allows the user to become immersed in the experience.

virus: [C-3] A small program, designed to alter the way a computer operates, that functions without the knowledge of the user.

web bug: [C-3] A small software program on a website that can record the actions users take while on the website.

wide area network (WAN): [C-3] A connection between computers or LANs over a large geographic area. They are commonly referred to as *backbone networks.*

wireless network: [C-3] A LAN or WAN that operates using wireless connections.

worm: [C-3] A program that moves from one computer to another automatically, usually through e-mail.

xenotransplantation: [C-2] The transplantation of animal cells, tissues, or organs into human beings.

zapping: [C-12] The modern day television watching habit of switching channels with a remote control to avoid watching commercials or to simply get a sampling of several different programs simultaneously.

zipping: [C-12] A technique enabling the viewer to fast-forward through a prerecorded tape to avoid watching any segments that do not interest him or her (such as commercials and time-outs during sporting events). With regard to DVDs, zipping refers to cueing the laser to a specific segment of the disc to begin watching the program at that point.

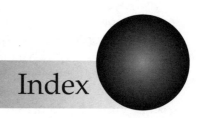

Index